WITHDR

Problems in the Behavioural Sciences
GENERAL EDITOR: Jeffrey Gray
EDITORIAL BOARD: Michael Gelder, Richard Gregory, Robert Hinde, Christopher Longuet-Higgins

Animal learning and cognition

This book describes several connectionist theories of animal learning and cognition. Starting with the simple assumption that psychological associations are represented by the strength of synaptic connections, Nestor Schmajuk offers mechanistic descriptions of complex cognitive behaviors. Part I presents neural network theories of classical conditioning and discusses the concepts of models of the environment, prediction of future events, reliable and salient predictors, redundancy reduction, competition for limited-capacity short-term memory, mismatch between predicted and observed events, stimulus configuration, inference generation, modulation of attention by novelty, storage and retrieval processes, and timing. Part II describes neural networks of operant conditioning, introduces the concept of response-selection mechanisms, and applies a model of operant conditioning to the description of animal communication. Part III discusses goal-directed mechanisms, spatial mapping, and cognitive mapping. Finally, Part IV shows how neural network models permit one to simultaneously develop psychological theories and models of the brain.

The book comes with an accompanying diskette containing a version of the original program for one of the neural network models described in the book, written for DOS computers. This permits the simulation on a personal computer of several classical conditioning paradigms and brain manipulations. It also allows users to simulate their experimental designs and examine the model's predictions.

All those interested in neural networks, from psychologists through neuroscientists to computer scientists working in artificial intelligence and robotics, will find this book an excellent guide to the subject.

Problems in the Behavioural Sciences

1. Contemporary animal learning theory
   A. DICKINSON
2. Thirst
   B. J. ROLLS & E. T. ROLLS
3. Hunger
   J. LE MAGNEN
4. Motivational systems
   F. TOATES
5. The psychology of fear and stress
   J. A. GRAY
6. Computer models of mind
   M. A. BODEN
7. Human organic memory disorders
   A. R. MAYES
8. Biology and emotion
   N. McNAUGHTON
9. Latent inhibition and conditioned attention theory
   R. E. LUBOW
10. Psychobiology of personality
    M. ZUCKERMAN
11. Frustration theory
    A. AMSEL
12. Genius
    H. EYSENCK
13. The psychology of associative learning
    D. SHANKS
14. Theoretical approaches to obsessive-compulsive disorder
    I. JAKES
15. Incentive relativity
    C. F. FLAHERTY
16. Animal learning and cognition: A neural network approach
    N. A. SCHMAJUK

Animal Learning and Cognition

A neural network approach

Nestor A. Schmajuk
Duke University

PUBLISHED BY THE PRESS SYNDICATE OF THE UNIVERSITY OF CAMBRIDGE
The Pitt Building, Trumpington Street, Cambridge CB2 1RP, United Kingdom

CAMBRIDGE UNIVERSITY PRESS
The Edinburgh Building, Cambridge CB2 2RU, United Kingdom
40 West 20th Street, New York, NY 10011-4211, USA
10 Stamford Road, Oakleigh, Melbourne 3166, Australia

© Cambridge University Press 1997

This book is in copyright. Subject to statutory exception
and to the provisions of relevant collective licensing agreements,
no reproduction of any part may take place without
the written permission of Cambridge University Press.

First published 1997

Printed in the United States of America

Typeset in Times Roman

Library of Congress Cataloging-in-Publication Data

Schmajuk, Nestor A.
Animal learning and cognition : a neural network approach/Nestor A. Schmajuk.
p. cm. – (Problems in the behavioral sciences)
Includes bibliographical references (p.).
ISBN 0-521-45086-1 (hc). – ISBN 0-521-45696-7 (pbk.)
1. Learning in animals. 2. Cognition in animals. 3. Neural networks (Neurobiology) I. Title II. Series.
QL785.S225 1996
591.51–DC20 95-37712
CIP

A catalog record for this book is available from the British Library.

ISBN 0 521 45086 1 hardback
ISBN 0 521 45696 7 paperback

QL
785
.S225
1997

101397-9240B1

To Mabel, Gabriela, and Mariana

Contents

Preface

This book is about how animals build models of the environment where they live and how they use these models to survive and flourish. It is about how animals learn to predict the future – what is going to happen, where it is going to happen, and when it is going to happen. It is about how animals select some of the profuse information around them and store only the most informative pieces in their internal models. It is about how animals compare their predictions about the world with observed events and sense the unrelenting novelty of the environment. It is about how animals navigate through open spaces using distal landmarks to guide them and steer their way through tortuous mazes using cognitive maps. It is also about how animals learn to perform certain responses in given circumstances and how they learn to communicate with each other.

Although important progress has been made in our theories of animal learning and cognition, our understanding of the field still lacks the completeness and precision that characterize other scientific domains. This book summarizes some of the recent progress in animal learning and cognition theorizing and, at the same time, it shows how much still remains for future efforts.

Acknowledgments

This book reflects the content of a course on animal learning and neural networks that I taught at the Institute of Psychiatry in London during the summer of 1992, with the support of a Guest Research Fellowship of the Royal Society. I want to thank Jeffrey Gray for suggesting that I write the book as part of the Problems in the Behavioural Sciences series, for his careful readings of the manuscript, and for his helpful criticisms. I also thank Peter Holland, John Moore, Alan Pickering, and John Staddon for their comments on the book. My students Chris Lai and Stephan Kasian were extremely helpful with the references.

The book reflects the influence of different advisors over many years: Enrique Segura, Bob Isaacson, John Moore, and Steve Grossberg. Although I do not expect them to agree completely with the contents of the book, I am sure they will recognize some of their ideas at different points.

This book incorporates material from several previous publications that is reprinted here with kind permission from the publishers listed:

Grossberg, S. and Schmajuk, N. A. (1989). Neural dynamics of adaptive timing and temporal discrimination during associative learning. *Neural Networks, 2:* 79–102. Copyright Elsevier Science Ltd., The Boulevard, Langford Lane, Kidlington OX5 1GB, UK.

Schmajuk, N. A. and Blair, H. T. (1993). Stimulus configuration, place learning, and the hippocampus. *Behavioral Brain Research, 59*:103–17. Copyright Elsevier Science.

(1993). Place learning and the dynamics of spatial navigation: An adaptive neural network. *Adaptive Behavior, 1*: 355–87. Copyright MIT Press.

(1995) Time, space, and the hippocampus. In N. E. Spear, L. P. Spear, and M. Woodruff (eds.), *Neurobehavioral plasticity: Learning, development, and response to brain insult*. Hillsdale, NJ: Lawrence Erlbaum Associates, pp. 35–56. Copyright Lawrence Erlbaum Associates.

Schmajuk, N. A. and DiCarlo, J. J. (1991). A neural network approach to hippocampal function in classical conditioning. *Behavioral Neuroscience, 105*:82–110. Copyright the American Psychological Association.

(1992). Stimulus configuration, classical conditioning, and the hippocampus. *Psychological Review, 99*: 268–305. Copyright the American Psychological Association.

Schmajuk, N. A. (1994). Behavioral dynamics of escape and avoidance: A neural network approach. In D. Cliff, P. Husbands, J.-A. Meyer, and S. W. Wilson (eds.), *From Animals to Animats 3*, Cambridge, MA: MIT Press, pp. 118–27. Copyright MIT Press.

Schmajuk, N. A., Lam, Y. W., and Gray, J. A. (1996) Latent inhibition: A neural network approach. *Journal of Experimental Psychology: Animal Behav-*

ior Processes, *22*: 321–349. Copyright the American Psychological Association.

Schmajuk, N. A. and Thieme, A. D. (1992). Purposive behavior and cognitive mapping: An adaptive neural network. *Biological Cybernetics*, *67*: 165–74. Copyright Springer-Verlag.

Glossary

$A(T)$: accuracy of theory T
$a(t)$: animal acceleration
an_j : output of hidden unit j
as_i : output of input unit i
B_k: aggregate prediction of event k by all CSs with τ_i's active at a given time
$\bar{B}_k$: average prediction of event k
B_{US}: aggregate prediction of the US by all CSs with τ_i's active at a given time
$\bar{B}_{US}$: average prediction of the US.
C: cortex
CL: cortical lesion
CR: conditioned response
CS: conditioned stimulus
CX: contextual stimuli
E_i^k: effective spatial stimulus i for stimulus S_k based on landmark L_i
EO: output error
γ: generalization factor
$G^k(x, y)$: generalization surface of the prediction of stimulus S_k at a point (x, y)
g_i: selective sampling signal
HF: hippocampal formation
HFL: hippocampal formation lesion
HP: hippocampus proper
HPL: hippocampus proper lesion
ISI: interstimulus interval
ITI: intertrial interval
λ_j: intensity of CS_j
$\bar{\lambda}_j$: average observed value of event j
λ_{US}: intensity of the US
$\bar{\lambda}_{US}$: average intensity of the US
LI: latent inhibition
LTD: long-term depression
LTM: long-term memory
LTP: long-term potentiation
M: animal's mass
NMR: nictitating membrane response
Novelty: absolute value of the difference between an event's predicted and observed magnitude
O_i : size of landmark L_i
$\Omega_i\,(d)$: visual angle of spatial landmark L_i at distance d
p_j: real-time prediction of view j
$f\,[d(p_i)/dt\,]$: fast-time prediction of place i
R: operant response
$\rho\,(T)$: explanatory efficiency of theory T
$r_{h,\mathrm{goal}}$: working memory of the prediction of the goal when examining alternative next place h
R_k : strength of alternative response k
r_k : random magnitude of response R_k
R_k' : strength of operant response R_k
S: discriminative stimulus or situation
STM: short-term memory
τ_i: short-term memory trace of stimulus i
T_i^k : memory of the value of the visual angle Ω_i at the place where stimulus S_k is encountered
Total novelty: sum of the individual novelties of all events
UR: unconditioned response
US: unconditioned stimulus
v : animal's velocity
v_t : animal's terminal velocity
$V_{i:}$ associations of sensory representation X_{i1} with the drive representation
$V_{i,k}$: association of neuron i with neuron k
V_i^k : association between effective stimulus $E_i^k(d)$ and stimulus S_k
$V_{i,US}$: association of CS_i with the US
VH_{ij} : association between input unit i and hidden unit j
VN_j : association between hidden unit j and the US
VS_i: association between input i and the US
X-OR: exclusive or
X_i : activity of neuron i
Y: activity in the drive representation node
y_i : amount of neurotransmitter in gate for input i
z_i : association of X_i with novelty in Chapter 5
z_i : association of sampling signal g_i with the US in Chapter 7
Z_{i1} : incentive motivation of sensory representation X_{i1} in Chapter 4
Z_{ik} : operant association of τ_i with the R_k in Chapter 8

Introduction

1 Neural networks and associative learning

Beyond the simple accumulation of data and the timid induction of general laws inferred from a wide collection of particular instances, the most exciting scientific endeavor is the conception of novel explanations, the creation of innovative models, and the formulation of fresh theories (Flew, 1979).

Scientific theories are formulated by generating a set of hypotheses that permit the deduction of already collected experimental data. After experimental data are integrated and compressed into a theory (see Waldrop, 1992), the theory can be rigorously tested by predicting the results of new experimental paradigms, different from those from which they were initially inferred.

Until recently, psychological theories were presented simply in verbal form. Therefore, the complexity and intricacies of some cognitive behaviors prevented these theories from providing a precise account of the data. With the advent of mathematical models, theoretical psychology entered a realm previously reserved for the more mature natural sciences.

Nowadays, complex behavioral interactions and complex brain mechanisms can be described in terms of the nonlinear dynamics of neural networks. The systems of differential equations that formalize the neural networks, although difficult to solve by hand, are easily solved by computers. In addition, scientists can rapidly experiment with computer models and predict the behavior of animals in different experimental situations (see Waldrop, 1992). Furthermore, neural network models allow scientists simultaneously to develop psychological theories and models of the brain.

This chapter summarizes the historical pathways leading from simpler, older concepts to the more sophisticated, recent learning theories.

Associationist theories of learning

Associationist theories of learning suggest that associations are the basis of learning. Aristotle (c. 350 B.C.), considered to be the first associationist, proposed that a thought can lead to other associated thoughts. Thoughts become associated (1) when they are presented in close succession (contiguity), (2) when they are similar, or (3) when they are opposites. In the nineteenth century, British associationist Thomas Brown (1820) suggested additional principles that regulate the formation of associations, among them: (1) overlap between sensations, (2) salience of the sensations, (3) frequency of the pairing, (4) recency of the pairing, and (5) the associations the sensations already have with other sensations. Also in line with the contiguity principle, William James (1890) submitted that "when two elementary brain processes have been active together or in immediate succession, one of them on reoccurring tends to propagate its excitement to the other."

The associationist approach was further extended by Thorndike (1898), who suggested that learning consisted of the association between stimuli S and responses R. When confronted with a given problem, animals find the correct response by applying a trial-and-error procedure. According to Thorndike's *law of effect,* S–R *connections* increase whenever the responses are followed by satisfaction and weakened whenever they are followed by discomfort. For Thorndike, satisfaction connotes a state the animal does not avoid, whereas discomfort indicates a state the animal tries to end.

Pavlov (1927) delineated the elements of classical conditioning, describing many of the experimental paradigms and theoretical concepts still in use today. In classical conditioning, a conditioned stimulus CS, which does not elicit a response, is presented in conjunction with an unconditioned stimulus US, which is able to elicit an unconditioned response UR. After several CS–US pairings, the CS comes to generate a conditioned response CR similar to the UR; that is, the CS becomes a substitute for the US (stimulus substitution theory). Pavlov assumed that a connection (an association) between a CS center and a US center was formed in cortical areas during acquisition of conditioning.

Tolman (1932a) proposed that during operant conditioning animals learn *expectancies* that the performance of response R1 in a situation S1 will be followed by a change to situation S2 (S_1–R_1–S_2 expectancies). In contrast to Thorndike's law of effect, Tolman suggested that no reward is needed for animals to learn this sequence of events; that is, associations are learned by simple temporal contiguity. During classical conditioning, animals learn to expect that a CS is followed by another CS or by the US. Tolman hypothesized that a large number of expectancies can be combined, through inferences, into a *cognitive map.*

Mathematical theories of learning

As described in the previous section, many associationist theories had been advanced by the first decades of the century. However, testing of the theories was frequently hindered by the fact that purely verbal theories do not provide unequivocal predictions of the effects of different experimental manipulations on learning. This deficiency results from the lack of rigorous specifications of the interaction between the different variables in the theory.

Contrasting with these nonrigorous, verbal theories, Hull (1951) described animal learning in the context of an accurate, falsifiable, mathematical model of conditioning. Hull's mechanistic model suggests that CS presentations trigger a short-term memory trace that activates excitatory and inhibitory associations, which are combined to generate the CR. Following Thorndike's law of effect, excitatory CS trace–CR associations (habits) are strengthened when the CR is followed by a satisfying US. Any CS trace could become associated with the CR independently of the associations of the other CS traces. Excitatory CS trace–CR associations are modulated by the internal motivational state of the animal. Hull (1951, p. 16) applied his theory to both classical and operant conditioning.

In 1972, Rescorla and Wagner proposed a computational model of classical conditioning that differed substantially from the association laws suggested by Hull. The model is a modification of a mathematical model closely related to the Hullian theory, the Bush and Mosteller (1955) linear model. First, Rescorla and Wagner assumed that rather than CS–CR associations, animals form CS–US associations used to generate predictions of the magnitude of the US based on the CS (see Rescorla, 1973). Second, Rescorla and Wagner suggested that CS–US associations change until the summed prediction of the US by all CSs present on a given trial equals the actual magnitude of the US; that is, until the US is no longer surprising. This means that the association of a given CS with the US depends on the associations of other CSs present simultaneously with the first CS.

In 1975, Mackintosh described a computational theory in which the conditioning rate associability of a given CS increases when it is the best predictor of the US, and decreases otherwise. Moore and Stickney (1980) and Schmajuk and Moore (1989) generalized Mackintosh's approach to include the predictions of other CSs, and assumed that CS associability increases when CS is the best predictor of other CSs or the US, but decreases otherwise.

Wagner (1978) suggested that CS–US associations are determined not only by the surprisingness of the US, as in the Rescorla–Wagner model, but also by the surprisingness of the CS.

According to Pearce and Hall (1980), when a CS is followed by an unconditioned stimulus US, a CS–US association is formed. This association can generate a prediction of the US by the CS. CS–US associations are changed until the CS associability is zero. Associability is defined as the absolute difference between the US intensity and the "aggregate prediction" of the US computed on all CSs present on a given trial.

Neural networks and connectionism

As mentioned, associationist theories advance the idea that associations are the basis of learning. Neural network theories suggest that associations can be stored by modifying the synaptic connections between elements that capture some of the most important information-processing aspects of neurons. Importantly, neural networks provide a mechanistic description that implements the psychological idea of association.

As early as 1905, Freud proposed that a neuron can activate other neurons through a contact barrier and that associations are stored by changing the value of this connection. In 1949, Hebb suggested that changes in the synaptic strength connecting neurons i and k, w_{ik}, can be described by $\Delta w_{ik} = x_i y_k$, where x_i represents the presynaptic activity and y_k the postsynaptic activity.

In 1943, McCulloch and Pitts suggested that, given the "all-or-none" properties of axonal activity in the neurons, the behavior of neural networks can be described in terms of propositional logic. McCulloch and Pitts assumed that a certain number of synapses must be excited for a neuron to fire; that is, a neuron is active when the sum of its excitatory and inhibitory inputs exceeds a given threshold ($\Sigma_i x_i \geq \theta$), and that acquisition and extinction of learning consist of changes in threshold θ. Since McCulloch and Pitts's formal neurons are able to represent propositional connections (negation, disjunction, conjunction, implication, and equivalence), the logical function of any neural network can be described by applying two-valued logical calculus.

BOX 1.1 The "Delta" Rule

The delta rule describes changes in the synaptic connections w_i between a set of input neurons i and the output neuron in a two-layered network. According to the rule, connection weights w_i change according to $\Delta w_i \sim x_i\ (\lambda - \Sigma_j w_j x_j)$, where x_i is the activity of input i, λ the desired value of the output, and $\Sigma_j w_j x_j$ is the actual value of the output.

Rosenblatt (1962) presented a neural network, called the perceptron, capable of learning a logical function by changing the weight of the synaptic connections w_i between inputs and outputs. In the perceptron, the output y of a neuron is active when the sum of its excitatory and inhibitory inputs exceeds a given threshold, $y = 1$ if $\Sigma_i w_i x_i \geq \theta$ and $y = -1$ otherwise, where $\Sigma_i w_i x_i$ is a linear discriminant function. When different input patterns should be classified in two categories, $\lambda = 1$ and $\lambda = -1$, they are linearly separable if for one subset of patterns $\Sigma_i w_i x_i \geq \theta$ and thus $y = 1$, but $\Sigma_i w_i x_i < \theta$ for the other subset and thus $y = -1$. To compute the values of w_i, Rosenblatt proposed an error-correcting procedure in which the different training patterns are presented several times and, each time, the desired response of the system λ is compared with the actual response y. Changes in w_i are given by $\Delta w_i = x_i f(\lambda - y)$, where f is an error function. It is $f = 0$ if $\lambda = y$, $f = 1$ if $(\lambda - y) > 0$, and $f = -1$ if $(\lambda - y) < 0$.

Widrow and Hoff (1960) applied their seminal paper to the analysis and synthesis of logical networks. When a logical function is to be implemented in a network, first a truth table including all input possibilities and desired outputs is defined and a logical function is constructed. Since it is desirable that the physical realization of the logical function be done in the simplest possible way (to avoid a combinatorial explosion), the logical function is algebraically simplified. On the basis of the simplified function, the network is then implemented. Widrow and Hoff (1960) argued that although there are systems in which the truth table must be followed precisely and reliably, such as in a digital computer, there are other situations in which errors are acceptable. In the perceptron, Widrow and Hoff (1960) proposed a neural network (adaline) that is active when the sum of its excitatory and inhibitory inputs exceeds a given threshold ($\Sigma_i w_i x_i \geq \theta$), and acquisition and extinction of learning consist of changes in weights w_i. They proposed a rule (least-mean-squares algorithm, or delta rule) that generates a linear discriminant function by minimizing the sum of squared errors between the desired output and actual output. According to the delta rule, connection weights w_i change according to $\Delta w_i \sim x_i\ (\lambda - \Sigma_j w_j\ x_j)$, where $(\lambda - \Sigma_j w_j x_j)$ is the error function (see Box 1.1).

Minsky and Papert (1969) pointed out the limitations of two-layered networks that resort to linear discriminant functions such as the perceptron or adalines. For example, such systems are unable to describe "exclusive or" (X-OR) functions, in which $y = 1$ when x_1 or x_2 are presented separately, but $y = 0$ when x_1 or x_2 are presented together. One solution is to add to a two-layered network a hidden layer placed between its inputs and outputs. Units in the hidden layer can be trained through a procedure called backpropagation (see Box 1.2), which was independently suggested by Bryson and Ho (1969), Werbos (1974), Parker (1985), and Rumelhart, Hinton, and Williams (1986). In backpropagation, the error sig-

BOX 1.2 Backpropagation

Two-layered networks that resort to linear discriminant functions, such as the perceptron or adalines, are unable to describe "exclusive or" (X-OR) functions, in which $y = 1$ when x_1 or x_2 are presented separately, but $y = 0$ when x_1 or x_2 are presented together. One solution is to add to a two-layered network a hidden layer placed between its inputs and outputs. Units in the hidden layer can be trained through a procedure called backpropagation. In backpropagation, the error signal that regulates the weights connecting hidden unit j with a single output is proportional to $d(o)e_o$, where $d(o)$ is the derivative of the output of hidden unit j, and $e_o = (\lambda - \Sigma_j w_j x_j)$ is the output error. The error signal that regulates the associations of input x_i with hidden unit j is proportional to $d(o_j)w_j e_o$, where $d(o_j)$ is the derivative of the output of hidden unit j, w_j the association of hidden unit j with the output, and e_o the output error.

nal that regulates the weights connecting hidden unit j with the output is proportional to $d(o)e_o$, where $d(o)$ is the derivative of the output with respect to its total input and $e_o = (\lambda - \Sigma_j w_j x_j)$ is the output error. The error signal that regulates the associations of input x_i with hidden unit j is proportional to $d(o_j)w_j e_o$, where $d(o_j)$ is the derivative of the output of hidden unit j with respect to its total input, w_j the association of hidden unit j with the output, and e_o the output error. Different biologically plausible versions of backpropagation have been offered, including the one described in Chapter 4.

Although backpropagation overcomes the problems inherent in linear separability and solves X-OR functions, it still contains the problems intrinsic to the error-minimizing delta rule. For instance, McCloskey and Cohen (1989) and Ratcliff (1988) have found that when backpropagation networks are trained in paradigms in which input patterns are presented one at a time and not further trained, well-learned patterns are forgotten as new patterns are learned.

In contrast to backpropagation, Carpenter and Grossberg (1985) proposed an adaptive resonance theory (ART) to describe stable self-organization of sensory patterns. In the ART system, top–down expectations are matched against bottom–up sensory signals. When a mismatch occurs, an orienting arousal-burst acts to reset the sensory representation of all cues currently being stored in short-term memory (STM). Representations with high STM activation tend to become less active; representations with low STM activation tend to become more active; and the novel event that caused the mismatch tends to be more actively stored than it would have been had it been expected.

Steinbuch (1961) described an associative network (which he named a *learning matrix*), in which two input patterns, X and Y, become associated by modifying the connections between the neurons representing their elements according to $\Delta w_{ij} = x_i y_j$. Outputs j are defined by $\text{out}_j = \Sigma_i w_{ij} x_i$. As pointed out by Willshaw, Buneman, and Longuet-Higgins (1969), these associative networks operate in parallel, nonlocally, and are damage-resistant. Kohonen (1977) classified associative networks as autoassociative and heteroassociative (see Box 1.3). Hopfield (1982) proposed that, for the special case of an autoassociative network with symmetrical connections ($w_{ij} = w_{ji}$), the system will move toward a stable state as the Lyapunov or energy of the system ($-\frac{1}{2} \Sigma_{i=j}\, x_i\, y_j\, w_{ij}$) decreases or remains constant.

BOX 1.3. Associative Networks

Associative networks are classified as autoassociative and heteroassociative. Autoassociative networks are those in which the input pattern is associated with itself in such a way that the whole pattern is retrievable on the basis of an arbitrary fragment of it (content addressable memory). Heteroassociative networks are those in which two different input patterns are associated in such a way that one of the patterns is retrievable on the basis of an arbitrary fragment of the other. Autoassociative recurrent networks are autoassociative networks in which the outputs either (1) constitute additional inputs or (2) are reinjected to the inputs of the network.

BOX 1.4 Real-Time Neural Networks

Real-time networks describe the unbroken, continual temporal dynamics of behavioral variables and, therefore, their output can be compared to behavior as it unfolds in real time. Furthermore, the dynamics of their intervening variables can be contrasted with neural activity, providing a basis for the study of the physiological foundations of behavior.

The network dynamics are formalized by a set of differential equations that depict changes in the values of neural activities and connectivities as a continuous function of time. However, computer simulations of the network generate values of the variables at discrete time instants.

Neural network theories of animal learning

Although mathematical theories provide unequivocal predictions of the effects of different experimental manipulations on learning, their equations are not constrained by neurophysiological data. Biologically constrained neural networks apply the principles of connectionist theories to the study of animal learning and cognition.

Many connectionist models describe behavior on a trial-to-trial basis, but real-time neural networks describe behavior as a moment-to-moment phenomenon (see Box 1.4).

Figure 1.1 shows a simple real-time neural network. $CS_1(t)$, $C2_2(t)$, and $CS_3(t)$ represent stimuli (e.g., lights, tones, odors) presented as pulses for a given length of time (e.g., 100, 200, 1,000 msec). Presentation of $CS_i(t)$ activates a trace, $\tau_i(t)$, that increases following the onset of $CS_i(t)$ and decays back to zero following $CS_i(t)$ offset (Figure 1.2).

Figure 1.1 also shows that neurons 1, 2, and 3 excite or inhibit neuron k in proportion to $\tau_i V_{ik}$, where V_{ik} is the value of the connection of neuron i with neuron k. As with real neurons, neuron k computes the sum of inputs τ_i. V_{ik} values can adopt excitatory or inhibitory values. Output of neuron (or neural population) k, R_k, can be described by a sigmoid and interpreted as the rate of firing of neuron k.

Extending the ideas fostered by mathematical learning theories, neural network theories of associative learning assume that the association between neurons i and k is represented by the efficacy of the synapses, V_{ik}, that connect a presynaptic neuron excited by $\tau_i(t)$ with a postsynaptic neuron k. At the begin-

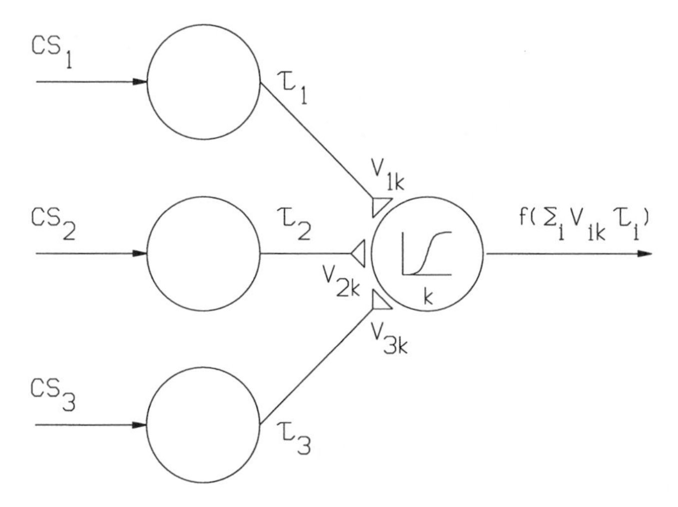

Figure 1.1. Diagram of a formal neuron (or neural population). $CS_1(t)$, $C2_2(t)$, and $CS_3(t)$: stimuli; $\tau_i(t)$: trace of CS_i; V_{ik}: connection weight between neurons *i* and *k*.

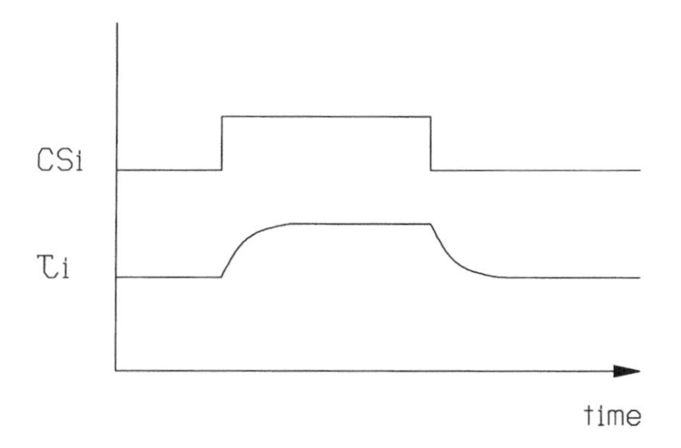

Figure 1.2. Short-term memory trace. Temporal courses of $CS_i(t)$ and $\tau_i(t)$.

ning of training, synaptic strength V_{ik} is small and, therefore, τ_i is incapable of exciting neuron *k*. As training progresses, synaptic strengths gradually increment and $\tau_i(t)$ comes to excite *k*. Different rules have been proposed to modify V_{ik}. Appendix 1.A presents a formal description of the basic building blocks employed in the real-time networks introduced in the following chapters.

Organization of the book

Starting with the rather simple assumptions that (1) neurons (or neural populations) generate a trace when activated, (2) neurons can excite or inhibit each other, and (3) associative learning reflects changes in the efficacy of synapses, this book provides mechanistic descriptions of simple and complex cognitive behaviors.

Part I describes several neural network theories applied to classical conditioning. Chapter 2 describes how, through classical conditioning, animals build a model of the world that predicts future events and how, thereby, they become able to minimize the impact of environmental perturbations, replacing a negative feedback system that partially decreases the environmental effect by a more powerful feedforward mechanism that substantially diminishes it. In addition, Chapter 2 succinctly describes the paradigms to be addressed later by different models.

Chapter 3 describes a neural network that builds and modifies the internal model of the world when predicted events differ from observed events. In consequence, more reliable CSs prevent less reliable or less salient ones from becoming associated with the US. Chapter 3 also describes a model of the world that combines classical conditioning associations (declarative representations of knowledge according to Dickinson, 1980), thereby inferring new knowledge from previously experienced situations in the framework of a cognitive map.

Chapter 4 presents an attentional neural network that assumes that CSs compete for a limited short-term memory capacity according to their salience and the value of their associations with the US. Consequently, the most reliable or salient predictors are selected to become associated with the US or other CSs. This shows that competition for a limited short-term memory is an alternative to the mismatch mechanism described in Chapter 3 to reduce redundancy in the stored model of the environment.

Chapter 5 introduces a neural model that builds on the cognitive map model portrayed in Chapter 3. The new model adds an attentional mechanism that increases the processing of those CSs associated with novelty. The addition of this attentional mechanism allows the model to describe not only association storage as in the previous models, but also association retrieval.

Chapter 6 describes a neural network that also builds on the real-time model presented in Chapter 3. The new model adds a hidden-unit layer that performs stimulus configuration. Associations between configural and simple stimuli are associated with the US to solve complex combinations of world events, thereby describing complex learning paradigms.

Chapter 7 introduces a neural network in which the model of the world predicts what events follow other events and also specifies the time at which they occur.

Part II describes a neural network applied to operant conditioning and animal communication. Chapter 8 introduces a neural network that learns different responses by combining classical conditioning with a "response-selection" mechanism. Chapter 8 also shows that communication between two animals can be described in terms of two interacting neural networks of operant conditioning.

Part III describes neural networks applied to animal cognition. Chapter 9 describes different cognitive paradigms to be addressed by the models. Chapter 10 describes a neural network that accomplishes place learning by assuming that animal behavior is controlled by a goal-seeking mechanism that approaches appetitive locations using a spatial map built with classical conditioning principles. Chapter 11 presents a neural network that depicts maze learning (and some problem-solving tasks) using a goal-seeking mechanism that approaches appetitive locations using a cognitive map such as those described in Chapters 3 and 6.

Part IV shows how neural network models permit one to develop simultaneously, not only behavioral theories, but also models of the brain. Chapter 12 summarizes the effects of different brain manipulations in different learning and cognitive tasks. Chapter 13 addresses the physiological basis of animal learning by showing how the network introduced in Chapter 5 can be mapped onto different regions of the brain, including the hippocampus, cortex, septum, and cerebellum.

As an overarching theme, the book uses Sokolov's (1960) suggestion that the brain constructs a *neural model of environmental events* to predict future events. In Parts I and II, most chapters (1) introduce a block diagram that describes the conceptual mechanisms used to build the model of the environment and (2) describe a neural network that implements the conceptual model. In Part III, chapters (1) introduce a block diagram that describes the conceptual mechanisms used to *build and employ* the model of the environment and (2) describe a neural network that implements the conceptual model.

The book includes computer software that allows the simulation of the model mentioned in Chapters 6 and 13. Computer simulations describe classical conditioning and the effect of different brain lesions on many classical paradigms.

Appendix 1.A

A brief introduction to neural networks

The present appendix describes some basic properties common to the brain-inspired networks presented in this book. The organization of the brain results from the interconnection of billions of blocks called neurons. When these relatively simple blocks are combined into a complex system, the system exhibits properties – called emergent properties – not exhibited by the blocks themselves. Similarly, neural networks incorporate simple neural elements that, when combined in different architectures, give rise to complex behaviors.

Typically, neurons consist of dendrites, cell body, axons, and synaptic terminals. Dendrites (and sometimes cell bodies) receive input from physical stimuli or neurotransmitters released by other cells. The cell body contains the nucleus and most organelles essential to maintain the living neuron. The axon is the output of the cell that emerges from the cell body. The portion next to the cell body is called the axon hillock. The axon ends at the synaptic terminals that contain synaptic vessicles storing neurotransmitters used to communicate with other neurons, muscles, or glands.

The properties of the cell membrane are critical for receiving and conducting information. Cell membranes exhibit a resting potential of approximately –70 mV. Excitatory input from physical stimuli or neurotransmitter released by other cells produces a decrease in the membrane potential (depolarization). Inhibitory input produces an increase in the membrane potential (hyperpolarization). Excitatory and inhibitory inputs received at the dendrites and cell body depolarize and hyperpolarize the membrane respectively. These changes in membrane potential become attenuated as they spread toward the axon hillock. The axon hillock integrates the incoming excitatory and inhibitory inputs through spatial and temporal summation. Spatial summation refers to the process of adding the effect of excitatory and inhibitory inputs received simultaneously from different places on the dendrites or cell bodies. Temporal summation refers to the process of adding the effect of excitatory and inhibitory inputs received at different times from the same place on the dendrites or cell bodies. If the integration of depolarizations and hyperpolarizations results in a potential of approximately –55 mV at the hillock (threshold), the axon generates an action potential. This action potential is conducted along the axon until it reaches the synaptic terminal. In general, because stronger physical inputs result in faster depolarization, they also result in higher frequency of axonal firing.

When the action potential reaches the synaptic terminal at the end of the axon, neurotransmitter is released into the synaptic cleft. The released neurotransmitter activates the receptors located on the dendrites of the following cell, producing depolarizations or hyperpolarizations that propagate toward the hillock of the cell. In general, due to temporal summation at the hillock, higher-input frequency results in higher-output firing.

Figure 1.1 shows a simple real-time neural network. $CS_1(t)$, $C2_2(t)$, and $CS_3(t)$ represent stimuli (e.g., lights, tones, odors) presented as pulses for a given length of time (e.g., 100, 200, 1,000 msec). Presentation of $CS_i(t)$ activates according to

$$d[\tau_i(t)]/dt = A\,[1 - \tau_i(t)]\,CS_i(t) - B\tau_i \quad (1.1)$$

According to Equation 1.1, $\tau_i(t)$ increases (with rate A) to a maximum of 1 following the onset of $CS_i(t)$ and decays (with rate B) back to zero following $CS_i(t)$ offset (see Grossberg, 1975; Hull, 1941).

An alternative equation describing $\tau_i(t)$ is

$$d[\tau_i(t)]/dt = A[CS_i(t) - \tau_i(t)] \quad (1.1')$$

According to Equation 1.1', $\tau_i(t)$ increases (with rate A) to a maximum of $CS_i(t)$ following the onset of $CS_i(t)$ and decays (also with rate A) back to zero following $CS_i(t)$ offset. $\tau_i(t)$ can be interpreted as the rate of firing of cell i when activated by $CS_i(t)$. $\tau_i(t)$ represents a short-term memory trace of CS_i. Figure 1.2 shows the temporal courses of $CS_i(t)$ and its corresponding $\tau_i(t)$.

Figure 1.1 also shows that neurons 1, 2, and 3 excite or inhibit neuron k in proportion to $V_{i,k}\,\tau_i$, where $V_{i,k}$ is the value of the connection of neuron i with neuron k. Activity of neuron k is given by

$$d[X_k(t)]/dt = A[1 - X_k(t)]\,(\Sigma_i V_{i,k}\tau_i) - BX_k \quad (1.2)$$

According to Equation 1.2, $X_k(t)$ increases over time to a maximum of 1 when activated by $\Sigma_i V_{i,k}\,\tau_i$ and decays back to zero when the inputs are absent. Equation 1.3 describes spatial and temporal summation of inputs τ_i.

Whereas the term $\Sigma_i V_{i,k}\,\tau_i$ describes the *additive* interaction between excitatory and inhibitory inputs, *shunting* inhibitory inputs are described by

$$d[X_k(t)]/dt = A[1 - X_k(t)]\,\Sigma_i X_i E_{i,k} - BX_j \Sigma_i X_i\, I_{i,k} \quad (1.3)$$

where $E_{i,k}$ represents excitatory and $I_{i,k}$ inhibitory connections. In contrast to inhibitory additive inputs, which subtract from the total amount of input excitation, shunting inhibitory inputs perform a division operation on the total excitatory input (see Levine, 1991).

Output of neuron k is described by

$$O_k(t) = R_k\,[X_k(t)] \quad (1.4)$$

R_k might be the sigmoid

$$R_k(t) = X_k(t)^n/[X_k(t)^n + \beta^n] \quad (1.5)$$

where n determines the slope of the function and β is the value of $X_k(t)$ for which $R_k(t)$ reaches half of its asymptotic value of 1.

As mentioned, V_{ik} is the value of the connection of neuron i with neuron k. This value can be fixed or variable. Fixed connections can assume excitatory or inhibitory values. Although numerous rules have been proposed for changing variable connections, in general they are of the form

$$\Delta V_{i,k} = f(\tau_i) f(X_k) \tag{1.6}$$

where τ_i represents the activity of the presynaptic neuron and X_k the activity of the postsynaptic neuron (Hebb, 1949).

The preceding equations represent the basic building blocks used in the following chapters.

I Classical conditioning

2 Classical conditioning: data and theories

Part I describes several neural network theories applied to classical conditioning. During conditioning, animals modify their behavior as a consequence of their exposure to the contingencies that hold between environmental events (Mackintosh, 1983). In classical conditioning, animals change their behavior as a result of the double contingency between CSs and the US.

Sokolov (1960) suggested that the brain constructs a neural model of environmental events. When there is coincidence between external events and the predictions generated by the brain model, animals respond without changing their internal model of the environment. However, when the external inputs do not coincide with the previously established internal model, an orienting response appears, and the neural model is modified. The model of the environment describes relationships between environmental events by storing classical conditioning associations (what follows what). As explained below, animals can use this model to predict future events, thereby minimizing the impact of adverse environmental perturbations or maximizing the effect of beneficial circumstances. When future conditions cannot be predicted, animals react to an environmental situation, thereby partially decreasing its adverse impact (negative feedback) or increasing its advantageous effects. When future conditions can be expected, animals act in anticipation of the environmental situation, thereby substantially diminishing its adverse impact (negative feedforward) or considerably increasing its advantageous effects.

Figure 2.1 shows how an animal interacts with the environment through both URs and CRs. Following Sokolov (1960) and Sutton and Barto (1981a), the model of the environment generates predictions of future events and controls the CR. The mismatch system compares observed and predicted events to compute novelty. Novelty is then used to modify the model of the environment. The reflex system controls the generation of the UR.

Classical conditioning and evolution

Animals have innate connections, called reflexes, between USs and their corresponding URs. Selected under evolutionary pressure, these mechanisms assist animals in adequately reacting to environmental stimuli (USs) that consistently have a significant behavioral consequence for all individuals in the species (e.g., an airpuff to the eye). Environmental stimuli (CSs) without an invariable significant behavioral consequence (e.g., a tone) are not selected by evolutionary mechanisms to establish innate connections with the URs.

Learning mechanisms have also been selected through evolution to enable animals to detect and store information about causal relationships (contingencies) in their environment (Dickinson, 1980; Revusky, 1971; Rozin and Kalat, 1971; Testa, 1974). For example, Rescorla (1968) showed how CS–US associa-

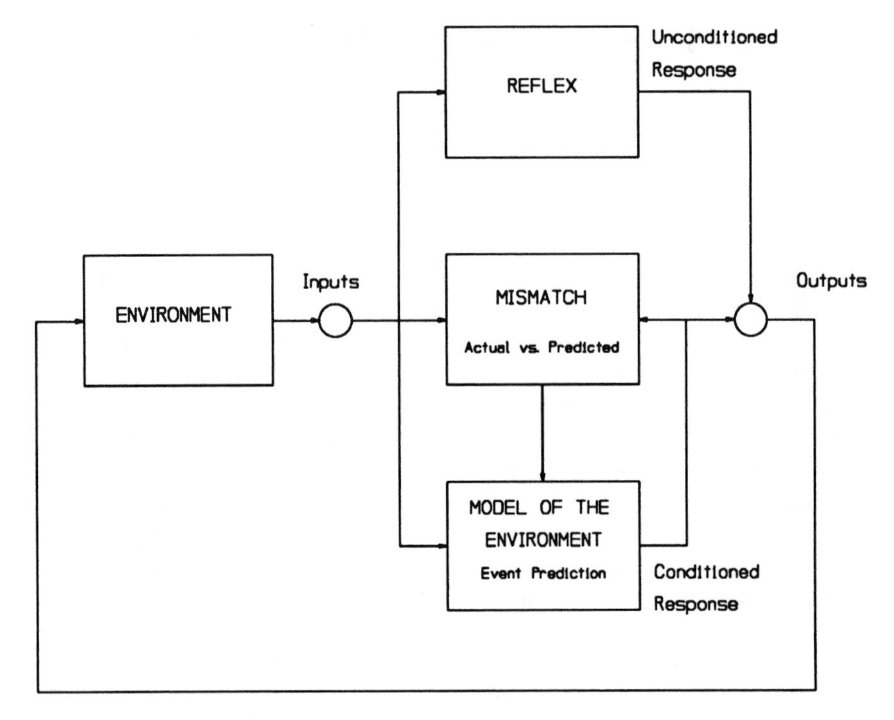

Figure 2.1. Block diagram describing the interactions between the environment and the animal during classical conditioning.

tions increase when the CS–US contingency is positive; that is, when the probability of receiving the US in the presence of the CS [P(US|CS)] is greater than the probability of receiving the US in the absence of the CS [P(US|notCS)].

The survival value of learning mechanisms can be quantitatively analyzed in terms of costs and benefits. Optimization theories (Krebs and Davies, 1981; Maynard Smith, 1978) propose that natural selection has chosen maximally efficient behavioral strategies; that is, those strategies governed by a particular trade-off between costs and benefits that give the maximum net benefit. According to optimization theories, the strategy of increasing CS–US associations with positive CS–US contingencies should be optimal for survival. As optimization theories predict, Schmajuk (1989) showed that the strategy of increasing CS–US associations with positive CS–US contingencies maximizes the expected benefit.

In the case of adverse environmental perturbations, animals react by activating negative feedback mechanisms that partially decrease the unfavorable impact of the environmental change. For instance, as shown in Figure 2.2, the reflexive closure (UR) of the eyelid in rats or the nictitating membrane in rabbits in response to a presentation of an airpuff US reduces the noxious effects of the US. Because of the very essence of negative feedback mechanisms, the system generates a UR after receiving the US, that is, after some degree of exposure to the US has already occurred (lined area in the middle panel of Figure 2.2). In contrast, generating a CR in response to a presentation of the CS substantially diminishes the deleterious consequences of the US (no exposure in top panel in

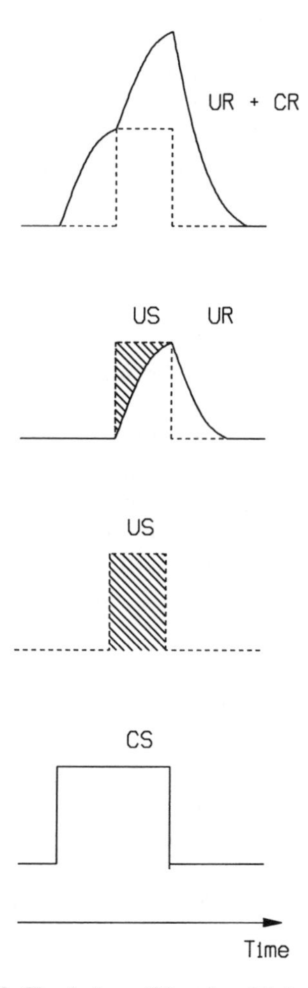

Figure 2.2. Classical conditioned eyeblink response. *Two lower panels:* A CS and an airpuff US are presented. The animal does not respond and receives the full effect of the airpuff US. *Middle panel:* The animal generates a UR, thereby decreasing the effect of the airpuff US. *Upper panel:* The animal generates a CR and a UR, thereby cancelling the effect of the airpuff US. The lined area under the US curve indicates the magnitude of the effect of an airpuff US in different situations.

Figure 2.2). Therefore, as shown in Figure 2.2, a model of the environment specifying that the CS is followed by the US can be used to act in anticipation of the US, thereby potentially cancelling its adverse impact. A system that acts in advance of the presentation of the perturbation, that is, before the animal is already affected, is called a feedforward system. It is important to notice that, although the closure of the eyelid or the nictitating membrane protects the eye, the formation of classical associations seems independent of the attenuation of the aversiveness of the US (see Mackintosh, 1974, p. 115). For instance, Gormezano and Coleman (1973) reported that rabbits trained in a classical conditioning paradigm, in which the US is always presented, learned faster than rabbits trained in an omission paradigm (see the section on avoidance in Chapter 8), in which the US is omitted when the animals produce a CR.

The feedforward argument offered for aversive USs might also be valid for the case of appetitive USs, in which classical associations increase the benefits of receiving the US.

In sum, the survival value of classical conditioning seems to reside in (1) the attenuation of the aversive, or enhancement of the favorable, effects of the US through the generation of a CR, (2) the control of the formation of stimulus–response operant associations (Chapter 8), and (3) the participation in spatial and maze navigation by prediction of the location of aversive and appetitive USs (Chapters 10 and 11).

Classical conditioning paradigms

During classical conditioning, animals change their behavior as a result of the contingency between CSs and USs. In different paradigms, contingencies may vary from very simple to extremely complex. This section describes in a succinct manner classical conditioning paradigms alluded to in Chapters 3–8. For an in-depth description of the paradigms, the reader is referred to any of the excellent books available on animal learning and cognition. It is important to notice that for many of the paradigms (e.g., latent inhibition), a large amount of data have been collected about the phenomenon (e.g., effects of changing the preexposure context during conditioning).

Acquisition

In simple acquisition, animals are exposed to CS(A) followed by the US. Although at the beginning of training animals only generate a UR when the US is presented, with an increasing number of CS(A)–US pairings, CS(A) presentations elicit a CR. CRs increase their amplitude over trials. In general, the CR is considerably analogous to the UR (see Gormezano, Kehoe, and Marshall, 1983).

Interstimulus intervals

Different CS–US interstimulus intervals (ISI) are usually employed. In simultaneous conditioning, both the CS and the US start at the same time; in delay conditioning, the CS onset precedes the US onset; in trace conditioning, the CS offset precedes the US onset; and in backward conditioning, the US offset precedes the CS onset. It has been consistently reported that CR amplitude and percentage are negligible with simultaneous conditioning, increase dramatically with delay conditioning, and gradually decrease with trace conditioning. Figure 2.3 depicts the temporal arrangement of CS and US in delay and trace conditioning.

Smith (1968) studied the effect of manipulating the CS–US interval and the US intensity on the acquisition of the rabbit's classically conditioned nictitating membrane response. The CS was a 50-msec tone and the US was a 50-msec electric shock. The ISI values were 125, 250, 500, and 1,000 msec. Smith (1968) found that the CR, measured as a percentage of responses and response amplitude, was determined by both ISI and US intensity, whereas response onset rate and peak time were determined by the ISI essentially independently of CS intensity. At ISIs of zero CRs are negligible, at ISIs of around 200 msec, CRs increase dramatically, and for longer ISIs, CRs gradually decrease (Schnei-

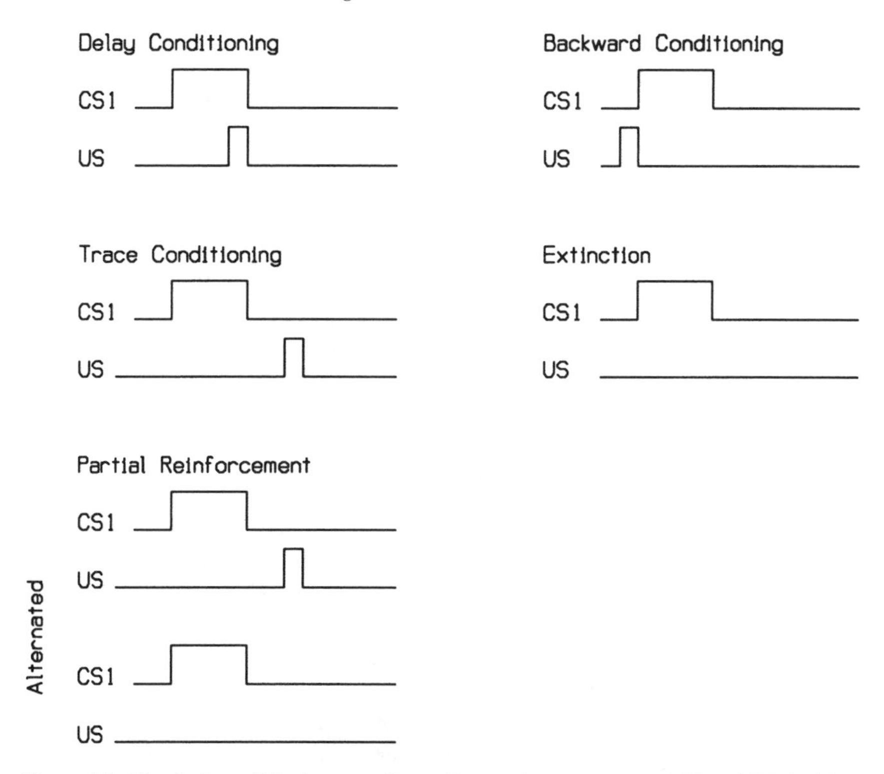

Figure 2.3. Classical conditioning paradigms. Temporal arrangement of CS and US in delay, trace, and backward conditioning, partial reinforcement, and extinction.

derman, 1966; Smith, 1968). In addition, an increase in the mean of the peak response time was correlated with an increase in the variance of the response curve for each ISI.

Analysis of Smith's (1968) data shows that as ISI increases, both CR peak times μ and CR temporal generalization σ increase according to Weber's law, that is, the ratio $W = \sigma/\mu$ is approximately constant. The Weber fraction is a measure of the acuity of discrimination. Evidence coming from the estimation of a rat's time in instrumental learning (Platt, Kuch, and Bitgood, 1973) also shows that CR peak times and CR temporal generalization are related by a Weber function.

It has been found that different response systems and species have different optimal ISIs. The nictitating mebrane conditioned response in rabbits has an optimal ISI of around 250 msec (Smith, 1968). Heart rate conditoning in rabbits is optimal with a 7-sec ISI (Schneiderman, 1972). Conditioned leg flexion in cats is optimal with a 500-msec ISI (McAdam, Knott, and Chiorini, 1965). Salivary conditioning in dogs is optimal with a 20-sec ISI (Konorski, 1948). Conditioned licking in rats is optimal with a 3-sec ISI (Boice and Denny, 1965). Heart rate conditioning in rats is optimal with a 5-sec ISI (Black and Black, 1967) (see Mackintosh, 1974, p. 64.)

Importantly, Millenson, Kehoe, and Gormezano (1977) found that when a tone CS was followed by a shock US at two different ISIs, rabbits acquired double-peak CRs.

Effect of increasing US duration

Burkhardt and Ayres (1978) found that conditioning increased as a function of US duration, as well as of CS–US overlap.

Backward conditioning

In backward conditioning, animals are exposed to the US followed by CS(A). Rescorla and LoLordo (1965) and Siegel and Domjan (1971) found that backward conditioning procedures yield inhibitory conditioning. Figure 2.3 depicts the temporal arrangement of CS and US in backward conditioning.

Extinction and reacquisition

When acquisition is followed by presentations of CS(A) alone, the CR becomes extinguished. Extinction of the CR proceeds with an orderly sequence of changes reversed from that shown during acquisition. Pavlov (1927) reported that a salivary CR extinguished after five trials with a 4-min intertrial interval (ITI), but did not completely extinguish after eight trials with a 16-min ITI. In general, the main effect of ITIs is that increases in ITI decrease the rate of extinction. Figure 2.3 depicts the temporal arrangement of CS and US in extinction.

Partial reinforcement consists of the alternated presentation of acquisition and extinction trials (Figure 2.3). Smith and Gormezano (1965) and Frey and Ross (1968) reported that following extinction, animals show faster reacquisition of CS(A)–US associations (savings effect).

Blocking and overshadowing

In blocking (Kamin, 1968, 1969), an animal is first conditioned to CS(A), and this training is followed by conditioning to a compound consisting of CS(A) and a second stimulus, CS(B). This procedure results in a weaker conditioning to CS(B) than would be attained if paired separately with the US. Although most available evidence suggested that blocking requires at least two compound trials to occur, some results suggest that one-trial blocking can occur (Balaz et al., 1982). Dickinson, Hall, and Mackintosh (1976) found that the surprising omission of a posttrial shock restored the ability to acquire associations between the to-be-blocked CS and the US (unblocking). Figure 2.4 depicts the temporal arrangement of the CS and US in the different phases of blocking and unblocking.

In overshadowing (Pavlov, 1927), an animal is conditioned to a compound consisting of CS(A) and CS(B). This procedure results in the animal's weaker conditioning to the less salient or less reinforced CS than it would achieve if it was independently trained. If both CSs are of similar salience and equally reinforced, both show weaker conditioning than each would achieve separately. If the CSs are of different salience and equally reinforced, the more salient CS overshadows the less intense CS, but not vice versa (Kamin, 1969). Figure 2.4 depicts the temporal arrangement of the CS and US in overshadowing.

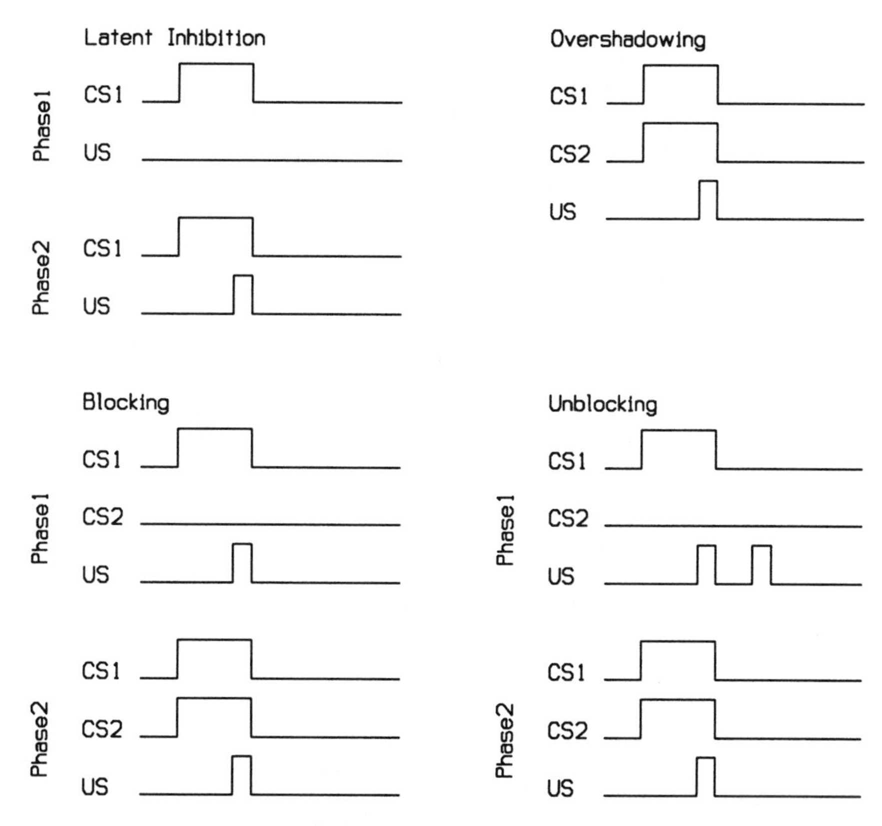

Figure 2.4. Classical conditioning paradigms. Temporal arrangement of CS and US in latent inhibition, blocking, unblocking, and overshadowing.

Overexpectation effects

Kremer (1978) reported that when separately reinforced CS(A) and CS(B) trials are followed by compound CS(A)–CS(B) reinforced trials, responding to CS(A) and CS(B) is weaker relative to controls that do not received compound training.

Generalization

Pavlov (1927) reported than when a CR is established to a CS, the same CR is elicited by other CSs in proportion to their similarity to the original CS.

The gradient of generalization is defined by the decline in responding to stimuli increasingly different from the training stimulus.

Latent inhibition

In latent inhibition (LI) (Lubow and Moore, 1959), preexposure to CS(A) retards the acquisition of CS(A)–US associations. Figure 2.4 depicts the temporal arrangement of CS and US in the different phases of latent inhibition.

Latent inhibition is characterized by a large number of properties, which we are going to address in Chapters 4 and 5.

Stimulus specificity

Lubow and Moore (1959, experiment 1) and Reiss and Wagner (1972, experiment 1) reported that preexposure to a given CS retards conditioning to that specific CS but not to a novel CS. Similar results were reported by Klein, Mikulka, and Hamel (1976) using solutions of different concentrations of sucrose. Furthermore, Schnur and Lubow (1976, experiment 2) showed that preexposure to the same stimulus intensity as that used in acquisition produced greater LI than when different intensities were used.

Latent inhibition of inhibitory conditioning

Rescorla's (1971a) and Reiss and Wagner's (1972, experiment 2) results demonstrate LI of conditioned inhibition.

Preconditioning to CS–weak US presentations

Hall and Pearce (1979) reported that CS–weak US presentations retard the acquisition of a CS–strong US association. Several studies were able to replicate the original result (see Lubow, 1989, p. 87).

Number of CS preexposures

Lantz (1973, experiment 1) varied the number of CS preexposure trials and found that LI increases with increasing number of CS preexposures. Recently, Fanselow (1990) reported that a small number of CS preexposures might result in facilitation of conditioning rather than in LI.

CS duration

According to Lubow (1989, p. 63), data from both conditioned suppression and conditioned taste aversion show that LI increases with increasing CS durations. For example, Westbrook, Bond, and Feyer (1981, experiment 3) reported that long preexposure to an odor CS results in a lasting decrement in odor conditionability.

Total duration of CS preexposure

Ayres et al. (1992) reported that the most important parameter of LI is the total CS-preexposure time (under different combinations of numbers of preexposure trials and CS duration), followed by the number of preexposure trials as a second significant parameter.

CS intensity

Schnur and Lubow (1976, experiment 2) found that LI is an increasing function of the intensity of the preexposed CS. Schnur and Lubow preexposed animals to a loud and a weak CS, and then animals in each group were conditioned with either a loud or a weak CS. Nonpreexposed animals showed faster conditioning with the loud than with the weak CS. When the louder CS was used in acquisition, preexposure with the more intense CS retarded conditioning more than preexposure with the weak CS. However, when the weaker CS was used in acquisition, preexposure with the more intense CS retarded conditioning less than preexposure with the weak CS. Schnur and Lubow (1976) interpreted the inconsistency in the results as the effect of the high specificity of CS preexposure and the consequent generalization decrement during conditioning. According to this interpretation, preexposure to the more (less) intense CS specifically retards conditioning to the more (less) intense CS. Nevertheless, Schnur and Lubow (1976) point out that the relative decrement in conditioning produced by CS preexposure still warrants the conclusion that LI varies proportionally with the intensity of the CS during preexposure.

Crowell and Anderson (1972, experiment 1) reported a direct relation between CS intensity during preexposure and the intensity of LI. Crowell and Anderson (1972) preexposed animals to a loud and a weak CS, and then animals in each group were conditioned with either a loud or a weak CS. Nonpreexposed animals showed faster conditioning with the loud than with the weak CS. In contrast to Schnur and Lubow (1976), Crowell and Anderson (1972) reported that preexposure with the louder CS retarded conditioning more than preexposure with the weak CS when either CS was used in acquisition. Also in contrast to Schnur and Lubow (1976), Crowell and Anderson (1972) did not find generalization decrement in their experiment.

In contrast with the results indicating that LI is an increasing function of the intensity of the preexposed CS, Solomon, Brennan, and Moore (1974, experiment 1) reported that LI is inversely related to CS intensity.

Intertrial interval duration

Lantz (1973, experiment 2) and Schnur and Lubow (1976, experiment 1) varied the ITI during CS preexposure and showed that LI is a positive function of the ITI. In contrast to these data, DeVietti and Barrett (1986) and Crowell and Anderson (1972) established that the ITI of the preexposure trials had no effect on the intensity of LI.

Preexposure to a compound CS

Mackintosh (1973), Rudy, Krauter, and Gaffuri (1976), and Holland and Forbes (1980) studied the effect of preexposing CS(A) and CS(B) in compound on the subsequent independent conditioning of CS(A) or CS(B) ("overshadowing" of LI). They reported little LI to CS(A) or CS(B). Honey and Hall (1988, experiment 1) found that LI was attenuated with simultaneous but not sequential presentations of CS(A) and CS(B). Honey and Hall (1989, experiment 3) established that, whereas preexposure to a compound with elements in the same modality (a click and a tone) overshadows LI to the elements, preexposure to a compound with elements in different modalities (a tone and a light) leaves LI to the elements intact. However, this result is in conflict with Rudy et al.'s (1976) reported overshadowing of LI using CSs in different modalities (a light and a noise).

Honey and Hall (1988, experiment 3) studied the effect of preexposing CS(B) followed by simultaneously preexposing CS(A) and CS(B) in compound on the subsequent conditioning of CS(A) ("blocking" of LI). They observed that with relatively few CS(B) preexposures, blocking and overshadowing of LI reduce LI by similar amounts. However, blocking with more extensive CS(B) preexposure tends to preserve LI. Interestingly, Rudy et al. (1976) observed that LI can be reduced simply by repeated presentations of CS(B) prior to preexposing CS(A).

Preexposure to a CS(A)–CS(B) compound or its elements

Holland and Forbes (1980) reported that preexposure to CS(A)–CS(B) is less effective than alternated preexposure to separate presentations of CS(A) and CS(B) in retarding conditioning (LI) of the CS(A)–CS(B) compound. In contrast, Baker, Haskins, and Hall (1990, experiment 1) reported that alternated preexposure to CS(A)–CS(B) and the context is more effective than alternated preexposure to separate presentations of CS(A) and CS(B) in yielding LI to the CS(A)–CS(B) compound.

Preexposure to CS(A) followed by the introduction of a surprising event

LI is disrupted by a surprising event following CS preexposure. For example, Lantz (1973, experiment 3) reported that the interpolation of a novel CS between CS preexposure and conditioning resulted in an augmentation of the subsequent conditioning. In the same vein, Rudy, Rosenberg, and Sandell (1977) reported that LI was attenuated when animals where placed in a novel black box preceding the conditioning phase. Similarly, Best, Gemberling, and Johnson (1979) found that when a preexposed CS was followed by a novel CS, LI was markedly reduced. As in the cases when a novel, unexpected event is presented, Hall and Pearce (1982) reported that the omission of an expected weak shock also decreases the magnitude of LI.

Preexposure to the apparatus prior to CS preexposure

Hall and Channel (1985a, experiments 1–3) reported that context preexposure prior to CS preexposure facilitates LI.

Context presentations following CS preexposure

Although sometimes CS preexposure in a given context (CX) followed by exposure to that context alone decreases LI (e.g., Baker and Mercier, 1982; Wagner, 1979), the extinction effect is small or nonexistent. Futhermore, Hall and Minor (1984, experiment 5) reported that a phase of exposure to the context alone, interposed between CS preexposure and conditioning in the same context, somewhat facilitates LI.

Context changes following CS preexposure

Wickens, Tuber, and Wickens (1983, experiment 3) and Hall and Channel (1985, experiment 3) showed that LI is disrupted by a change in the context from the CS preexposure phase to the conditioning phase. Attenuation of LI occurs even when conditioning occurs in a context already familiar (Hall and Channel, 1985; Hall and Minor, 1984; Lovibond, Preston, and MacKintosh, 1984).

Preexposure to the apparatus prior to CS preexposure combined with context changes following CS preexposure

McLaren et al. (1994) reported that, even when LI is disrupted when conditioning takes place in a context different from that of CS preexposure, LI is preserved if animals are exposed to the context of CS preexposure before the CS preexposure phase.

Restoration of LI in a novel context by a cueing treatment

As explained in the section on context changes following CS preexposure, LI is abolished when conditioning occurs in a context different from that used for CS preexposure. However, Gordon and Weaver (1989) showed that LI can be restored if animals are exposed to a background noise – present during preexposure – in the conditioning context, prior to the conditioning trials.

Stimulus and context changes

Lubow, Rifkin, and Alek (1976) carried out an experiment to study the effect of CS and context changes on LI. Animals received CS preexposure in a given environment, followed by conditioning (1) with the old CS in the old environment (SoEo or LI group), (2) with a new CS in the old environment (SnEo group), (3) with the old CS in a new environment (SoEn group), and (4) with a new CS in a new environment (SnEn group). Lubow et al. (1976) reported that the SoEn and SnEo groups showed the fastest rate of conditioning, the SoEo (LI) group showed the slowest rate of conditioning, and the SnEn group displayed an intermediate rate of conditioning.

Perceptual learning

Although CS preexposure typically yields LI, preexposure to a pair of stimuli might facilitate subsequent discrimination. Channel and Hall (1981, experiment 1) trained rats in a simultaneous discrimination task using vertically and horizontally striped objects. Whereas preexposure to these stimuli in the rats' home cage helped them to learn the discrimination (perceptual learning), preexposure to the stimuli in the training environment retarded the acquisition of the discrimination (LI). In contrast to these results, Trobalon, Chamizo, and Mackintosh (1992) reported that preexposure to intramaze cues resulted in perceptual learning only when these cues were presented in the same context during preexposure and discrimination training.

Orienting response and LI

Hall and Channel (1985b) and Hall and Schachtman (1987) found clear dissociations between the orienting response OR and LI. Hall and Channel (1985b) studied the relationship between the OR and LI. Hall and Channel (1985b, experiment 1) reported that the OR to a light CS presented in context A habituates over trials, but dishabituates when the CS is presented in a novel context B. Hall and Channel (1985b, experiment 2) also showed that CS presentations in context A alternated with exposure to context B prevented the dishabituation of the ORs when the CS is presented in context B. Finally, Hall and Channel (1985b, experiment 3) found that the context change that fails to produce dishabituation of the OR in experiment 2 is sufficient to prevent LI. Also suggesting a dissociation between changes in the OR and LI, Hall and Schachtman (1987) reported that a period of time that succeeds in producing dishabituation of the OR is insufficient to prevent LI. In sum, experimental data suggest a disconnection between changes in the OR and LI effects: LI can be impaired by contextual changes that fail to produce dishabituation of the OR, and can remain undisturbed after a period of time that produces dishabituation of the OR.

Attenuation of LI by extinction of the training context

Grahame et al. (1994) examined the effect of context extinction, following CS preexposure and conditioning, on LI. LI was obtained by preexposing a CS in the training context followed by CS–US pairing. Subsequently, LI was attenuated by extensive exposure to the training context in the absence of the US. Testing was done in a different context (1) to minimize differences in responding to the CS based on context–US associations and (2) to avoid the reactivation of the CS representations by context–CS associations, which might differentially affect CS preexposed and nonpreexposed groups.

Attenuation of LI by US presentations in another context

Kasprow et al. (1984) showed that LI following CS preexposure and CS–US pairings in a given context was reduced by US presentations in another context. Animals were preexposed to the CS in the training context or to the training context alone, alternated with preexposure to the testing context. Preexposure was followed by conditioning in the training context. Following conditioning, animals were presented with the US in a reminder context. The reminder context was similar in some aspects to the training context (e.g., illumination levels, grid floors), and dissimilar in other aspects (e.g., size, noise levels). After being exposed to the US, animals were tested in a radically different testing context to avoid the generalization of the associations formed between the reminder context with the US.

Attenuation of LI by delayed testing

Kraemer, Randall, and Carbary (1991) studied the interaction between CS preexposure and retention interval. Animals were exposed either to the CS in the training context or to the training context alone. Following conditioning, animals were tested 1, 7, or 21 days after conditioning. Animals were preexposed and conditioned in sets of three animals, but tested individually. The retention interval was spent in their home cages. LI was found at the 1-day but not at 7- or 21-day retention intervals.

Classical discrimination

In a discrimination paradigm (Pavlov, 1927), reinforced trials with CS(A) are alternated with nonreinforced trials with a second CS(B). During discrimination reversal, the original nonreinforced CS(B) is reinforced, whereas the CS(A) reinforced in the first phase is presented without the US. If animals are trained on a long series of independent discriminations, they gradually become more competent at solving the new discrimination. Figure 2.5 depicts the temporal arrangement of CS and US in discrimination acquisition.

Conditioned inhibition

In conditioned inhibition (Pavlov, 1927), CS(B) acquires inhibitory conditioning following CS(A) reinforced trials interspersed with CS(A)–CS(B) nonreinforced trials. In contrast to excitatory conditioning, presentations of CS(B) alone do not extinguish inhibitory conditioning. Inhibitory conditioning can be extinguished, however, by reinforced presentations of CS(B) (Zimmer-Hart and Rescorla, 1974). Baker (1974) found that a neutral stimulus CS(A) presented with a conditioned inhibitor CS(B) does not become excitatory. Figure 2.5 depicts the temporal arrangement of CS and US in the different phases of conditioned inhibition.

Interestingly, extinction of the excitatory conditioning of CS(A) seems to extinguish the inhibitory conditioning of CS(B) (conditioned inhibition as a "slave" process) (Lysle and Fowler, 1985). Presence of a conditioned inhibitor CS(A) during compound CS(A)–CS(B) reinforced presentations promotes superconditioning of CS(B) (Rescorla, 1971b).

Conditioned inhibitors can also be obtained during differential conditioning [alternated reinforced CS(A) trials and CS(B) nonreinforced trials] or by reducing the magnitude of the US on compound CS(A)–CS(B) trials that follow CS(A) reinforced trials (Cotton, Goodall, and Mackintosh, 1982).

Second-order conditioning

Second-order conditioning consists of a first phase in which CS(A) is paired with the US. In a second phase, CS(A) and CS(B) are paired together in the absence of the US. Finally, when CS(B) is presented alone, it generates a CR. Rashotte, Griffin, and Sisk (1977) and Rescorla (1978) found that extinction of the CS(A)–US association led to substantial reduction in responding to CS(B). Opposite results, however, were obtained by Rizley and Rescorla (1972) and Holland and Rescorla (1975). Figure 2.5 depicts the temporal arrangement of CS and US in second-order conditioning.

Secondary inhibitory conditioning consists of two phases. In phase 1, CS(A) is paired with a US. In phase 2, the offset of CS(A) can generate an off-response that can condition a subsequent CS(B) to become an inhibitory conditioned reinforcer (Rescorla, 1976a).

Sensory preconditioning

Sensory preconditioning (Brogden, 1939) consists of a first phase in which two CSs, CS(A) and CS(B), are paired together in the absence of the US. In a second

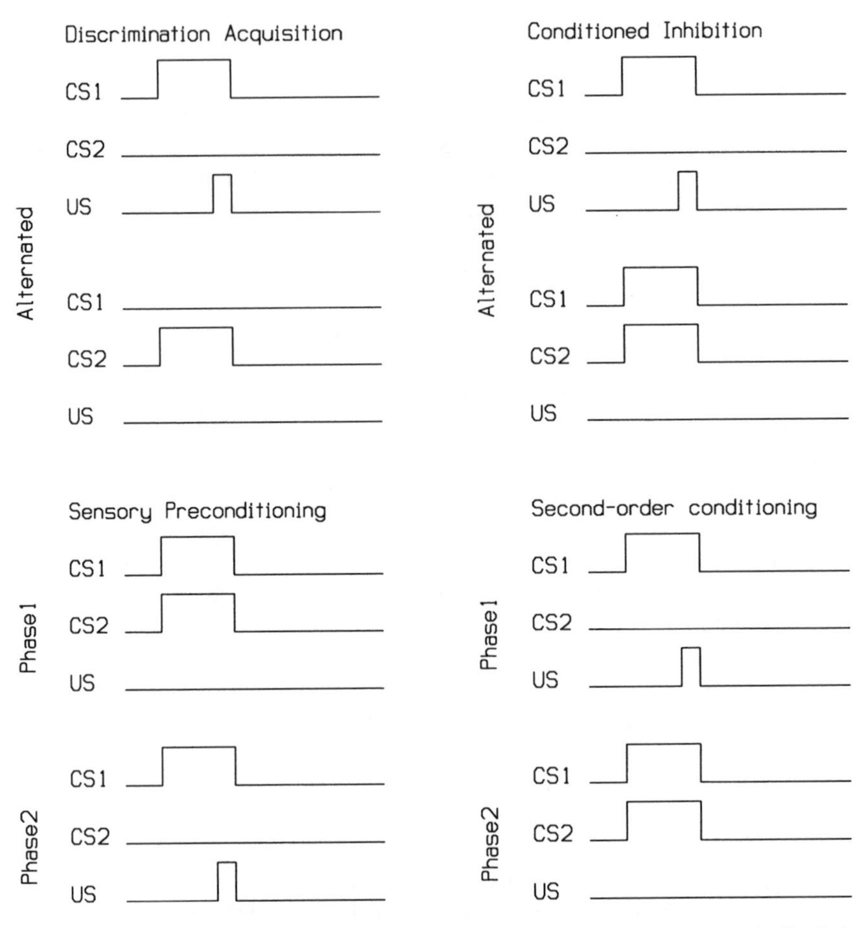

Figure 2.5. Classical conditioning paradigms. Temporal arrangement of CS and US in discrimination acquisition, sensory preconditioning, conditioned inhibition, and second-order conditioning.

phase, CS(A) is paired with the US. Finally, when CS(B) is presented alone, it generates a CR. Rescorla (1980) found that simultaneous presentations of CS(A) and CS(B) produced higher levels of sensory preconditioning than successive presentations. Figure 2.5 depicts the temporal arrangement of CS and US during sensory preconditioning.

Compound conditioning

In compound conditioning, two or more stimuli are presented together in the presence of the US. Kehoe (1986) compared the responding to a CS(A)–CS(B) compound and its components, CS(A) and CS(B), under different proportions of reinforced presentations of CS(A)–CS(B), CS(A), and CS(B), and found that responding to each component, CS(A) and CS(B), decreased with increasing proportion of CS(A)–CS(B) presentations.

In negative patterning, reinforced component presentations [CS(A) or CS(B)] are intermixed with nonreinforced compound [CS(A)–CS(B)] presentations. Negative patterning is attained when the response to the compound is smaller than the sum of the responses to the components. Bellingham, Gillette-

Bellingham, and Kehoe (1985) found that at the beginning of training, animals respond to both compound and components, and this is followed by a gradual decline in the response to the compound.

In positive patterning, reinforced compound [CS(A)–CS(B)] presentations are intermixed with nonreinforced component [CS(A) or CS(B)] presentations. Positive patterning is attained when the response to the compound is larger than the sum of the responses to the components. Bellingham et al. (1985) reported that at the beginning, rabbits respond to both compound and components, and this is followed by a gradual decline of the response to the components. Figure 2.6 depicts the temporal arrangement of CS and US in positive and negative patterning.

Simultaneous and serial feature-positive discrimination (occasion setting)

In a simultaneous feature-positive discrimination, animals receive reinforced simultaneous compound presentations [CS(A) overlapping with CS(B)] alternated with nonreinforced presentations of CS(B). Animals learn to respond to CS(A) but not to CS(B). In a serial feature-positive discrimination, animals receive reinforced serial compound presentations [CS(A) preceding CS(B)] alternated with nonreinforced presentations of CS(B). Animals learn to respond to the CS(A)–CS(B) compound but not to CS(A) or CS(B). Figure 2.6 depicts the temporal arrangement of CS and US in simultaneous and serial feature-positive patterning.

Holland (1983) suggested that different strategies can be used to solve feature-positive discriminations. When CS(A) and CS(B) are presented simultaneously, CS(A) acquires simple associations with the US. When CS(A) precedes CS(B) in compound CS(A)–CS(B) trials, CS(A) acquires occasion-setting properties. To the extent that responding during a compound CS(A)–CS(B) in simultaneous feature-positive discrimination is the consequence of simple conditioning of CS(A), repeated presentations of CS(A) alone after training extinguish that conditioning, and hence abolish any responding to the CS(A)–CS(B) compound that reflected CS(A)–US associations. After serial feature-positive training, repeated presentations of CS(A) alone after training extinguish CS(A)–US conditioning, but CS(A)'s ability to modulate behavior controlled by CS(B) is unaffected.

Theories of classical conditioning

Modern learning theories assume that the association between events CS_i and CS_k, $V_{i,k}$, can be used to generate the *prediction* that CS_i will be followed by CS_k (Dickinson, 1980). As mentioned in the Introduction, neural network or connectionist theories frequently assume that the association between CS_i and CS_k is represented by the efficacy of the synapses, $V_{i,k}$, that connect a presynaptic neural population excited by CS_i with a postsynaptic neural population that is excited by CS_k (event k might be another CS or the US). When CS_k is the US, this second population controls the generation of the CR. As Arbib, Kilmer, and Spinelli (1976) noted, in a neural network, a given CS_i might become associated with some CS_k more rapidly and easily than with others – a concept akin to that of "preparedness" proposed by Seligman (1970).

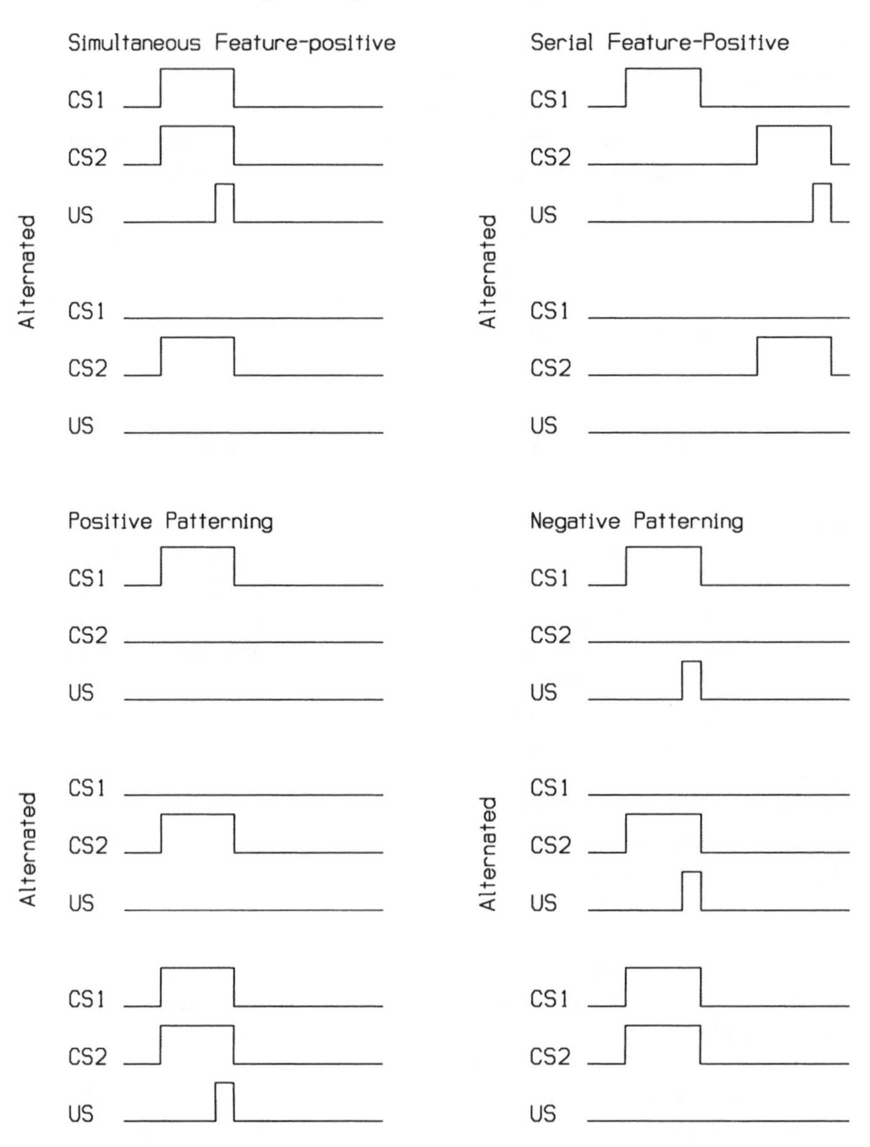

Figure 2.6. Classical conditioning paradigms. Temporal arrangement of CS and US in simultaneous and serial feature-positive patterning, and positive and negative patterning.

Storage processes

In neural networks, storage of CS_i–CS_k associations is achieved by changing the synaptic strength $V_{i,k}$. Changes in $V_{i,k}$ are described by $\Delta V_{i,k} = f(CS_i)\, f(CS_k)$, where $f(CS_i)$ represents a function of the presynaptic activity and $f(CS_k)$ the postsynaptic activity. Different $f(CS_i)$ and $f(CS_k)$ functions have been proposed. Learning rules for $V_{i,k}$ either assume variations in the effectiveness of the US, $f(CS_k)$, the CS_i, $f(CS_i)$ (Dickinson and Mackintosh, 1978), or both.

Variations in the effectiveness of the US (CS_k)

Rescorla and Wagner (1972) proposed a psychological equivalent of the "delta" rule (see Boxes 1.1 and 2.1). According to the simple delta rule, CS_i–US associ-

BOX 2.1 Delta rules in classical conditioning

Rescorla and Wagner (1972) described classical conditioning with a two-layer network trained by the equivalent of a simple delta rule (see Box 1.1). According to the rule, CS_i–US associations are changed until the difference between the US intensity and the "aggregate prediction" of the US computed on all CSs present at a given moment ($US - \Sigma_j V_{j,US} CS_j$) is zero.

Schmajuk and DiCarlo (1992) described classical conditioning with a three-layer network in which the hidden layer is trained by the equivalent of a "generalized" delta rule or backpropagation (see Box 1.2). The network is able to solve exclusive-or problems and, hence, negative patterning.

ations are changed until the difference between the US intensity and the "aggregate prediction" of the US computed on all CSs present at a given moment ($US - \Sigma_j V_{j,US} CS_j$) is zero. The term ($US - \Sigma_j V_{j,US} CS_j$) can be interpreted as the effectiveness of the US to become associated to the CS. Real-time refinements of the Rescorla and Wagner model were offered by Sutton and Barto (1981b) and Klopf (1988).

Recently, Schmajuk and DiCarlo (1992) introduced a model that, by employing a generalized delta rule or backpropagation (see Boxes 1.2 and 2.1) to train a layer of hidden units that configure simple CSs, is able to solve exclusive-or problems and, hence, negative patterning. Importantly, the use of the generalized delta rule allows the model to describe multiple combinations of input–output (CS, US) relationships employed in classical conditioning paradigms, including occasion setting (see Schmajuk and Blair, 1993).

Theories assuming variations in the effectiveness of the US are presented in Chapters 3 and 6.

Variations in the associability of the CS (CS_i)

Attentional theories assume that the associability or effectiveness of CS_i to form CS_i–US associations depends on the magnitude of the "internal representation" of CS_i. In neural network terms, attention may be interpreted as the modulation of the CS representation that activates the presynaptic neuronal population involved in associative learning. For instance, Mackintosh's (1975) attentional theory suggests that the associability of CS_i increases when it is the best predictor of the US and decreases otherwise. Moore and Stickney (1980) and Schmajuk and Moore (1989) generalized Mackintosh's approach to include the predictions of other CSs, and assumed that CS_i associability increases when CS_i is the best predictor of other CSs or the US, but decreases otherwise. In contrast to Mackintosh's (1975) view, Pearce and Hall (1980) suggested that CS_i associability increases when CS_i is a poor predictor of the US; that is, when CS_i has been followed recently by the unexpected presentation of the US.

According to Grossberg's (1975) neural attentional theory, pairing of CS_i with a US causes both an association of the sensory representation of CS_i with the US (conditioned reinforcement learning) and an association of the drive representation of the US with the sensory representation of CS_i (incentive motivation learning). Sensory representations compete among themselves for a limited-capacity short-term memory activation that is reflected in the CS_i–US associations. CSs with larger incentive motivation associations accrue CS–US associations faster than those with smaller incentive motivation.

Theories assuming variations in the effectiveness of the CS are presented in Chapter 4.

Variations in the effectiveness of both the CS (CS_i) and the US (CS_k)

Some classical conditioning theories have combined variations in the effectiveness of both the CS and the US. For example, Frey and Sears (1978) proposed a model of classical conditioning that assumed variations in the effectiveness of both the CS and the US: The internal representation of CS_i is modulated by its association with the US, $V_{i,\mathrm{US}}$. More recently, Ayres, Albert, and Bombace (1987) proposed a real-time model of conditioning that combines the Frey and Sears (1978) attentional rule with the Rescorla and Wagner (1972) model.

Wagner (1978) suggested that CS_i–US associations are determined by (1) the effectiveness of the US ($US - \Sigma_j V_{j,\mathrm{US}} CS_j$), as in the Rescorla–Wagner model, and (2) the effectiveness (associability) of the CS_i ($CS_i - V_{\mathrm{CX},i}$ CX), where CX represents the context and $V_{\mathrm{CX},i}$ the strength of the CX–CS_i association.

Schmajuk and Moore (1988; Schmajuk, 1986) offered a real-time attentional-associative model of classical conditioning that incorporates CS–CS as well as CS–US associations. In the model, CS salience (assumed to modulate the rate of CS–CS and CS–US associations) is determined by CS novelty; that is, the absolute value of the difference between its predicted and observed amplitude, $|CS_i - \Sigma_j V_{j,i} CS_j|$. McLaren et al. (1989) proposed a similar model in which the formation of CS_i–CS_j and CS_i–US associations is modulated by the associability of CS_i ($CS_i - \Sigma_j V_{j,i} CS_j$).

Gluck and Myers (1993) presented a computational theory of classical conditioning that includes three three-layer networks working in parallel. One of the networks represents the hippocampal region, another one the cortex, and a third one the cerebellum. The output and hidden layers of the hippocampal network are trained using backpropagation to associate CSs with those same CSs and the US. The output layers of both cortical and cerebellar networks are also trained by the US. However, hidden units of both cortical and cerebellar networks are trained by the hidden units of the hippocampal network. Weights in the cortical hidden layer represent CS salience. In the cortical and cerebellar networks, elements of the hidden layers are initialized according to a uniform distribution, U(–0.3 to + 0.3), except for two random weights from each CS that are initialized from U(–3.0 to + 3.0). In contrast, the hippocampal hidden layer is initialized according to a uniform distribution U(–0.3 to + 0.3).

A theory assuming variations in the effectiveness of both the CS and the US is presented in Chapter 5.

CS_i–CS_k associations and inference generation

During classical conditioning, animals learn to expect (predict) that a CS is followed by another CS or by the US. Tolman (1932a) proposed that multiple expectancies (predictions) can be integrated into larger units through a reasoning process called inference. One simple example of inference formation is sensory preconditioning (see Bower and Hilgard, 1981, p. 330). As mentioned previ-

BOX 2.2 Cognitive maps

Tolman (1932a, b) proposed that multiple expectancies (predictions) can be integrated into larger units through a reasoning process called inference. Tolman hypothesized that a large number of expectancies can be combined into a cognitive map.

ously, sensory preconditioning consists of a first phase in which two conditioned stimuli, CS(A) and CS(B), are paired together in the absence of the US. In a second phase, CS(A) is paired with the US. Finally, when CS(B) is presented alone, it generates a CR: The animal has inferred that CS(B) predicts the US. Tolman hypothesized that a large number of expectancies can be combined into a cognitive map (see Box 2.2 and Chapters 5 and 11).

Dickinson (1980) suggested that knowledge can be represented in declarative or procedural form. Whereas in the declarative form knowledge is represented as a description of the relationships between events (knowing that), in the procedural form knowledge is represented as the prescription of what should be done in a given situation (knowing how). Examples of declarative knowledge are classical CS–CS associations [CS(A) precedes CS(B)] or CS–US [CS(B) precedes the US] associations. Examples of procedural knowledge are operant S–R associations (if S is present, then do R). Dickinson indicates that declarative, but not procedural, knowledge can be integrated through inference rules.

By including CS–CS associations, some models of classical conditioning are able to generate inferences and therefore describe sensory preconditioning. For instance, Gelperin, Hopfield, and Tank (1985; Gelperin, 1986) proposed an autoassociative recurrent network capable of describing stimulus–stimulus associations during classical conditioning. The network can simulate first- and second-order conditioning, extinction, sensory preconditioning, and blocking in the terrestrial slug *Limax maximus.* Schmajuk (1986, 1987; Schmajuk and Moore, 1988, 1989) proposed networks that separately store CS–CS and CS–US associations and combine them to generate new expectancies. For instance, if CS(A) predicts (is associated with) CS(B) and CS(B) predicts the US, the network *infers* that CS(A) also predicts the US.

Theories describing inference generation are described in Chapters 3 and 5.

Storage and retrieval processes

As noted by Spear, Miller, and Jagielo (1990), current theories of conditioning tend to underscore the process of acquisition of CS–US associations but neglect the process of expression of the association during performance. In contrast, the network model introduced in Chapter 5 combines both acquisition and expression processes and offers a quantitative description of both operations and their interactions. In the model, storage and retrieval processes are two different aspects of single neural mechanism. In a sense, the model reconciles Spear's (1981) suggestion that many phenomena typically attributed to storage processes are a consequence of retrieval processes with more traditional views that stress storage operations.

Opponent processes

In agreement with Pavlov's "stimulus substitution theory," all the aforementioned theories suggest that repeated CS–US pairing leads the CS to substitute for the US in generating the UR. Although in many cases the CR is considerably analogous to the UR, there are some cases in which the CR might be the opposite of the UR.

To predict whether a CR will be analogous or opposite to the UR, Wagner (1981) proposed a theory called a sometimes opponent process (SOP). The theory proposes that presentation of the US drives the US node into primary activity A1 that rapidly grows and decays back to zero, followed by secondary activity A2 that grows and decays at a slower rate. A CS paired with the US can trigger only activity A2 in the US node. For instance, rats receiving a brief footshock US develop a UR that consists of hyperactivity (A1) followed by freezing (A2). A CS paired with the footshock US elicits a CR that consists of freezing. In contrast, rabbits receiving a corneal airpuff US develop a UR that consists of an eyeblink (A1 and A2). A CS paired with the airpuff US elicits an eyeblink CR. According to the SOP model, excitatory CS–US associations increase whenever the CS and the US node are in the A1 state. Conversely, inhibitory CS–US associations increase whenever the CS node is in state A1 and the US node is in state A2. In a conditioned inhibition paradigm, CS_1–US presentations will cause the US node to assume state A2 during CS_1–CS_2 presentations. Because during CS_1–CS_2 presentations node CS_2 is in state A1 and the US node in state A2, inhibitory CS_2-US associations are formed. In sum, SOP is able to predict when the CR will resemble and when it will differ from the UR, and when excitatory or inhibitory conditioning is obtained.

Grossberg and Schmajuk (1989) presented a real-time neural network model, READ (REcurrent Associative Dipole) circuit, that combines a mechanism of associative learning with an opponent-processing circuit. CSs form excitatory associations with direct activations of a dipole and inhibitory associations with the antagonistic rebounds of a previously habituated dipole. Conditioning can be actively extinguished, and associative saturation prevented, by a process called opponent extinction, even if no passive memory decay occurs. Opponent extinction exploits a functional dissociation between read-in and read-out of associative memory, which may be achieved by locating the associative mechanism at specific locations in the neuronal dendrites. The READ architecture is able to explain conditioning and extinction of a conditioned excitor, second-order conditioning, conditioning and nonextinction of a conditioned inhibitor, and properties of conditioned inhibition as a "slave" process.

3 Cognitive mapping

As mentioned, Sokolov (1960) suggested that the brain constructs a neural model of environmental events. When there is coincidence between external events and the predictions generated by the brain model, animals respond without changing their internal model of the environment. However, when the external inputs do not coincide with the previously established internal model, an orienting response appears and the neural model is modified. Figure 3.1 shows a block diagram of a model that incorporates Sokolov's ideas of a model of the environment and a mismatch or novelty system.

Sokolov's (1960) view that the internal model of the environment is updated whenever observed events differ from predicted events is well captured by the "delta" rule (see Boxes 1.1 and 2.1). Because according to the "simple" delta rule, CS_i–US associations are changed until $f(\text{US}) = \text{US} - \Sigma_j V_{j,\text{US}} \text{CS}_j$ is zero, earlier, more reliable, or more salient CSs prevent later, less reliable, or less salient CSs from becoming associated with the US.

Sutton and Barto (1981b) presented a temporally refined version of the Rescorla–Wagner model. In the model, the effectiveness of CS_i is given by the temporal trace $\tau_i = f[\text{CS}_i(t)] = Af[\text{CS}_i(t)] + B\text{CS}_i(\text{t})$, which does not change over trials. The effectiveness of the US changes over trials according to $f[\text{US}(t)] = [y(t) - y'(t)]$, where the output of the model is $y(t) = f[\Sigma_j V_{j,\text{US}} f(\text{CS}_j) + f(\text{US})]$, $f(\text{US})$ is the temporal trace of the US, and $y'(t) = Cy'(t) - (1 - C)\, y(t)$. Computer simulations show that the model correctly describes acquisition, extinction, conditioned inhibition, blocking, overshadowing, primacy effects, and second-order conditioning. More recently, Barto and Sutton (1990) proposed a new rendering of the Sutton and Barto (1981b) model, designated the temporal difference model, in which $f(\text{US}) = [\text{US} + \gamma y'(t + 1) - y'(t)]$. The temporal difference model correctly describes ISI effects, serial-compound conditioning, no extinction of conditioned inhibition, second-order conditioning, and primacy effects.

Klopf (1988) introduced an interesting extension of the Sutton–Barto model, termed a drive-reinforcement model of single neuron function. According to this neuronal model, changes in pre- and postsynaptic activity levels are correlated to determine changes in synaptic efficacy. Changes in presynaptic signals, $p(\text{CS}_i) = f(\Delta\text{CS}_i)$, are correlated with changes in postsynaptic signals, $p(\text{US}) = \Delta y$. Changes in the efficacy of a synapse are also proportional to the current efficacy of the synapse. The model can describe delay and trace conditioning, conditioned and unconditioned stimulus duration and amplitude effects, partial reinforcement effects, intertrial interval effects (ITI), second-order conditioning, conditioned inhibition, extinction, reacquisition effects, backward conditioning, blocking, overshadowing, compound conditioning, and discriminative stimulus effects.

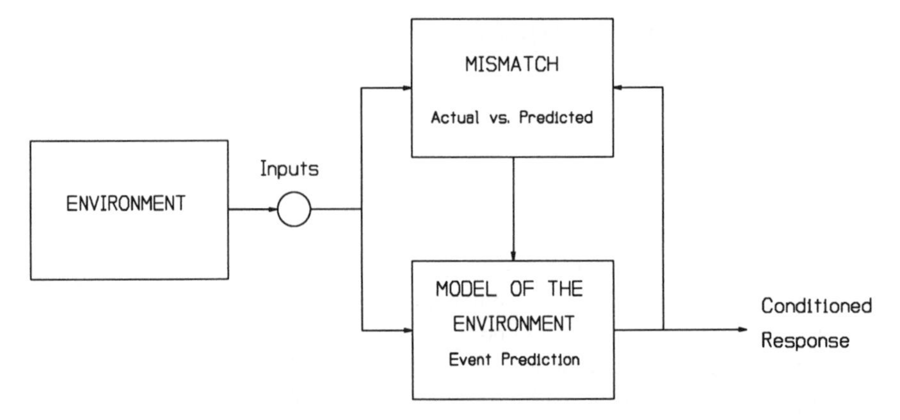

Figure 3.1. A block diagram of a simple model of the environment. The model of the environment generates predictions of future events and controls the conditioned response. The mismatch system compares observed and predicted events to compute novelty. Novelty is then used to modify the model of the environment.

A real-time model of classical conditioning

Rescorla and Wagner (1972) showed that their model correctly describes paradigms such as acquisition, extinction, conditioned inhibition, blocking, and overshadowing. Their model, however, cannot describe other paradigms, such as trace conditioning, CS and US duration effects, and ITI effects, in which time is an important independent variable. The circuit represented in Figure 3.2 can be regarded as a real-time neural network rendition of the Rescorla and Wagner (1972) model. Figure 3.2 shows that in neural network terms, *f*(US) can be construed as the modulation of the US signal that activates the postsynaptic neural population involved in associative learning.

In contrast to models of classical conditioning that describe behavior on a trial-to-trial basis (e.g., Frey and Sears, 1978; Mackintosh, 1975; Pearce and Hall, 1980; Rescorla and Wagner, 1972; Wagner, 1978), the real-time network depicted in Figure 3.2 describes behavior as a moment-to-moment phenomenon (see Box 1.4). Box 3.1 summarizes the main properties of the network. A formal description of the model is presented in Appendix 3.A.

The network incorporates two types of memory: (1) a trace short-term memory (STM) of CS_i and (2) CS_i–US associative long-term memories (LTM).

Trace STM

Trace STM $\tau_i(t)$ increases over time to a maximum following CS_i presentation and then decays back to zero (see Grossberg, 1975; Hull, 1943). By simultaneously reaching a critical locus of learning, $\tau_i(t)$ allows CS_i to build associations with the US even when they are separated by a temporal gap, such as in the case of trace conditioning. Figure 3.3 illustrates the time courses of the CS_i, τ_i, and US in a delay conditioning experiment.

CS–US associative LTM

Associative LTM, $V_{i,US}$ represents the past experience that CS_i is followed by the US. Changes in $V_{i,US}$ are given by $d(V_{i,US})/dt \sim \tau_i(\lambda_{US} - B_{US})$, where λ_{US} rep-

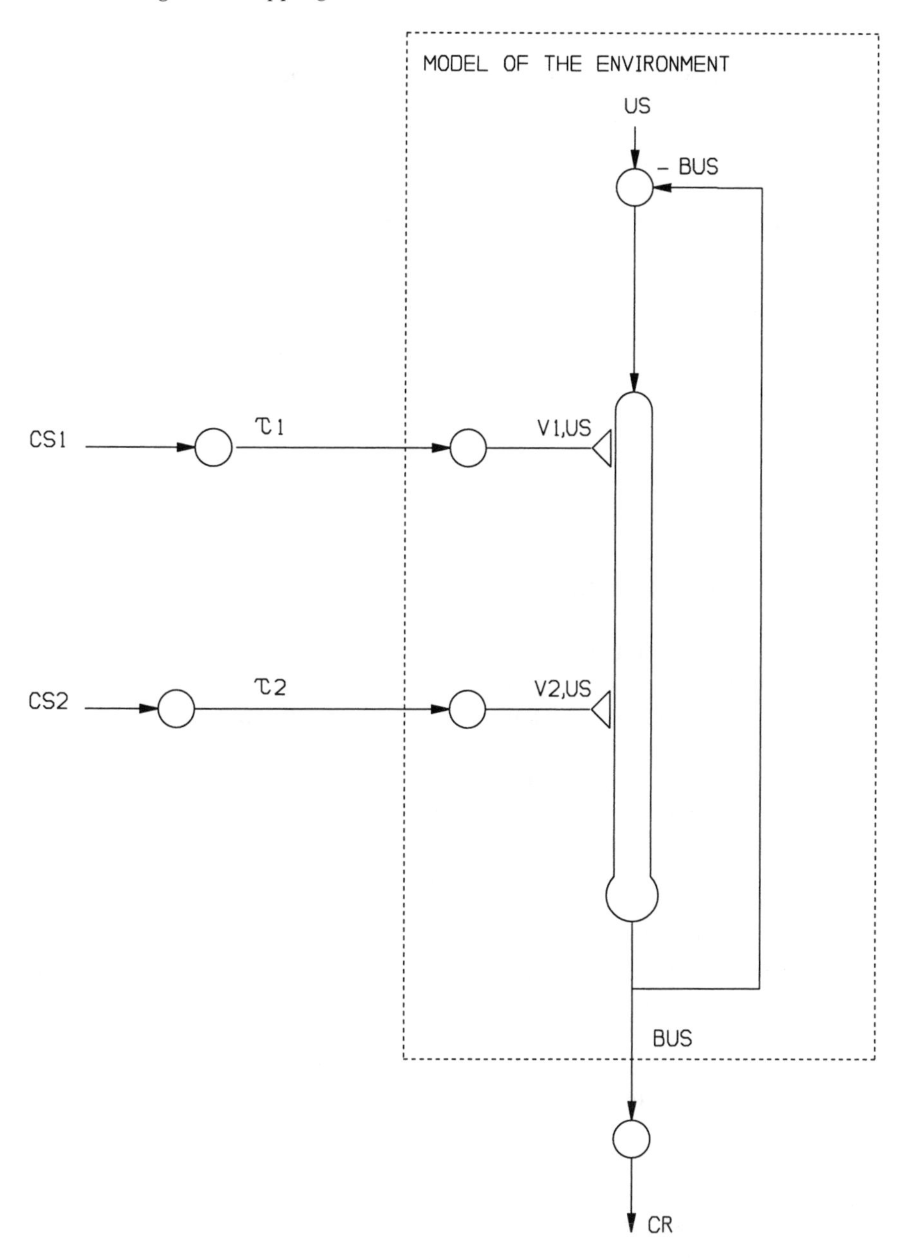

Figure 3.2. Neural implementation of the delta rule. CS_i: conditioned stimulus; τ_i: trace STM of CS_i; $V_{i,US}$: CS_i–US association; B_{US}: aggregate prediction of the US; CR: conditioned response. Arrows represent fixed synapses, open triangles variable synapses.

resents the US intensity. $V_{i,US}$ increases when $\tau_i(t) > 0$ and the observed intensity of the US exceeds the predicted intensity, that is, $\lambda_{US} > B_{US}$. $V_{i,US}$ decreases when $\tau_i(t) > 0$ and the predicted intensity of the US exceeds the observed intensity, that is, $\lambda_{US} < B_{US}$. Note that, in a real-time network, periods of acquisition and extinction occur within the same "acquisition" trial. Asymptotic learning is reached when the amounts of acquisition and extinction are similar within an acquisition trial.

Box 3.1 Real-time version of Rescorla and Wagner's (1972) model

The important aspects of the network are:

1. CS_i generates a STM trace, τ_i.
2. B_{US} is the aggregate prediction of the US by all CSs with $\tau_i(t)$ greater than zero.
3. Changes in CS_i–US associations, $V_{i,US}$, are proportional to the value of τ_i at the time of US presentation and controlled by a simple delta rule (see Boxes 1.1 and 2.1).
4. The CR is proportional to B_{US}.

Aggregate predictions and learning

B_{US} is the aggregate prediction of the US by all CSs with representations active at a given time, $\tau_i(t) > 0$, $B_{US} = \Sigma_i V_{i,US} \tau_i$.

Computer simulations

By applying the neural network described in Figure 3.2, this section contrasts the experimental results in classical conditioning with computer simulations of the following paradigms: (1) acquisition of trace conditioning, (2) extinction, (3) partial reinforcement, (4) overshadowing, (5) blocking, (6) conditioned inhibition, (7) discrimination acquisition, and (8) discrimination reversal. Appendix 3.B shows the parameters used in the simulations.

Acquisition and extinction of delay conditioning

Figure 3.4 shows real-time simulations of acquisition and extinction of delay conditioning with a 200-msec CS, 50-msec US, and 150-msec ISI. As CS–US and CX–US associations, $V_{i,US}$, increase over 50 acquisition trials, $d(V_{i,US})/dt \sim \tau_i(\lambda_{US} - B_{US})$, so does CR amplitude. During the following 50 extinction trials, CS–US associations decrease, $d(V_{i,US})/dt \sim \tau_i(0 - B_{US})$, and CX–US associations become slightly inhibitory (see later section on conditioned inhibition).

ISI curve

Figure 3.5 shows an ISI curve representing peak CR amplitude after 50 conditioning trials with a 200-msec CS, 10-msec US, and 0-, 100-, 200-, 300-, and 400-msec ISIs. In simultaneous conditioning the ISI is 0 msec, in delay conditioning the ISI is less than or equal to 200 msec, and in trace conditioning the ISI is greater than 200 msec.

Simulated results resemble the "inverted-U" curve obtained by Smith (1968) and Schneiderman and Gormezano (1964) with a peak CR at the 100-msec ISI. In agreement with experimental data, peak CR amplitude is zero with simultaneous conditioning, increases with delay conditioning, and decreases with trace conditioning. In terms of the model, larger peak CR amplitudes are obtained at ISIs that correspond to larger values of τ_i (see Equation 3.3 in Appendix 3.A). Therefore, as noted by Gormezano, Kehoe, and Marshall (1983), the ISI curve approximately reflects the shape of τ_i, which has a small initial value, increases

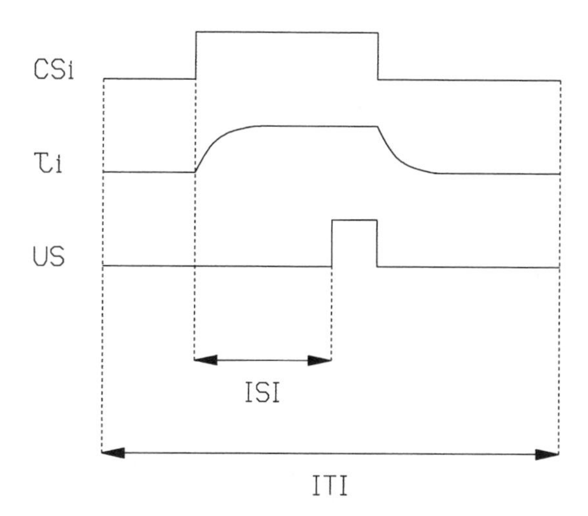

Figure 3.3. Time courses of the CS_i, τ_i, and the US in a delay conditioning experiment. ISI: interstimulus interval; ITI: intertrial interval.

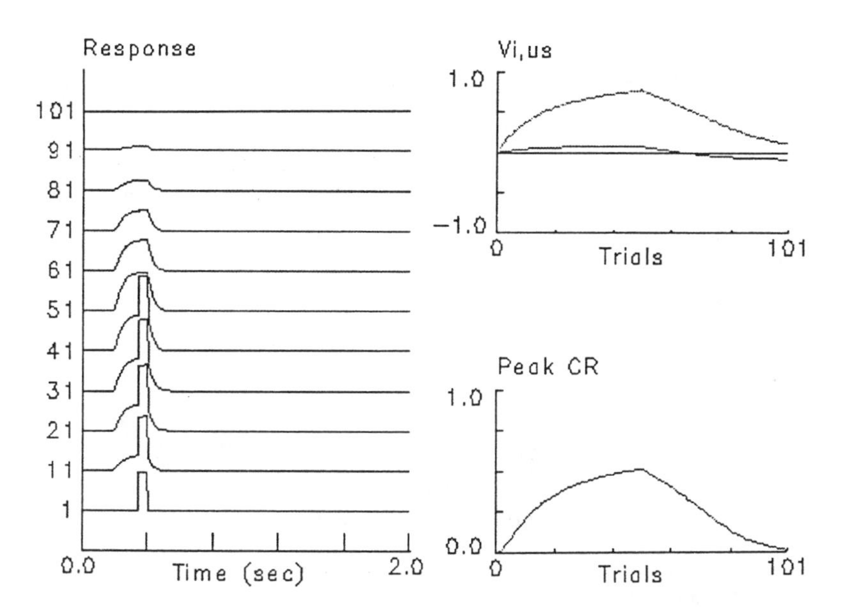

Figure 3.4. Acquisition and extinction of classical conditioning. *Left panels.* Real-time simulated conditioned and unconditioned responses for a CS presented between 200 and 400 msec and a US presented between 350 and 400 msec. Trial 1 is presented at the bottom of the panel. Trials 1–50 are reinforced and 51–101 are not reinforced. *Right panels.* Peak CR: peak CR as a function of trials; output weights: $V_{i,\mathrm{US}}$ as a function of trials.

over time to a maximum following CS_i presentation, and then decays back to zero (see Figure 3.3).

Partial reinforcement

Figure 3.6 shows peak CR amplitude after 100 trials with different proportions (1/1, 1/2, 1/3, 1/4, 1/5, and 1/8) of CS(A)–US and CS(A) alone trials. In agreement with data on partial reinforcement (see Gormezano and Moore, 1969, for a

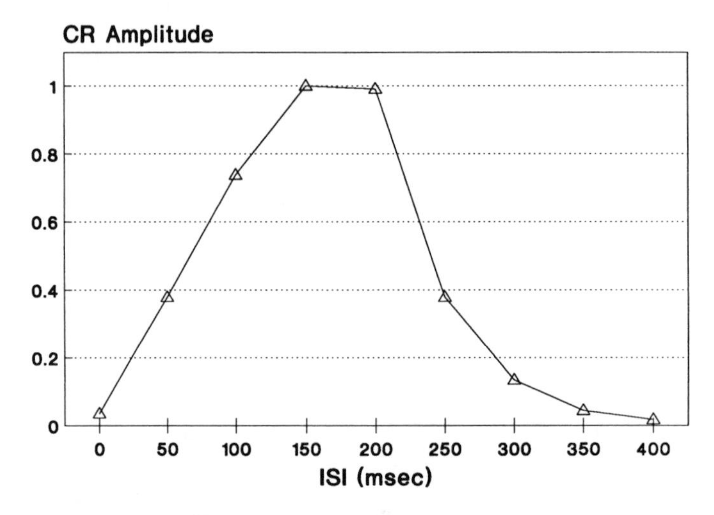

Figure 3.5. ISI curves. Peak CR amplitude after 50 conditioning trials with a 200-msec CS, 50-msec US, and 0-, 100-, 200-, 300-, and 400-msec ISIs.

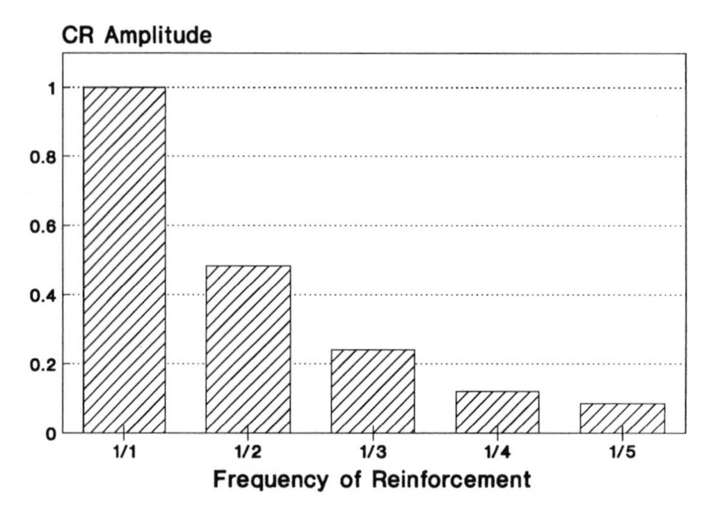

Figure 3.6. Partial reinforcement. Peak CR amplitude after 100 trials with different proportions (1/1, 1/2, 1/3, 1/4, and 1/5) of CS(A)–US and CS(A) alone trials.

review), increasing conditioning is obtained with increasing percentages of reinforced trials.

In terms of the model, because CS–US associations increase during reinforced trials, $d(V_{i,\mathrm{US}})/dt \sim \tau_i(\lambda_{\mathrm{US}} - B_{\mathrm{US}})$, and decrease during nonreinforced trials, $d(V_{i,\mathrm{US}})/dt \sim \tau_i(0 - B_{\mathrm{US}})$, conditioning increases with increasing percentages of reinforced trials.

Overshadowing and blocking

Figure 3.7 shows peak CR amplitude evoked by CS_2 in acquisition, overshadowing, and blocking. Acquisition consisted of CS_2 presentations paired together with the US during 20 trials. Overshadowing consisted of CS_1 and CS_2 presen-

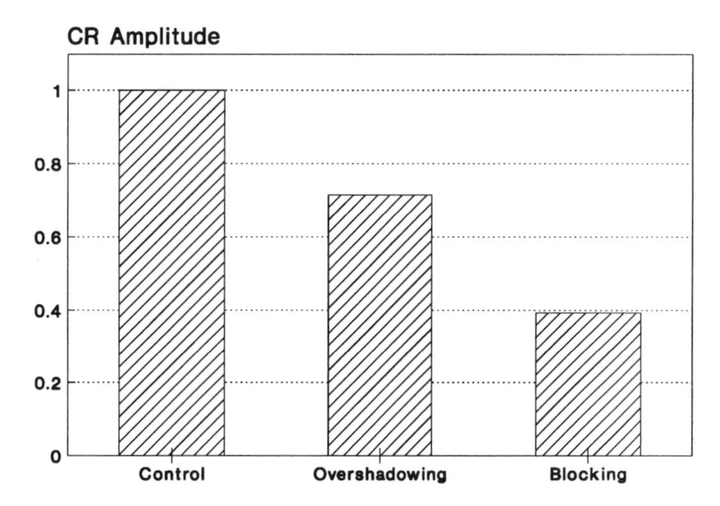

Figure 3.7. Blocking and overshadowing. Peak CR amplitude evoked by CS_2 after 20 reinforced CS_1–CS_2 trials following 20 reinforced CS_1 trials in the case of blocking and 20 CS_1–CS_2 reinforced trials in the case of overshadowing. Peak CR amplitude is expressed as a percentage of the peak CR to CS_2 after 20 reinforced CS_{21} trials.

tations paired together with the US during 20 trials. Blocking consisted of 20 CS_1–CS_2 acquisition trials following 20 CS_1 acquisition trials. Figure 3.7 shows that the model exhibits overshadowing because the CR for CS_2 is smaller in the overshadowing condition than in simple acquisition. In addition, Figure 3.7 indicates that the model exhibits blocking because the CR for CS_2 is smaller in the blocking than in the overshadowing case.

In terms of the network, both overshadowing and blocking are the consequence of the competition between stimuli to gain association with the US. During overshadowing, the presence of two CSs causes the aggregate prediction of the US, $B_{US} = \Sigma_j V_{j,US}\tau_j$, to grow faster than during simple conditioning. As B_{US} grows faster, error $(\lambda_{US} - B_{US})$ approaches zero faster, and $V_{1,US}$ and $V_{2,US}$ reach lower asymptotic values. During blocking, the aggregate prediction of the US, $B_{US} = \Sigma_j V_{j,US}\tau_j$, approaches the asymptotic level λ_{US}, and error $(\lambda_{US} - B_{US})$ approaches zero during CS_1 reinforced trials. During CS_1–CS_2 reinforced trials, little additional learning occurs as error $(\lambda_{US} - B_{US})$ is close to zero. Importantly, because CS_2–US associations depend on the value of the error at the beginning of the CS_1–CS_2 phase, the model can describe blocking in one trial.

Conditioned inhibition

Figure 3.8 shows peak CR amplitude evoked by CS_1 and CS_1–CS_2 after 50 reinforced CS_1 trials alternated with 50 nonreinforced CS_1–CS_2 trials. Peak CR amplitude is expressed as a percentage of the peak CR to CS_1. In agreement with experimental data, responding to CS_1 is stronger than responding to the CS_1–CS_2 compound.

In terms of the model, $V_{1,US}$ increases during reinforced CS_1 trials and decreases on nonreinforced CS_1–CS_2 trials. During nonreinforced CS_1–CS_2 trials, $\lambda_{US} = 0$, $V_{1,US} > 0$, and therefore error $(0 - B_{US}) < 0$. In consequence, $V_{2,US}$ becomes negative (inhibitory) until the aggregate prediction of the US, $B_{US} = \Sigma_j V_{j,US}\tau_j$, and error $(0 - B_{US})$ become zero.

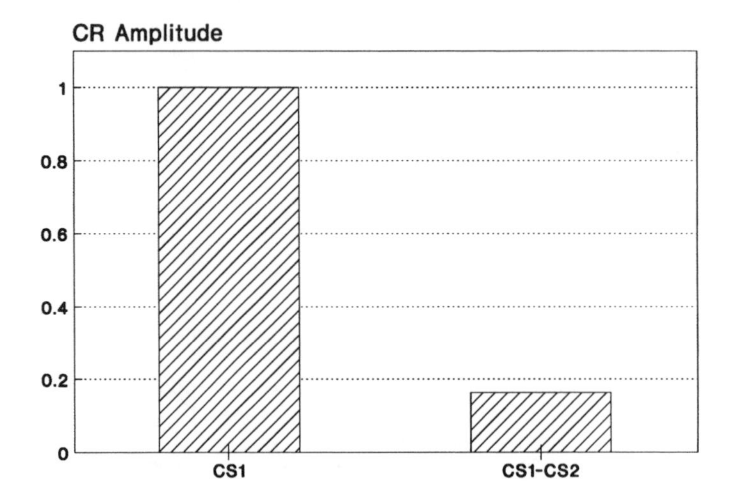

Figure 3.8. Conditioned inhibition. Peak CR amplitude evoked by CS_1 and CS_1–CS_2 after 50 reinforced CS_1 trials alternated with 50 nonreinforced CS_1–CS_2 trials. Peak CR amplitude is expressed as a percentage of the peak CR to CS_1.

Importantly, because B_{US} cannot become negative, the model correctly predicts that presentations of CS_2 alone do not produce extinction of the inhibitory $V_{2,US}$. When CS_2 is presented alone, the prediction of the US is computed as $B_{US} = -V_{2,US}\tau_2$, and therefore B_{US} and error $(0 - B_{US})$ equal zero. In the absence of error, no change is produced in the value of $-V_{2,US}$, $d(V_{2,US})/dt \sim \tau_2(0 - 0)$. However, inhibitory association $-V_{2,US}$ can become zero or even excitatory by pairing CS_2 with the US, $d(V_{i,US})/dt \sim \tau_i (\lambda_{US} - 0)$.

Discrimination acquisition and reversal

Figure 3.9 shows simulations of discrimination acquisition and extinction. During discrimination acquisition, 30 reinforced CS_1–US trials alternated with 30 nonreinforced CS_2 trials. Figure 3.9 also shows simulations of the CR amplitude elicited by CS_1 in a discrimination reversal paradigm. During reversal, the original nonreinforced CS_2 is reinforced for 30 trials and these trials alternate with 30 nonreinforced CS_1 trials.

Generalization

Generalization is measured, for instance, by the responses elicited by tones of different frequencies following conditioning to a tone of a given frequency. Blough (1975) proposed that associations between the different tone components and the US are regulated by a Rescorla–Wagner (delta) rule. By incorporating Blough's (1975) ideas, the present network is able to describe generalization. Blough proposed that training to a tone of given frequency generalizes to tones of other frequencies according to a generalization factor γ, $0 \leq \gamma \leq 1$. Blough assumed that γ increases with stimulus similarity and, therefore, γ is proportional to the ordinate of a Gaussian density distribution centered on the stimulus being presented. Generalization was simulated by first pairing a CS (representing a tone of a given training frequency) together with 10 CSs (representing tones of frequencies separated $\pm$ 1, $\pm$ 2, $\pm$ 3, $\pm$ 4, and $\pm$ 5 arbitrary steps from

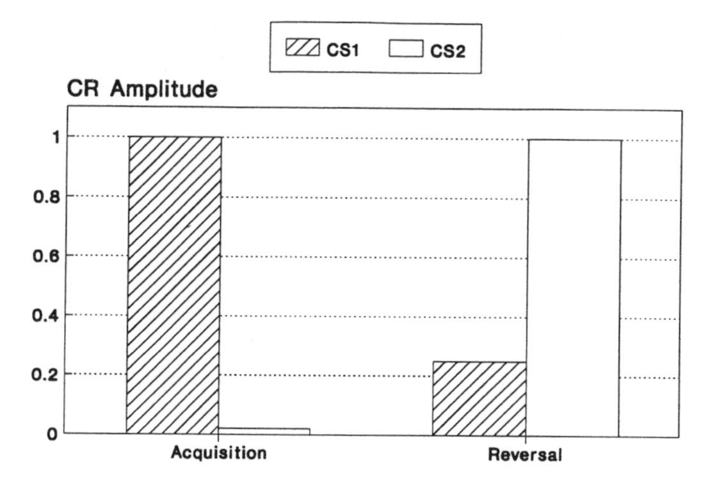

Figure 3.9. Discrimination. *Discrimination acquisition:* peak CR amplitude evoked by CS_1 and CS_2 after 30 reinforced CS_1 trials alternated with 30 nonreinforced CS_2 trials. Peak CR amplitude is expressed as a percentage of the peak CR to CS_1. *Discrimination reversal:* peak CR amplitude evoked by CS_1 after 30 nonreinforced CS_1 trials alternated with 30 reinforced CS_2 trials. Peak CR amplitude is expressed as a percentage of the peak CR to CS_2.

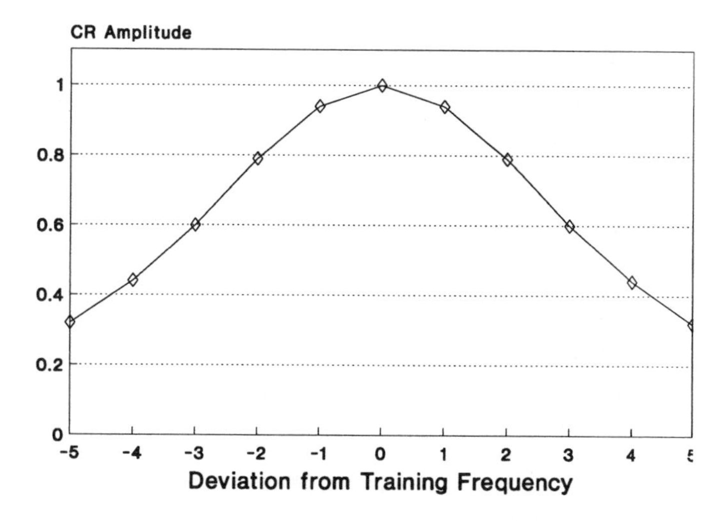

Figure 3.10. Generalization. Simulated peak CR amplitude cases generated by five simulated tones with frequencies that deviate from the training frequency by ±1, ±2, ±3, ±4, and ±5 arbitrary steps. In each case, CR amplitude is expressed as a percentage of the peak CR to the original tone.

the training frequency) with the US for 10 trials. Generalization curves were generated by simulating the CR elicited by a CS (representing a tone of a testing frequency) together with 10 CSs (representing tones of frequencies separated ± 1, ± 2, ± 3, ± 4, and ± 5 arbitrary steps from the testing frequency).

Figure 3.10 shows a generalization curve in which CR amplitude has been normalized to the value of the CR amplitude elicited by the training frequency. In agreement with experimental data (e.g., Solomon and Moore, 1975), simulated animals respond maximally to the training frequency and gradually less with increasing differences between the training and testing frequencies.

Table 3.1. *Simulations obtained with the real-time version of the Rescorla–Wagner (1972) model and a simplified version of the Schmajuk and Moore (1988) cognitive map model.*

Paradigm	Real-Time Delta Rule	Cognitive Map
Delay conditioning	Y	Y
Trace conditioning	Y	Y
Partial reinforcement	Y	Y
Extinction	Y	Y
Acquisition series		
Extinction series		
Latent inhibition		
Blocking	Y	Y
Unblocking		
Overshadowing	Y	Y
Discrimination	Y	Y
Conditioned inhibition	Y	Y
Inhibition by differential conditioning	Y	Y
Inhibition by decrease in US	Y	Y
Counterconditioning	Y	Y
Superconditioning	Y	Y
Overprediction effects	Y	Y
Generalization	Y	Y
Compound conditioning		
Simultaneous feature-positive discrimination	Y	Y
Simultaneous feature-negative discrimination	Y	Y
Serial feature-positive discrimination		
Serial feature-negative discrimination		
Negative patterning		
Positive patterning		
Sensory preconditioning		Y
Second-order conditioning		Y
Context switching		
External inhibition		

Note. Y: The model correctly describes the experimental data.

Discussion

The simple model depicted in Figure 3.2 can describe (1) acquisition and extinction of delay conditioning, (2) acquisition and extinction of trace conditioning, (3) ISI curves, (4) partial reinforcement, (5) overshadowing, (6) blocking, (7) conditioned inhibition, (8) counterconditioning, (9) simultaneous feature-positive and feature-negative discriminations, (10) superconditioning, (11) overprediction effects, (12) discrimination, and (13) generalization (see Table 3.1).

Although a simple, real-time delta rule describes many paradigms with few parameters, more complexity is needed to explain other classical conditioning protocols. For example, the next section introduces an extension of the model that allows the characterization of second-order conditioning and sensory preconditioning. Chapter 5 shows that a "generalized" delta rule model is able to describe more complex paradigms, such as negative patterning and occasion setting. Finally, Chapter 6 introduces a neural network of timing that, by introducing multiple CS-activated traces, generates better ISI curves than the present model.

A real-time model of cognitive mapping

Although the model presented in the previous section is able to portray CS–US associations, this is not sufficient for a complete description of environmental

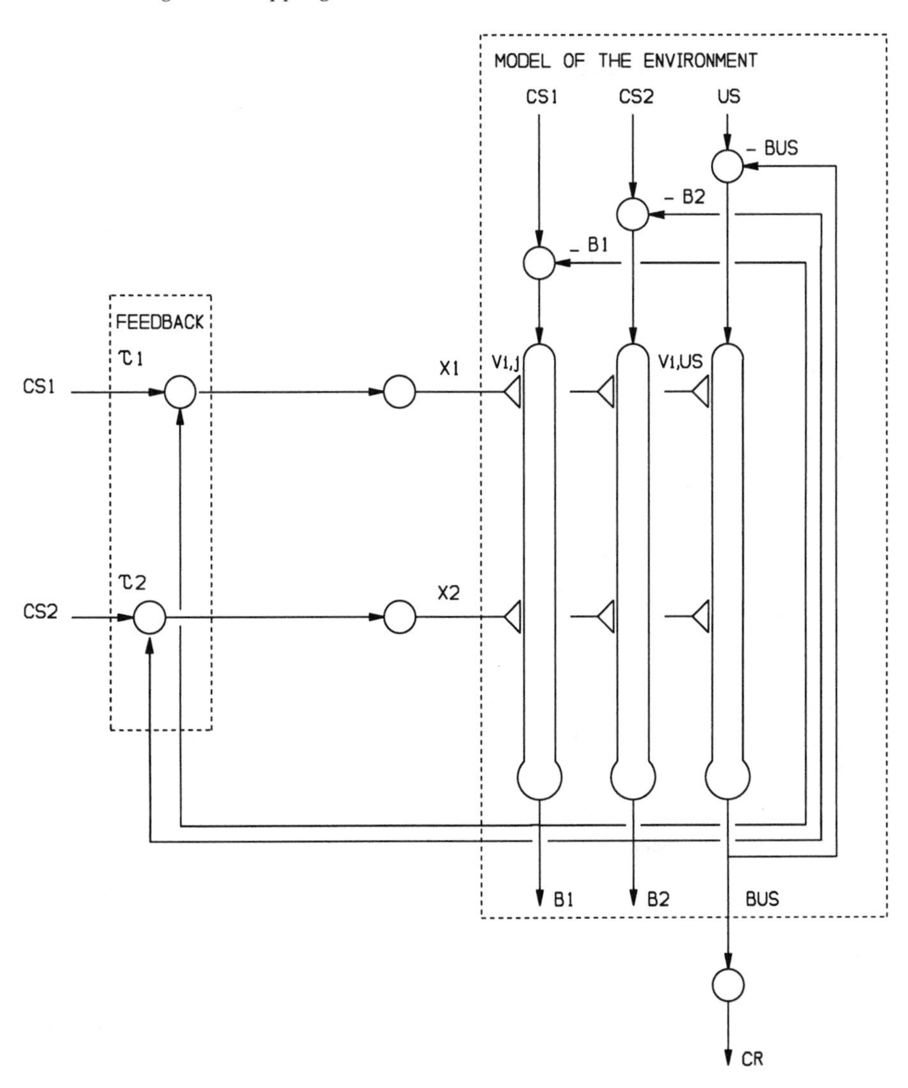

Figure 3.11. A neural network for cognitive mapping. CS_i: conditioned stimulus; τ_i: trace STM of CS_i; X_i: internal representation of CS_i; $V_{i,US}$: CS_i–US association; $V_{i,j}$: CS_i–CS_j association; B_k: aggregate prediction of event k; B_{US}: aggregate prediction of the US; CR: conditioned response. Arrows represent fixed synapses, open triangles variable synapses.

events. Therefore, this section introduces a model that, by combining CS–CS and CS–US associations, is able to describe second-order conditioning and sensory preconditioning. This integration of the predictions generated through CS–CS and CS–US associations into a new prediction is similar to the process that Tolman (1932a) called inference. This view is compatible with Dickinson's (1980) notion that classical conditioning associations are declarative representations of knowledge that can be combined in a cognitive map (see Box 2.2).

Figure 3.11 shows a diagram of a network that describes second-order conditioning and sensory preconditioning. The theory is a rendition of a model of classical conditioning described by Schmajuk and Moore (1988) and Schmajuk (1986, 1989). The network is an important part of the more powerful network introduced in Chapter 5. Box 3.2 summarizes the main properties of the network. A formal description of the model is presented in Appendix 3.C.

Box 3.2 Simplified version of Schmajuk and Moore's (1988) cognitive map model

The important aspects of the network are:

1. CS_i generates a STM trace, τ_i.
2. B_k is the aggregate prediction of event k by all CSs with $\tau_i(t)$ greater than zero.
3. B_{US} is the aggregate prediction of the US by all CSs with $\tau_i(t)$ greater than zero.
4. Changes in CS_i–CS_k associations, $V_{i,k}$, are proportional to the value of τ_i at the time of the presentation of event k and controlled by a simple delta rule (see Boxes 1.1 and 2.1).
5. Changes in CS_k-US associations, $V_{k,US}$, are proportional to the value of τ_k at the time of US presentation and controlled by a simple delta rule (see Boxes 1.1 and 2.1).
6. The network is a recurrent autoassociative network (see Box 1.3). This network makes it possible to combine different associative CS_i–CS_k and CS_k–US values into a cognitive map (see Box 2.2).
7. The CR is proportional to the aggregate prediction B_{US}.

In addition to (1) a trace short-term memory (STM) of CS_i and (2) associative CS_i–US LTMs, used in the preceding model, the network incorporates (3) associative CS_i–CS_k LTMs.

CS–CS associative LTM

Associative LTM $V_{i,k}$ represents the past experience that CS_i is followed by CS_k. Changes in $V_{i,k}$ are given by $d(V_{i,k})/dt \sim \tau_i(\lambda_k - B_k)$, where λ_k represents the observed intensity of CS_k and B_k, its predicted intensity. $V_{i,k}$ increases when $\tau_i(t) > 0$ and the observed intensity of the CS_k exceeds the predicted intensity, that is, $\lambda_k > B_k$. $V_{i,k}$ decreases when $\tau_i(t) > 0$ and the predicted intensity of CS_k exceeds the observed intensity, that is, $\lambda_k < B_k$.

Aggregate predictions and learning

B_k is the aggregate prediction of the CS_k by all CSs with representations active at a given time, $\tau_i(t) > 0$, $B_k = \Sigma_j V_{j,k} \tau_j$.

Cognitive mapping

The network storing CS–CS and CS–US associations in Figure 3.11 is a recurrent autoassociative network (see Box 1.3). This network makes it possible to combine different associative values to generate predictions of an event that never temporally overlapped with the internal representation of CS_i. This overlap might be absent because (1) the length of the CS_i–CS_k interstimulus interval exceeds the duration of the trace STM, $\tau_i(t)$, as in some cases of serial compound conditioning, or (2) CS_i and CS_k were never presented together, as in the case of second-order conditioning or sensory preconditioning. Sensory preconditioning is a simple example of inference (see Bower and Hilgard, 1981,

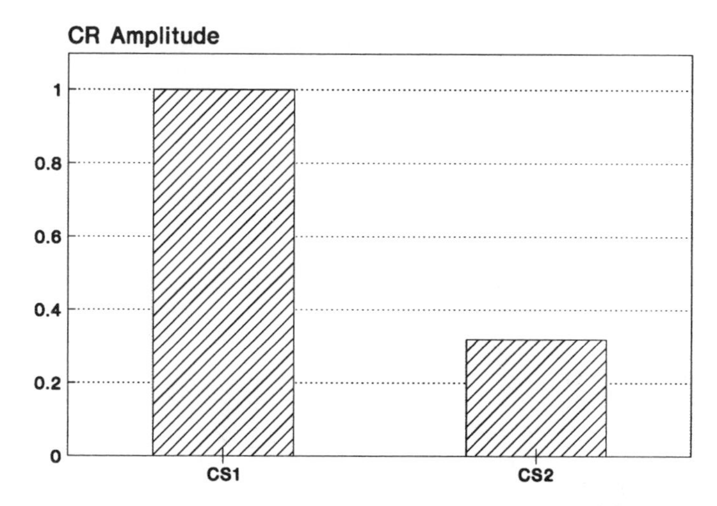

Figure 3.12. Second-order conditioning. Peak CR amplitude evoked by CS_2 after (1) 100 reinforced CS_1 trials and (2) 5 CS_1–CS_2 nonreinforced trials.

p. 330). Tolman (1932a) hypothesized that a large number of associative values can be combined into a *cognitive map* (see Box 2.2).

Computer simulations

By applying the neural network described in Figure 3.11, this section compares experimental results regarding second-order conditioning and sensory preconditioning with computer simulations. Appendix 3.D shows the parameters used in the simulations.

Second-order conditioning

Figure 3.12 shows peak CR amplitude evoked by CS_2 after (1) 100 reinforced CS_1 trials and (2) 5 CS_1–CS_2 nonreinforced trials. In agreement with experimental data, CS_2 is able to generate a CR after being paired with the previously reinforced CS_1. Also in agreement with experimental data, as the number of CS_1–CS_2 nonreinforced trials increases, conditioned inhibition takes precedence over second-order conditioning and responding to CS_2 decreases.

In terms of the network, second-order conditioning is described as follows. During CS_2 reinforced trials, $V_{2,US}$ associations grow, and $V_{1,2}$ associations increase during CS_1–CS_2 trials. When CS_1 (never presented with the US before) is presented by itself on test trial, it activates the prediction of CS_2, B_2, which in turn activates the$V_{2,US}$ association, thereby generating a CR.

The model description of second-order conditioning as the combination of CS_1–CS_2 and CS_2–US_2 associations agrees with Rashotte, Griffin, and Sisk's (1977) and Rescorla's (1978) finding that extinction of the CS_2–US association leads to substantial reduction in responding to CS_1. Opposite results, however, were obtained by Rizley and Rescorla (1972) and Holland and Rescorla (1975).

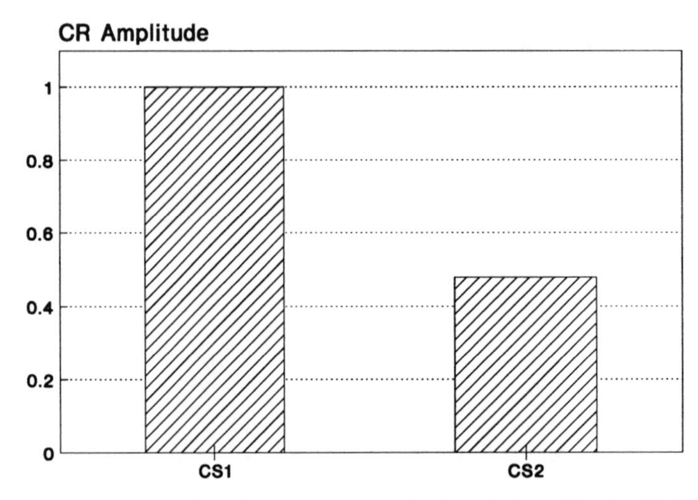

Figure 3.13. Sensory preconditioning. Peak CR amplitude evoked by CS_1 after (1) 10 nonreinforced CS_1–CS_2 trials and (2) CS_2 100 reinforced trials.

The model also correctly predicts Ross and Holland's (1981) results showing that CS_1 presentations lead to the extinction of CS_2–US associations.

Sensory preconditioning

Figure 3.13 shows peak CR amplitude evoked by CS_1 after (1) 10 nonreinforced CS_1–CS_2 trials and (2) 100 CS_2 reinforced trials. The results agree with Brogden's (1939) experimental results.

In terms of the network, sensory preconditioning is described as follows. During CS_1–CS_2 trials, $V_{1,2}$ associations increase, and $V_{2,US}$ associations grow during CS_2 reinforced trials. When CS_1 (never presented with the US before) is presented by itself on test trial, it activates the prediction of CS_2, B_2, which in turn activates the $V_{2,US}$ association, thereby generating a CR.

Discussion

The present section introduces a neural network that, like the previous one, describes (1) acquisition and extinction of delay conditioning, (2) acquisition and extinction of trace conditioning, (3) ISI curves, (4) partial reinforcement, (5) overshadowing, (6) blocking, (7) conditioned inhibition, (8) counterconditioning, (9) simultaneous feature-positive and feature-negative discriminations, (10) superconditioning, (11) overprediction effects, (12) discrimination, and (13) generalization. In addition, the present model describes (14) second-order conditioning and (15) sensory preconditioning (see Table 3.1).

Chapter 4 shows that most features of latent inhibition can be described by a network that is a simple extension of that represented in Figure 3.11. Similarly, Chapter 11 shows that latent learning and detour tasks in complex mazes can be described by a neural cognitive map organized under the same principles described in the present chapter.

Appendix 3.A

A formal description of the model

This section formally describes the model depicted in Figure 3.2.

Short-term memory trace of the CS

CS_i generates a STM trace τ_i according to

$$d(\tau_i)/dt = K_1(CS_i - \tau_i) \quad (3.1)$$

where K_1 represents the rate of increase and decay of τ_i.

Aggregate predictions

The *aggregate prediction* of the US by all CSs with τ_i's active at a given time, B_{US}, is given by

$$B_{US} = \Sigma_i V_{i,US}\tau_i \quad (3.2)$$

where $V_{i,US}$ represents the association of τ_i with the US.

Long-term memory CS–US associations

Changes in $V_{i,US}$, are given by

$$d(V_{i,US})/dt = K_2\, \tau_i(\lambda_{US} - B_{US})\,(1 - |V_{i,US}|) \quad (3.3)$$

where τ_i is the STM of CS_i, λ_{US} the intensity of the US, B_{US} the aggregate prediction of the US by all CSs active at a given time. By Equation 3.3, $V_{i,US}$ increases whenever τ_i is active and $\lambda_{US} > B_{US}$ and decreases when $\lambda_{US} < B_{US}$. In order to prevent the extinction of conditioned inhibition or the generation of an excitatory CS by presenting a neutral CS with an inhibitory CS, we assume that when $B_{US} < 0$, then $B_{US} = 0$.

Performance rules

The amplitude of the CR is given by

$$CR = B'_{US} \quad (3.4)$$

where B'_{US} is given by $B'_{US} = B^2_{US}/(K^2_3 + B^2_{US})$. According to Equation 3.4, the magnitude of the CR increases with increasing predictions of the US.

The amplitude of the total response is given by

$$R = CR + K_4\, US \quad (3.5)$$

Appendix 3.B

Learning with weights that assume only positive values

Because real synapses do not change from excitatory to inhibitory or vice versa (but see Alkon et al., 1992), connectionist models have been criticized for al-

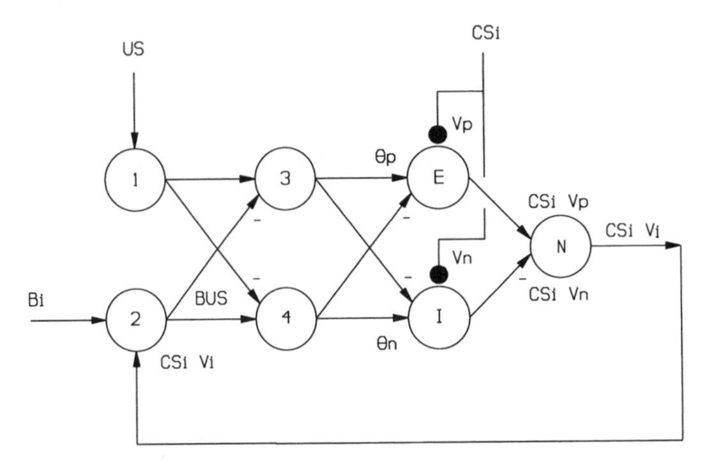

Figure 3.14. Neural implementation of excitatory and inhibitory associations using exclusively positive plastic connections. US: unconditioned stimulus; CS_i: conditioned stimulus; B_{US}: aggregate prediction of the US; B_i: aggregate prediction of the US by all CSs different from CS_i; V_i: net CS_i–US association; V_p: positive excitatory CS_i–US association; V_n: positive inhibitory CS_i–US association; θ_p: neural population active when US > B_{US}; θ_n: neural population active when US < B_{US}. Arrows represent fixed synapses, solid circles variable synapses.

lowing weights to change from positive and negative and, conversely, from negative to positive. Pearce and Hall (1980) proposed a mathematical model of classical conditioning that incorporates CS–US associations that are only positive. Pearce and Hall proposed that *positive excitatory* CS–US associations increase whenever CS and US are presented together, whereas *positive inhibitory* CS–US associations increase whenever the CS is presented in the absence of the US. The net CS–US association is obtained by subtracting the positive inhibitory from the *positive excitatory* association. In a connectionist context, the Pearce–Hall (P–H) scheme avoids the use of weights that change from positive to negative values and vice versa. However, because in the P–H model positive excitatory and inhibitory associations increase but never decrease, they may reach unrealistically high values.

Figure 3.14 shows a network that also employs *positive excitatory and inhibitory* associations, but these always positive associations can *increase and decrease.* In the network illustrated in Figure 3.14, node 1 receives input from the US. Node 2 receives one input proportional to $CS_i\ V_i$ from node *N* and another input proportional to B_i. B_i represents the aggregate prediction of the US, $B_i = \Sigma_{j \neq i} V_j CS_j$, based on all CSs other than CS_i. The aggregate prediction of the US, B_{US}, is given by $B_{US} = B_i + CS_i V_i$.

The output of node 3 is proportional to $f\,[US - B_{US}]$, where $f[x] = (US - B_{US})$ if $x > 0$, and $f\,[x] = 0$ if $x \leq 0$. The output of node 4 is proportional to $f[B_{US} - US]$, where $f\,[x] = (B_{US} - US)$ if $x > 0$, and $f\,[x] = 0$ if $x \leq 0$. Therefore, node 3 is active when the actual US is underpredicted $f\,[US > B_{US}]$ and node 4 is active when the actual US is overpredicted (B_{US} > US). At a given time, either node 3 or 4 is active, but both cannot be active at the same time. We call the output of node 3, $\theta_p = (US - B_{US})$, and the output of node 4, $\theta_n = B_{US} - US$. Notice that θ_p is similar to the error expression given in Equation 3.3.

Node E receives excitatory input from θ_p and inhibitory input from θ_n. Node I receives excitatory input from θ_n and inhibitory input from θ_p. The CS_i reaches both nodes E and I through modifiable synapses, with strengths V_p and V_n, respectively. V_p and V_n are always *positive* and vary between zero and 1. V_p increases whenever CS_i and θ_p are active, and decreases whenever CS_i and θ_n are active. V_n increases whenever CS_i and θ_n are active, and decreases whenever CS_i and θ_p are active. Node N computes the net output, CS_iV, by combining the output from node E, CS_iV_p, with the output from node N, CS_iV_n, according to $CS_i(V_p - V_n) = CS_iV_i$.

Similar results are obtained with the equations presented in Appendix 3.A and the equations describing the network illustrated in Figure 3.14. For simplicity and speed in the simulations, the present chapter made use of the equations presented in Appendix 3.A that assume that weights change from positive to negative and vice versa.

Appendix 3.C

Simulation parameters

In our computer simulations, each trial is divided into 200 time units. Each time unit represents approximately 10 msec. CSs of amplitude 1 are presented in general between 200 and 400 msec. A 50-msec US of intensity 1.0 is applied at 350 msec for all simulations with the exception of those for the ISI curve.

Parameters values used in all simulations are $K_1 = 0.1$, $K_2 = 0.05$, $K_3 = 0.15$, and $K_4 = 0.5$. Generalization factors are $\gamma_{\pm 1} = 0.9$, $\gamma_{\pm 2} = 0.6$, $\gamma_{\pm 3} = 0.3$, $\gamma_{\pm 4} = 0.15$, $\gamma_{\pm 5} = 0$, where subindices indicate the number of arbitrary steps separating different frequencies from the training frequency.

Appendix 3.D

A formal description of cognitive mapping

This section formally describes the model depicted in Figure 3.11.

Short-term memory trace of the CS

CS_i generates a STM trace τ_i according to

$$d(\tau_i)/dt = K_1(CS_i - \tau_i) \tag{3.6}$$

where K_1 represents the rate of increase and decay of τ_i.

Internal representation of the CS

The internal representation of CS_i, X_i, is given by

$$X_i = K_2(\tau_i + K_3B_i) \tag{3.7}$$

where K_3 represents a reinjection coefficient for the prediction of $CS_i(B_i)$ and K_2 a constant. By Equation 3.7, X_i is active either when (1) CS_i is present or (2) CS_i is predicted by other CSs. Increasing CS_i–CS_i associations also increase the magnitude of X_i.

Aggregate predictions

The *aggregate prediction* of event *k* by all CSs with representations active at a given time, B_k, is given by

$$B_k = \Sigma_i V_{i,k} X_i \tag{3.8}$$

where $V_{i,k}$ represents the association of X_i with CS_k.

The *aggregate prediction* of the US by all CSs with representations active at a given time, B_{US}, is given by

$$B_{US} = \Sigma_i V_{i,US} X_i \tag{3.9}$$

where $V_{i,US}$ represents the association of X_i with the US.

Long-term memory CS–US associations

Changes $V_{i,US}$ are given by

$$d(V_{i,US})/dt = K_4 X_i (\lambda_{US} - B_{US})\ (1 - |V_{i,US}|) \tag{3.10}$$

where X_i is the internal representation of CS_i, λ_{US} the intensity of the US, B_{US} the aggregate prediction of the US. By Equation 3.10, $V_{i,US}$ increases, with $K_4 = K'_4$, whenever X_i is active and $\lambda_{US} > B_{US}$, and decreases, with $K_4 = K''_4$, when $\lambda_{US} < B_{US}$. In order to prevent the extinction of conditioned inhibition or the generation of an excitatory CS by presenting a neutral CS with an inhibitory CS, we assume that when $B_{US} < 0$, then $B_{US} = 0$.

Long-term memory CS–CS associations

Changes $V_{i,j}$ are given by

$$d(V_{i,j})/dt = K_4 X_i (\lambda_j - B_j)\ (1 - |V_{i,US}|) \tag{3.11}$$

where X_i is the internal representation of CS_i, and B_j represents the aggregate prediction of event *j*. By Equation 3.11, $V_{i,j}$ increases, with $K_4 = K'_4$, whenever X_i is active and $\lambda_j > B_j$, and decreases, with $K_4 = K''_4$, when $\lambda_j < B_j$. When $B_j < 0$, then $B_j = 0$. $V_{i,i}$ represents the association of CS_i with itself.

Performance rules

The amplitude of the CR is given by

$$\text{CR} = B_{US} \tag{3.12}$$

According to Equation 3.12, the magnitude of the CR increases with increasing predictions of the US.

The amplitude of the total response is given by

$$\text{R} = \text{CR} + K_5\text{US} \tag{3.13}$$

Appendix 3.E

Simulation parameters

Parameters values used in all simulations are $K_1 = 0.1$, $K_2 = 0.2$, $K_3 = 2$, $K'_4 = 0.05$, $K''_4 = 0.01$, and $K_5 = 0.5$.

4 Attentional processes

Chapter 3 showed how Sokolov's (1960) view that the internal model of the environment is updated whenever observed events differ from predicted events is well captured by the "delta" rule (Rescorla and Wagner, 1972) that assumes CS_i–US associations are changed until the US effectiveness, $f(\text{US}) = \text{US} - \Sigma_j V_{j,\text{US}} \text{CS}_j$, is zero. In consequence, earlier, more reliable, or more salient CSs prevent later, less reliable, or less salient CSs from becoming associated with the US. The delta rule describes many classical paradigms, including (1) acquisition and extinction of delay conditioning, (2) acquisition and extinction of trace conditioning, (3) ISI curves, (4) partial reinforcement, (5) overshadowing, (6) blocking, (7) conditioned inhibition, (8) counterconditioning, (9) simultaneous feature-positive and feature-negative discriminations, (10) superconditioning, (11) overprediction effects, (12) discrimination, and (13) generalization (see Table 3.1). However, the delta rule is unable to describe latent inhibition (LI), a phenomenon by which the association of a CS with the US is retarded by preexposing the CS (Lubow and Moore, 1959).

The present chapter presents Grossberg's (1975) attentional neural network. In addition to the description of (1) acquisition of trace conditioning, (2) extinction, (3) partial reinforcement, (4) overshadowing, (5) blocking, (6) discrimination acquisition, (7) discrimination, but not conditioned inhibition, the network is able to describe (8) LI.

Storage and retrieval theories of latent inhibition

Theorizing about LI has taken two opposite views, one suggesting that CS preexposure disturbs the storage of CS–US associations, another proposing that it hinders the retrieval of CS–US associations (see Chapter 2). According to the storage position, CS preexposure disrupts the formation of CS–US associations during conditioning by either (1) decreasing the associability of the CS (e.g., Frey and Sears, 1978; Lubow, Weiner, and Schnur, 1981; Mackintosh, 1975; Moore and Stickney, 1980; Pearce and Hall, 1980; Schmajuk and DiCarlo, 1991a, b; Wagner, 1978), or (2) fostering the formation of CS–no consequence associations that interfere with the subsequent establishment of CS–US associations (Revusky, 1971; Testa and Ternes, 1977). According to the retrieval position, CS preexposure disrupts the subsequent retrieval of CS–US associations (e.g., Kasprow et al., 1984; Kraemer, Randall, and Carbary, 1991). That is, whereas the first approach suggests that LI can be explained in terms of a mechanism operating during memory storage, the second approach proposes that LI is the result of a mechanism operating during memory retrieval (Bouton, 1993; Spear, 1981; Spear, Miller, and Jagielo, 1990).

Modulation of storage of CS–US associations

Most theories of classical conditioning assume that temporal contiguity between CS_i and the US leads to the formation of CS_i–US associations. Different rules have been proposed to describe changes in CS_i–US associations. As mentioned in Chapter 2, these rules either assume variations in the effectiveness of the US or variations in the associability or effectiveness of CS_i.

Variations in the associability of the CS

As mentioned in Chapter 2, Mackintosh's (1975) attentional theory suggests that the associability of CS_i increases when it is the best predictor of the US and decreases otherwise. CS_i-preexposed animals show LI because both the CS_i and the context are equally poor predictors of the US. Moore and Stickney (1981) and Schmajuk and Moore (1989) generalized Mackintosh's approach to include the predictions of other CSs, and assumed that CS_i associability increases when CS_i is the best predictor of other CSs or the US, but decreases otherwise. In contrast to Mackintosh's (1975) view, Pearce and Hall (1980) suggested that CS_i associability increases when CS_i is a poor predictor of the US, that is, when CS_i has been recently followed by the unexpected presentation of the US. The model describes LI by assuming that during preexposure, the (initially large) associability of CS_i declines.

Schmajuk and DiCarlo (1991a, b) applied Grossberg's (1975) model to describe LI. They assumed that the initial value of incentive motivation decreases during CS preexposure, thereby retarding the acquisition of CS–US associations. Schmajuk and DiCarlo's (1991a, b) version of Grossberg's (1975) model is fully described in the present chapter.

Lubow, Weiner, and Schnur (1981; Lubow, 1989) presented a conditioned attention theory (CAT) of LI. According to CAT, attention to CS_i is a response R_i that occurs when CS_i is presented. The same laws of conditioning govern the acquisition of CR and R_i. The theory assumes that (1) R_i declines with repeated nonreinforced CS_i presentations and increases with reinforced CS_i presentations, (2) R_i may become conditioned to other CSs, and (3) R_i is correlated with CS_i associability. Levels of R_i that are higher than the level of R_i on the first presentation of CS_i represent *attention* to CS_i, whereas levels that are lower represent *inattention* to CS_i. Lubow (1989) showed that the model generates numerous predictions regarding (1) the conditioning of inattention to CS_i when CS_i is presented in isolation, and (2) the modulation of attention to CS_i when CS_i is presented with other CSs.

To the extent that LI involves CS preexposure, it can be succesfully described by models that depict variations in CS effectiveness. However, in order to describe arbitrary input–output functions in classical conditioning, rules that assume variations in the effectiveness of the US are necessary.

Variations in the effectiveness of both the CS and the US

As mentioned in Chapter 2, Wagner (1978) suggested that CS_i–US associations are determined by (1) the effectiveness of the US ($US - \Sigma_j V_{j,US} CS_j$), as in the Rescorla–Wagner model, and (2) the effectiveness (associability) of the CS_i ($CS_i - V_{CX,i}\ CX$), where CX represents the context and $V_{CX,i}$ the strength of the CX–CS_i association. CS preexposure causes the CS to be predicted by the context ($V_{CX,i}$ increases) and, therefore, CS_i associability decreases.

As mentioned in Chapter 2, Schmajuk and Moore (1988; Schmajuk, 1986) offered a real-time attentional-associative model of classical conditioning that incorporates CS–CS as well as CS–US associations. In the model, CS salience (assumed to modulate the rate of CS–CS and CS–US associations) is determined by CS novelty, that is, the absolute value of the difference between its predicted and observed amplitude, $|CS_i - \Sigma_j V_{j,i} CS_j|$. The model describes LI because $|CS_i - \Sigma_j V_{j,i} CS_j|$ and CS salience decrease during CS preexposure. McLaren, Kaye, and Mackintosh (1989) proposed a similar model that describes perceptual learning and LI, in which the formation of CS_i–CS_j and CS_i–US associations are modulated by the associability of CS_i ($CS_i - \Sigma_j V_{j,i} CS_j$). In a similar vein, Chapter 5 describes an extension of the cognitive map model introduced in Chapter 3, in which CS salience (assumed to modulate the rate of acquisition of CS–CS and CS–US associations) is determined by the total novelty, that is, the sum of the absolute values of the difference between the predicted and observed amplitudes of all environmental events (CSs and USs) detected at the time of the CS presentation. The model describes LI because novelty and CS salience decrease during CS preexposure.

As mentioned in Chapter 2, Gluck and Myers (1993) presented a computational theory of classical conditioning that describes LI in terms of changes in the CS internal representation stored in a hidden-unit layer. LI seems to reflect the fact that, during CS preexposure, weights in the cortical hidden layer decrease to the smaller value of the hippocampal hidden layer. So far, Gluck and Myers (1993) have shown that the model predicts that conditioning is slower following CS preexposure than following no preexposure either to the CS or the context, and that changes in the context produce release from LI. Unfortunately, the lack of an adequate control group (preexposure to the context alone) makes it difficult to evaluate the success of this network in describing LI.

Recently, Schmajuk, Lam, and Gray (1996) offered a formal theory of LI in the context of a real-time, neural network model of classical conditioning such as that described by Schmajuk and Moore (1988) or presented in Chapter 3. The model (fully described in Chapter 5) incorporates an attentional system that enhances internal representations of events active at a time when the total environmental novelty is large (by increasing attention), and decreases internal representations of those events active at a time when the total novelty is small (by decreasing attention or increasing inattention). The magnitude of the internal representations controls the storage of information into the model of the environment (associability) and the retrieval of information from the model (retrievability). The amplitude of the conditioned response (1) increases with increasing values of the prediction of the unconditioned stimulus, and (2) decreases with increasing values of total novelty. The orienting response is assumed to be proportional to total novelty. According to the model, LI reflects the decreased internal representation of a CS as a consequence of decreased attention after preexposure to that CS. Computer simulations demonstrate that, due to its combined storage and retrieval attributes, the neural network correctly describes the effects on LI of numerous experimental manipulations preceding CS preexposure, during CS preexposure, or during conditioning. Most intriguingly, the network correctly describes attenuation of LI after different postconditioning manipulations.

Interference with the formation of CS–US associations

As an alternative to the notion that LI results from a decrease in CS associability following CS preexposure, it has been proposed that CS preexposure fosters the formation of CS–no consequence associations that interfere with the subsequent establishment of CS–US associations.

Revusky (1971) has suggested that the formation of an association between the CS and any other event interferes with the ability of the CS to form other associations. In the case of LI, the CS becomes associated with other events during preexposure, and these associations reduce the ability of the CS to form associations with the US. Gordon and Weaver (1989) suggested that CS–no consequence associations are encoded along with the preexposure context. When animals are conditioned in a novel context, they fail to retrieve the CS–no consequence association and, therefore, LI is attenuated.

Testa and Ternes (1977) contended that conditioning is determined by the conditional occurrence of the US in the presence of the CS, p(US/CS). CS preexposure decreases p(US/CS) and, therefore, results in weaker conditioning.

Recently, Hall (1991) presented a hybrid theory of LI that extends the original Pearce and Hall (1980) model. According to Hall (1991, p. 137), CS preexposure not only produces a loss of CS associability, but also allows the formation of potentially interfering associations. Interference and low associability retard the formation of CS–US associations during conditioning. Interfering associations depend on the preexposure context and, therefore, LI is attenuated by changes in the context.

Modulation of retrieval of the CS–US associations

Theories explaining LI as the consequence of impaired acquisition find difficulty in explaining data showing that responding to the CS can be increased by postconditioning procedures such as extinction of the training context or testing after long retention intervals. Presumably, these results are better explained by theories that assume CS preexposure produces a deficit in performance despite the fact that CS–US associations had been adequately learned.

LI as a retrieval failure is well addressed by Miller and Schachtman's (1985) comparator hypothesis. According to this hypothesis, conditioned responding is assumed to be the result of a comparison between CS–US associations and CX–US associations. The context is the context in which the CS was trained. During testing, presentation of the CS activates the US directly and indirectly through the combination of CS–CX and CX–US associations. When the direct activation of the US representation is stronger than the indirect activation of the US representation, excitatory responding is expected. When the indirect activation is stronger than the direct activation, inhibitory responding is expected. Therefore, excitatory responding to a CS decreases with increasing CS–CX and CX–US associations.

In terms of the comparator hypothesis, LI is the result of increased CS–CX associations generated during CS preexposure. Although CS–US associations develop during conditioning, increased CS–CX associations will activate CX–US associations, thereby decreasing conditioned responding. Extinction of the training context will decrease the indirect activation of the US representation, thereby increasing conditioned responding and decreasing LI. The hypothesis predicts recovery from LI after a retention interval (Kraemer et al., 1991) by assuming that CX–US associations (but not CS–US associations) decrease during the interval.

Table 4.1 presents a taxonomy of the different models, summarizing and comparing their mechanisms. The present chapter concentrates on Schmajuk

Table 4.1 *Comparison between different theories of classical conditioning and latent inhibition.*

A. MODULATION OF THE STORAGE OF CS–US ASSOCIATIONS

1. Modulation of CS associability

1.1. CS_i associability is modulated by comparisons between observed and predicted events

Model	Associability Is Modulated by	Computed as	Effect of CS Preexposure
Wagner (1978)	Efficacy of CS_i	Difference between the observed and the context-predicted CS_i	Decreases associability of CS_i
Pearce and Hall (1980)	Novelty of the US	Absolute value of the difference between the observed value and the aggregate prediction of the US	Decreases associability of CS_i
Schmajuk and Moore (1988)	Novelty of CS_i	Absolute value of the difference between the observed value and the aggregate prediction of CS_i	Decreases associability of CS_i
McLaren, Kaye, and Mackintosh (1989)	Efficacy of CS_i	Difference between the observed value and the aggregate prediction of CS_i	Decreases associability of CS_i
Schmajuk, Lam, and Gray (1993)	Total novelty of US and CSs	Sum of absolute values of the differences between the observed values and the aggregate predictions of the US and CSs	Decreases associability of CS_i

1.2. CS_i associability is modulated by the quality of CS_i predictions of future events

Model	Associability Is Modulated by	Computed as	Effect of CS Preexposure
Mackintosh (1975)	CS_i relative predictive value	Differences between CS_i–US associations and context–US associations	Decreases associability of CS_i
Grossberg (1975)	CS_i incentive motivation	US–CS_i association	Decreases associability of CS_i
Frey and Sears (1978)	CS_i predictive value	CS_i–US association	Decreases associability of CS_i
Lubow, Weiner, and Schnur (1981)	CS_i predictive value	Not a computational model	Decreases associability of CS_i
Moore and Stickney (1980)	CS_i relative predictive value	Difference between CS_i–US and CS_i–CS_k associations and CS_j–US and CS_j–CS_k associations	Decreases associability of CS_i
Ayres, Albert, amd Bombace (1987)	CS_i predictive value	CS_i–US association	Decreases associability of CS_i
Schmajuk and Moore (1989)	CS_i relative predictive value	Difference between CS_i–US and CS_i–CS_k associations and CS_j–US and CS_j–CS_k associations	Decreases associability of CS_i
Schmajuk and DiCarlo (1991a, b)	CS_i incentive motivation	CS_i–US association	Decreases associability of CS_i
Gluck and Myers (1993)	CS_i internal representation	Input-hidden unit weights	Decreases in CS_i internal representation

2. Interference with the formation of CS–US associations

Model	Effect of CS Preexposure	Computed as	Effect on CS–US Associations
Revusky (1971)	Increases CS–CX associations	Not a computational model	Interference with CS–US associations
Testa and Ternes (1977)	Decreases p(US/CS)	Not a computational model	Conditioning proportional to p(US/CS)

B. MODULATION OF THE RETRIEVAL OF THE CS–US ASSOCIATIONS

Model	Effect of CS Preexposure	Computed as	Effect on CS–US Associations
Miller and Schachtman (1985)	Increases CS–CX associations	Not a computational model	Increases activation of US representation by CX, decreases CR magnitude
Schmajuk, Lam, and Gray (1993)	Decreases total novelty	See above	Decreases retrieval of CS–US associations

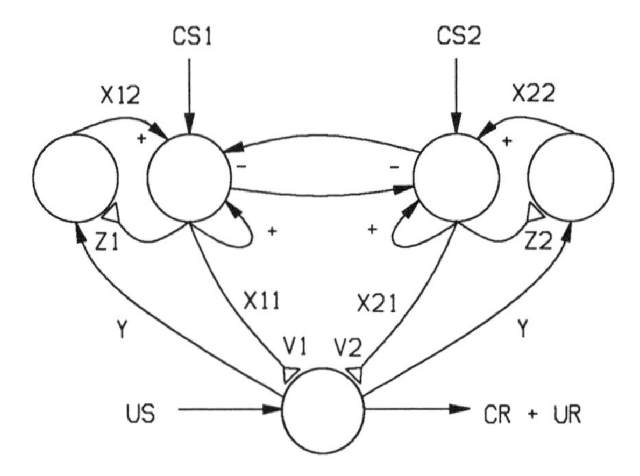

Figure 4.1. Diagram of Grossberg's (1975; Grossberg and Levine, 1987) attentional network for classical conditioning. Conditioned stimuli CS_1 and CS_2 activate sensory representations X_{11} and X_{21} that compete among themselves for a limited-capacity short-term memory. Sensory representations X_{11} and X_{21} become associated with the US by changing synaptic weights V_1 and V_2 (conditioned reinforcement learning). X_{11} and X_{21} also become associated with output Y by changing synaptic weights Z_1 and Z_2 (incentive motivation learning). Arrows represent fixed connections, triangles modifiable connections.

and DiCarlo's (1991a, b) rendition of Grossberg's (1975) network. Chapter 5 presents the Schmajuk, Lam, and Gray (1996) neural network.

The Grossberg attentional model

Briefly, Grossberg's (1975) real-time theory (see Box 1.4) suggests that a CS activates neural populations whose activity constitutes a sensory representation, or short-term memory (STM), of the CS. A US unconditionally activates neural populations of the drive representation of the US. Sensory representations compete among themselves for a limited-capacity STM activation that is reflected in a long-term memory (LTM) storage. Pairing of a CS with a US has two consequences: (1) The sensory representation of the CS becomes associated with the drive representation of the US (conditioned reinforcement learning in Grossberg's terms), and (2) the drive representation of the US becomes associated with the sensory representation of CS (incentive motivation learning in Grossberg's terms). Although "incentive motivation" has mainly an appetitive connotation, in the Grossberg model the term applies to both appetitive and aversive USs. Incentive motivation associations reflect the association of the US with a CS representation and mediate the enhancement of the sensory representation of the CS.

A network based on Grossberg's (1975) architecture is shown in Figure 4.1. A CS activates sensory representation nodes X_{i1}. In agreement with Hebb (1949), STM is sustained by a positive feedback loop. A US activates neural populations of the drive representation Y. Simultaneous activation of the drive representation and of CS sensory representations causes X_{i1} to become associated with the output of the drive representation. The LTM association is implemented by an increase in the synaptic weight V_i. After X_{i1} becomes associated with the drive representation Y, it becomes a second-order reinforcer for other

Box 4.1 Grossberg's (1975) real-time attentional theory

The most relevant aspects of the network are:

1. CS_i activates a STM sensory representation X_{i1}.
2. The US activates neural populations of the drive representation of the US, Y.
3. Sensory representations X_{i1} compete among themselves for a limited-capacity STM activation that is reflected in LTM. Stimuli with large X_{i1}'s change their V_i and Z_i (see the remarks that follow) faster than stimuli with weak X_{i1}'s.
4. Sensory representations X_{i1} become associated with the output of the drive representation Y. The LTM association is implemented by an increase in the synaptic weight V_i.
5. After X_{i1} becomes associated with the drive representation Y, it becomes a secondary reinforcer for other CSs.
6. Sensory representations X_{i1} also become associated, at a different rate, with the output of the drive representation Y. This second LTM association, interpreted as incentive motivation, is implemented by an increase in the synaptic weight Z_i.
7. Incentive motivation associations mediate the enhancement of the sensory representations X_{i1}.
8. CR amplitude is proportional to the sum of all sensory representations X_{i1} gated by their associations with the drive representation V_i.

CSs. In the original Grossberg model, simultaneous activation of the drive representation and of a sensory representation causes the drive representation Y to become associated with X_{i1}. In the Schmajuk and DiCarlo (1991a, b) rendering of the Grossberg network, simultaneous activation of the drive representation and of a sensory representation causes X_{i1} to become associated with the drive representation Y. This LTM association is implemented by an increase in the synaptic weight Z_i. Conditioning of the X_{i1}–X_{i2} pathway increases X_{i1} sensory representation by incentive motivation. A sensory cue with large V_i and Z_i can augment the STM activity of its sensory representation. Sensory representations compete among themselves for a limited STM capacity. This limited STM capacity is implemented by letting STM activity X_{i1} be excited by CS_i and inhibited by the sum of all other STM activities, $\Sigma_{j \neq i} X_{j1}$. Through this competition, CSs with strong V_i and Z_i inhibit CSs with weak V_i and Z_i.

STM activity translates into permanent LTM traces. Stimuli with strong STM (strong sensory representations) change their V_i and Z_i faster than stimuli with weak STM. Therefore, stimuli with strong STM acquire stronger V_i associations than stimuli with weak STM, when presented with the US. The way STM activation is reflected in LTM storage in different classical conditioning paradigms will be illustrated throughout the text. Box 4.1 summarizes the most important aspects of the model. A formal description of the model is presented in Appendix 4.A.

Computer simulations

The following section contrasts experimental results with computer simulations using the neural model in the following paradigms: (1) acquisition of delay and

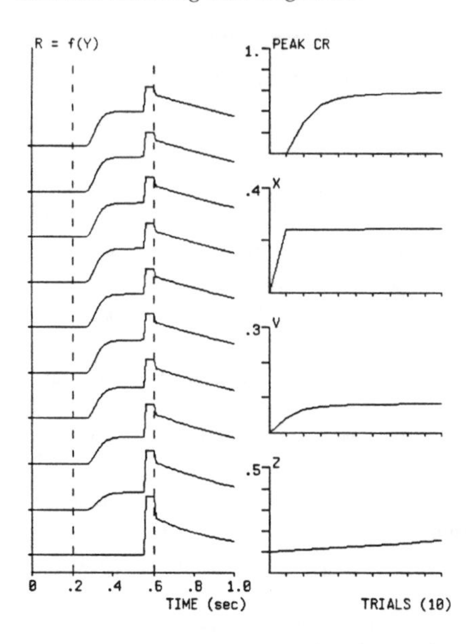

Figure 4.2. Acquisition of classical conditioning. *Left panels.* Real-time simulated conditioned and unconditioned response in 10 reinforced trials. First trial is represented at the bottom of the panel. *Right panel.* Peak CR: peak CR as a function of trials; *X:* average X_{11} as a function of trials; *V:* V_1 as a function of trials; *Z:* Z_1 as a function of trials.

conditioning, (2) extinction of delay conditioning, (3) acquisition–extinction series of delay conditioning, (4) latent inhibition, (5) blocking, (6) overshadowing, (7) discrimination acquisition, (8) discrimination reversal, and (9) secondary reinforcement.

Acquisition of delay conditioning

Figure 4.2 shows real-time simulations of 10 trials in a delay conditioning paradigm with a 400-msec CS, 50-msec US, and 350-msec ISI. CR amplitude, X_1, V_1, and Z_1 increase over trials.

ISI curve

Figure 4.3 shows an ISI curve displaying peak CR amplitude after 20 acquisition trials with 200-msec CSs; 0-, 100-, 200-, 300-, 400-, 500-, 600-, 700-, and 800-msec ISIs; and a 50-msec US. In simultaneous conditioning the ISI is 0 msec, in delay conditioning the ISI is less than or equal to 200 msec, and in trace conditioning the ISI is greater than 200 msec.

In agreement with experimental data (Schneiderman and Gormezano, 1964; Smith, 1968), peak CR amplitude is zero with simultaneous conditioning, increases with delay conditioning, and decreases with trace conditioning. In terms of the model, larger peak CR amplitudes are obtained at ISIs that correspond to larger values of X_{i1} (see Equation 4.8 in Appendix 4.A). Therefore, as noted by Gormezano, Kehoe, and Marshall (1983), the ISI curve approximately reflects the shape of τ_i, which has a small initial value, increases over time to a maximum following CS_i presentation, and then decays back to zero (see Figure 4.4).

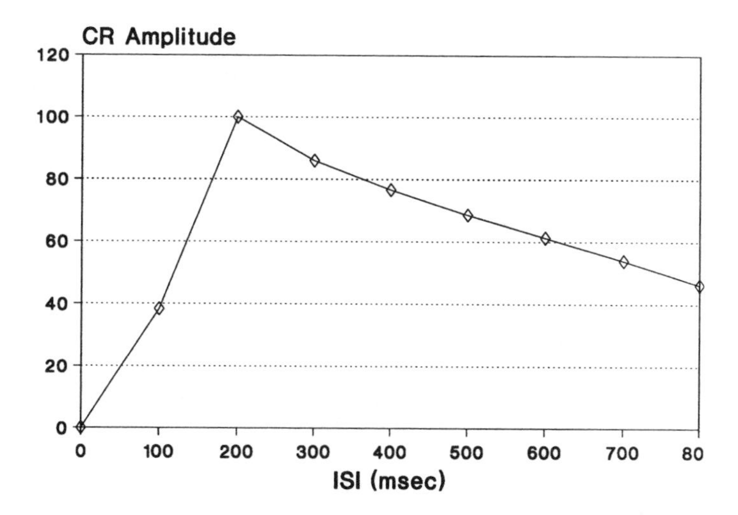

Figure 4.3. ISI curves. Peak CR amplitude after 20 reinforced trials for different interstimulus intervals, 200-msec CS, and 50-msec US. Peak CR amplitude is expressed as a percentage of the maximum peak CR over all ISIs.

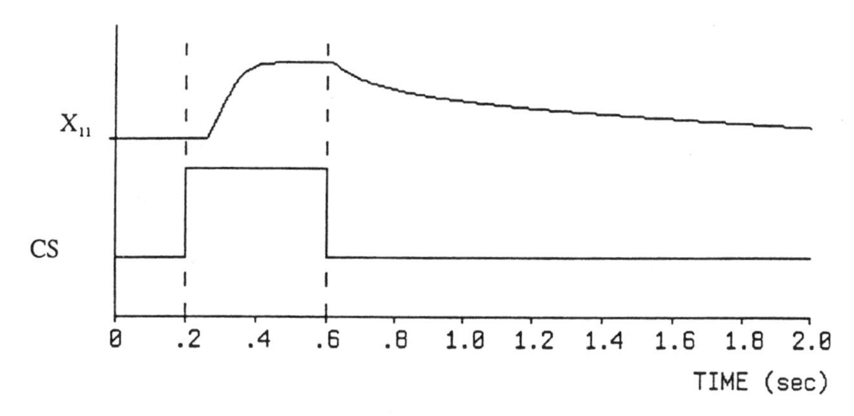

Figure 4.4. Sensory representations of a CS presented in isolation. Sensory representations X_{11} generated by a 400-msec CS.

The ISI curve shown in Figure 4.3 is similar to, but more accurate than, that displayed in Figure 2.5.

Acquisition–extinction series: savings effects

Figure 4.5a shows simulations of peak CR amplitude after 3 alternating acquisition–extinction series, each one with 50 acquisition and 10 extinction trials. Consistent with Schmaltz and Theios (1972), animals show savings. In the model, X_{11} receives an increasing amount of incentive motivation Z_1 over acquisition trials. As a result, after the first session acquisition proceeds faster (savings effect).

Figure 4.5b shows peak CR amplitude after one extinction trial expressed as the percentage of the peak CR amplitude on the last of 50 acquisition trials. As a consequence of the increasing value of Z_1 over extinction trials, the model

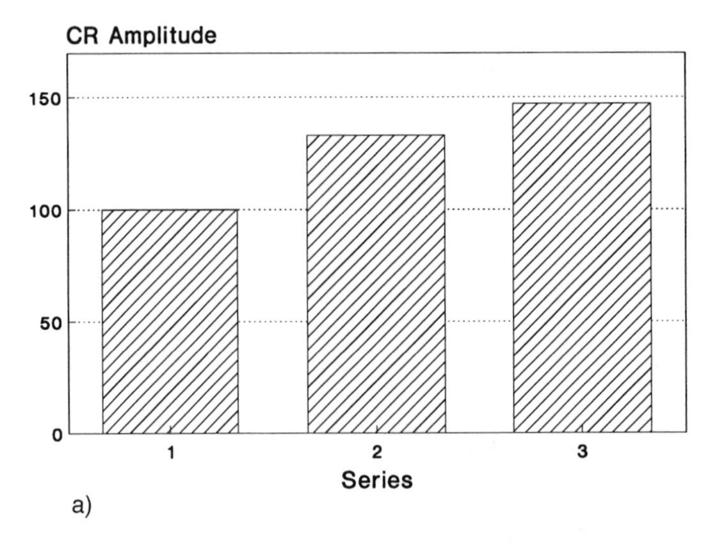

Figure 4.5. Acquisition–extinction series. (a) *Acquisition:* peak CR amplitude on three series of acquisition trials. Peak CR amplitude is expressed as a percentage of the peak CR on the fourth acquisition trial of the first series. (b) *Extinction:* percentage of peak CR amplitude on the first non-reinforced trial following 50 acquisition trials. Percentage is computed with respect to the peak CR at the end of each acquisition series.

predicts no improvement in extinction over series. These results are in disagreement with the acquisition–extinction series of the Schmaltz and Theios (1972) study showing that animals decreased the number of trials to extinction criterion.

Latent inhibition

Figure 4.6 shows peak CR amplitude after 4 acquisition trials following 200 CS preexposure trials. Schmajuk and DiCarlo (1991a, b) assumed that animals have an initial value of incentive motivation. Because the initial value of incentive motivation Z_1 decreases during CS_1 preexposure, the latent inhibition group

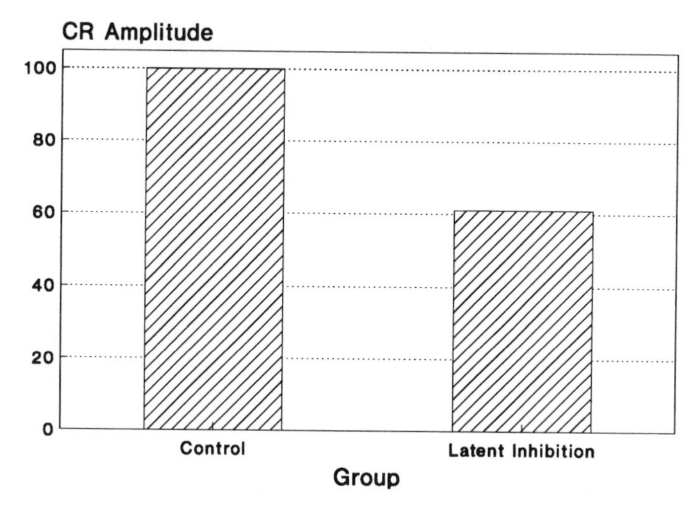

Figure 4.6. Latent inhibition. Peak CR amplitude after 4 acquisition trials following 200 CS preexposure trials or 200 control trials. Peak CR amplitude is expressed as a percentage of the peak CR.

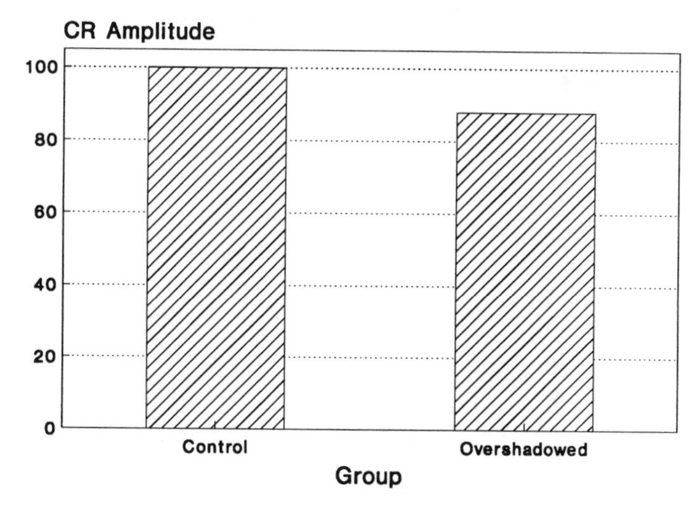

Figure 4.7. Overshadowing. Peak CR amplitude evoked by CS_1 after 10 reinforced CS_1–CS_2 trials. Peak CR amplitude is expressed as a percentage of the peak CR to CS_1 after 10 reinforced CS_1 trials.

shows slower acquisition than a control group that received preexposure to the context alone.

Overshadowing

Figure 4.7 shows peak CR amplitude after 10 CS_1–CS_2 acquisition trials. Simulated training consisted of 400-msec CSs, 50-msec US, and 400-msec ISI. In the model, the CR elicited by CS_1 is smaller than CRs generated by CS_1 when it had been reinforced alone, that is, the model yielded overshadowing.

Figure 4.8 illustrates how the network describes overshadowing. When CS_1 and CS_2 are paired with the US, competition between X_{11} and X_{21} reduces X_{11} and X_{21}, thereby reducing the rate of association of CS_1 and CS_2 with the US. Figure

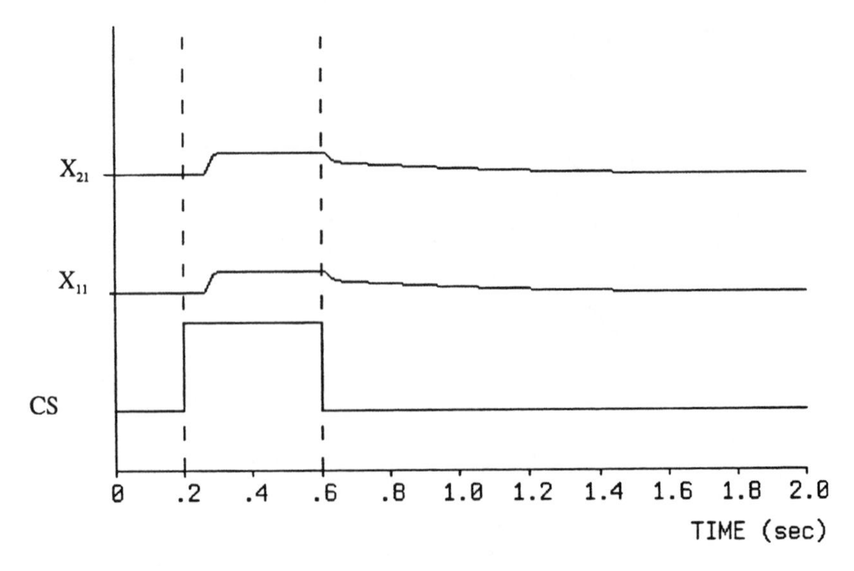

Figure 4.8. Interaction between sensory representations during overshadowing. Sensory representations generated by CS_1 and CS_2 active for 400 msec. X_{11} and X_{21} inhibit each other, thereby achieving less association with the US than when presented separately.

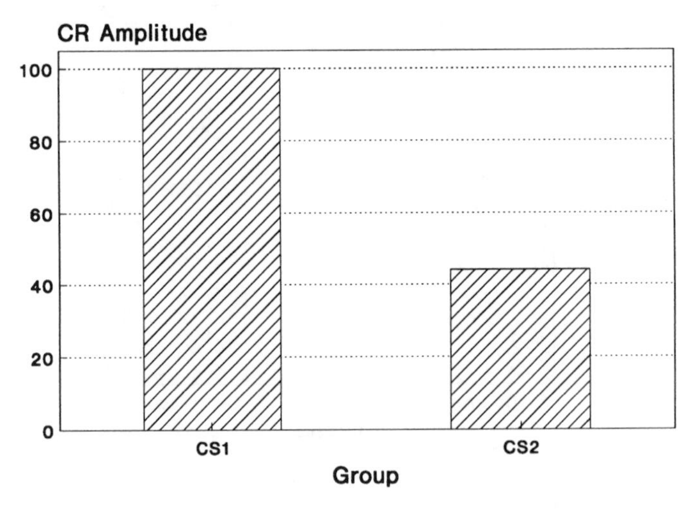

Figure 4.9. Blocking. Peak CR amplitude evoked by CS_2 after 10 reinforced CS_1–CS_2 trials following 10 reinforced CS_1 trials. Peak CR amplitude is expressed as a percentage of the peak CR to CS_1 after 10 reinforced CS_1 trials.

4.8 shows sensory representations X_{11} and X_{21} in a trial in which CS_1 and CS_2 are presented together. In the model, overshadowing is produced because sensory representations X_{11} and X_{21} are smaller than when each CS is presented separately (see Figure 4.4) and therefore they accrue less association with the US.

Blocking

Figure 4.9 shows peak CR amplitude after 10 CS_1–CS_2 acquisition trials following 10 CS_1 acquisition trials. Simulated training consisted of 400-msec CSs, 50-msec US, and 400-msec ISI. Figure 4.9 indicates that the model simulated

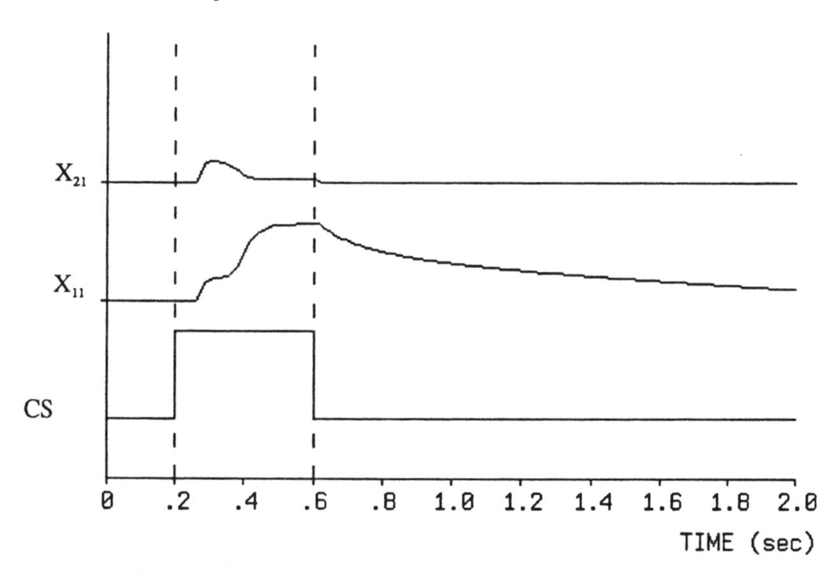

Figure 4.10. Interaction between sensory representations during blocking. Sensory representations generated by CS_1 and CS_2, active for 400 msec, after 10 reinforced CS_1 trials. X_{11} is enhanced by previous training and inhibits X_{21}, thereby blocking its association with the US.

blocking because the CR for the CS_2 was smaller in the experimental condition than in the control condition, in which CS_1 and CS_2 were paired together with the US during 10 trials (overshadowing in Figure 4.7).

Figure 4.10 illustrates how the model describes blocking. It shows sensory representations X_{11} and X_{21} in the second phase of blocking, after 10 CS_1–US trials. In the first phase of blocking, CS_1 is paired with the US and associations V_1 and Z_1 are created. During the second phase of blocking, incentive motivation Z_1 enhances the value of X_{11}, causing X_{11} to inhibit X_{21} and therefore preventing X_{21} from acquiring association with the US.

Discrimination acquisition and reversal

Figure 4.11 shows simulations of discrimination acquisition and reversal. During discrimination acquisition, five reinforced CS_1–US trials alternated with five nonreinforced CS_2 trials. CS duration was 850 msec, US duration 50 msec, and ISI 800 msec. Figure 4.11 also illustrates simulations of a discrimination reversal paradigm. During reversal, the original nonreinforced CS_2 is reinforced for three trials and these trials alternate with three nonreinforced CS_1 trials. Again, CS duration was 850 msec, US duration 50 msec, and ISI 800 msec.

Second-order conditioning

Figure 4.12 shows simulations of second-order conditioning. During second-order conditioning, 40 reinforced CS_1–US trials alternate with 40 CS_2–CS_1 trials. CS_1 increases its incentive motivation association during reinforced trials. When presented with CS_2 on the nonreinforced trials, CS_1 blocks CS_2 (see Figure 4.9) and, therefore, CS_2 acquires only some association with the drive representation activated by CS_1.

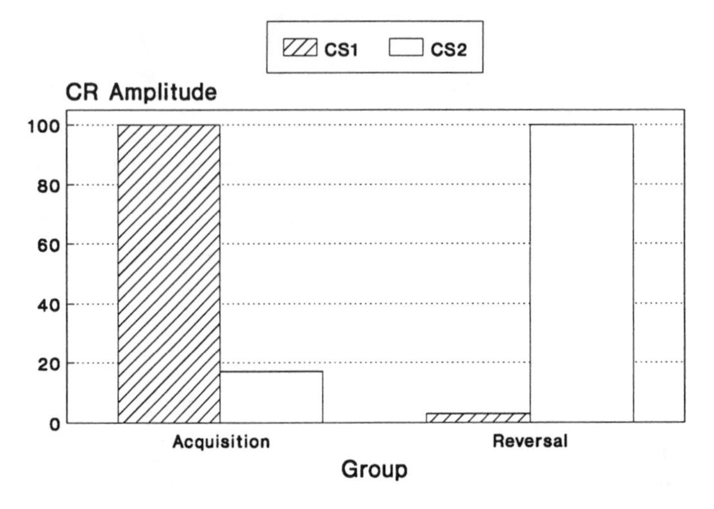

Figure 4.11. Classical discrimination. *Discrimination acquisition:* peak CR amplitude evoked by CS_1 and CS_2 after five reinforced CS_1 trials alternated with five nonreinforced CS_2 trials. Peak CR amplitude is expressed as a percentage of the peak CR to CS_1. *Discrimination reversal:* peak CR amplitude evoked by CS_1 and CS_2 after three nonreinforced CS_1 trials alternated with three reinforced CS_2 trials. Peak CR amplitude is expressed as a percentage of the peak CR to CS_2.

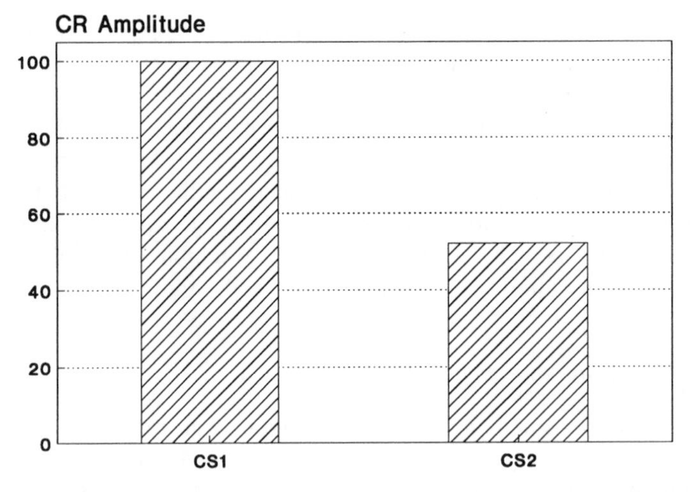

Figure 4.12. Second-order conditioning. Peak CR amplitude evoked by CS_1 and CS_2 after 40 reinforced CS_1 trials alternated with 40 nonreinforced CS_1–CS_2 trials. Peak CR amplitude is expressed as a percentage of the peak CR to CS_1.

Discussion

The present chapter introduces Grossberg's (1975) network. According to Grossberg's (1975) neural attentional theory, pairing of CS_i with a US causes both an association of $f(CS_i)$ with the US and an association of the US with $f(CS_i)$. Sensory representations $f(CS_i)$ compete among themselves for a limited-capacity short-term memory activation that is reflected in CS_i–US associations. Table 4.2 shows that the Schmajuk and DiCarlo (1991a, b) version of Grossberg's (1975) model correctly describes (1) delay and trace conditioning, (2) extinction, (3) partial reinforcement, (4) saving effects in acquisition but not ex-

Table 4.2. *Simulations obtained with the Schmajuk and DiCarlo (1991a, b) version of Grossberg's (1975) attentional model.*

Paradigm	Model
Delay conditioning	Y
Trace conditioning	Y
Partial reinforcement	Y
Extinction	Y
Acquisition series	Y
Extinction series	
Latent inhibition	Y
Blocking	Y
Unblocking	
Overshadowing	Y
Discrimination	Y
Conditioned inhibition	
Inhibition by differential conditioning	
Inhibition by decrease in the US	
Counterconditioning	
Superconditioning	
Overprediction effects	
Generalization	
Compound conditioning	
Simultaneous feature-positive discrimination	
Simultaneous feature-negative discrimination	
Serial feature-positive discrimination	
Serial feature-negative discrimination	
Negative patterning	
Positive patterning	
Sensory preconditioning	
Second-order conditioning	Y
Context switching	
External inhibition	

Note. Y: The model correctly describes the experimental data.

tinction series, (5) latent inhibition, (6) blocking, (7) overshadowing, (8) discrimination acquisition and reversal, and (9) second-order conditioning.

Importantly, the model shows that a limited-capacity STM mechanism can describe classical conditioning paradigms, such as overshadowing and blocking, also described by the delta rule.

Appendix 4.A

A formal description of the model

The model described in this chapter is depicted in Figure 4.1. Short-term memory variables (X_{i1}, X_{i2}, and Y) and long-term variables (V_i and Z_i) have been introduced in the text. The model used here is that presented by Schmajuk and DiCarlo (1991b) and differs in some aspects from those offered by Grossberg and Levine (1987) and Schmajuk and DiCarlo (1991a).

The sensory representation of CS$_i$, X_{i1}, is defined by

$$d(X_{i1})/dt = -K_1 X_{i1} + K_2(K_3 - X_{i1})\, I_{i1} - K_4 X_{i1} J_{i1} \qquad (4.1)$$

where $-K_1 X_{i1}$ represents the passive decay of STM, K_2 represents the rate of increase of X_{i1}, constant K_3 is the maximum possible value of X_{i1}, I_{i1} is the total excitatory input, and J_{i1} represents the total inhibitory input. K_3 can be regarded as the number, or the percentage, of cells or membrane active sites that can be excited. Therefore, $(K_3 - X_{i1})$ represents the number, or the percentage, of inactive sites that can be excited, and X_{i1} the number of active sites that can be inhibited.

Grossberg and Levine (1987) assumed that positive feedback signals trigger a process of habituation that steadily attenuates the size of the feedback signals. For simplicity, the present chapter assumes no attenuation, and the total excitatory input I_{i1} is given by

$$I_{i1} = R_1(XCS_i) + K_5 X_{i1}^m + K_6 R_1(XCS_i)\, X_{i2} \tag{4.2}$$

where $R_1(XCS_i)$ represents the activity of a node excited by CS_i, and $K_5 X_{i1}^m$ a positive feedback from X_{i1} to itself. $K_6 R_1(XCS_i)\, X_{i2}$ represents a signal from X_{i2} to X_{i1} that is active only if CS_i is present. This last term, which can be regarded as a presynaptic modulation of X_{i2} by CS_i, avoids the activation of sensory representations by drive representations in the absence of CS_i.

We assume that CS_i activates node XCS_i according to

$$XCS_i = -K_7 XCS_i + K_8(K_9 - XCS_i)\, CS_i \tag{4.3}$$

The output of the XCS_i node is a sigmoid given by

$$R_1(XCS_i) = XCS_i^n / \beta_1^n + XCS_i^n \tag{4.4}$$

Like Grossberg and Levine (1987), we assume that the total inhibitory input to X_{i1}, J_{i1}, is the sum of the activities of all other nodes, $\Sigma_{j \neq i} X_{j1}$. Therefore, J_{i1} is given by

$$J_{i1} = \Sigma_{j \neq i} R_2(X_{j1}), \tag{4.5}$$

where $R_2(X_{j1})$ equals 0 if $X_{j1} < 0.2$, and $R_2(X_{j1}) = X_{j1}$ if $X_{j1} \geq 0.2$.

Activity in the drive representation node is given by

$$dY/dt = -K_{10} Y + K_{11}\, (\Sigma_i V_i X_{i1} + US) \tag{4.6}$$

where $-K_{10} Y$ represents the passive decay of drive representation activity, $\Sigma_{j \neq i} V_i X_{i1}$ is the sum of sensory representations gated by the corresponding LTM traces, and US is the US input to the drive representation node. Notice that whereas Grossberg and Levine (1987) assumed that the US acts on the sensory input, thereby explaining ISI effects, we assume that the US acts only on the drive node and the CS is delayed in XCS_i. We assume that Y is the output of the system and is proportional to the sum of the conditioned response (CR) and the unconditioned response (UR).

Whereas Grossberg and Levine (1987) assumed that the signal from Y activates the LTM of Z_i, the present chapter assumes that the sensory representation X_{i1} activates LTM trace Z_i and determines the activity of X_{i2}:

$$d(X_{i2})/dt = -K_{12}\, X_{i2} + K_{13} X_{i1}^m Z_i \tag{4.7}$$

Whereas Grossberg and Levine (1987) assumed that associations of sensory representation X_{i1} with the drive representation V_i undergo extinction even when X_{i1} is zero, we assume that changes in V_i are possible only when X_{i1} is active. Also in contrast to Grossberg and Levine (1987), we assume that V_i has a maximum value. Changes in V_i are given by

$$d(V_i)/dt = -K_{14}\, V_i X_{i1} + K_{15}(K_{16} - V_i)\, Y X_{i1} \tag{4.8}$$

where $-K_{14}V_iX_{i1}$ is the active decay in V_i when X_{i1} is active, and $K_{15}(K_{16} - V_i)YX_{i1}$ is the increment in V_i when Y and X_{i1} are active together. K_{16} can be regarded as the number, or percentage, of cells or membrane patches that can be modified by learning. Therefore, $(K_{16} - V_i)$ represents the number, or percentage, of unmodified sites that can increase the efficacy of their connection, and V_i the number of already modified sites that can decrease their connectivity. Notice that Equation 4.8 is a Hebbian rule (V_i increases with concurrent pre- and postsynaptic activity) with extinction (V_i decreases with *presynaptic* activity alone).

As in Grossberg and Levine (1987), Equation 4.8 generates second-order conditioning. This is so because, by Equation 4.6, Y is activated either by presentation of the US or a CS already associated with the US. Therefore, if CS_1 is associated with the US, it generates activity Y that becomes associated with X_{21} by Equation 4.8.

For simplicity, Grossberg and Levine (1987) assumed that Z_i was identical to V_i. In order to explain the savings effect seen in acquisition–extinction series, we conjectured that Z_i varies at a slower rate than V_i. Therefore, we have computed Z_i with

$$d(Z_i)/dt = -K_{17}Z_iX^m_{i1} + K_{18}(K_{19} - Z_i)\,YX^m_{i1}, \tag{4.9}$$

where $-K_{17}Z_iX^m_{i1}$ is the active decay in Z_i when X^m_{i1} is active, and $K_{18}(K_{19} - Z_i)YX^m_{i1}$ is the increment in Z_i when Y and X_{i2} are active together. If we assume that X_{i1} represents presynaptic activity and Y characterizes postsynaptic excitation, Equation 4.9 states that synapse Z_i is potentiated with concurrent presynaptic and postsynaptic activities ($X_{i1} > 0$ and $Y > 0$) and depotentiated when the presynaptic neuron is active ($X_{i1} > 0$) but the postsynaptic neuron is inactive ($Y = 0$).

The output of the system, that is, the CR and UR, is another sigmoid given by

$$R_3(Y) = Y^2/(Y^2 + \beta_2^2) \tag{4.10}$$

Appendix 4.B

Simulation parameters

Simulations assume that one computer time step is equivalent to 10 msec. Each trial consisted of 500 steps, equivalent to 5 sec. Unless otherwise specified, the simulations assumed 200-msec CSs, the last 50 msec of which overlaps the US. CS onset was at 200 msec. Parameters were selected so that simulated asymptotic values of V_i were reached in approximately 10 acquisition trials. Since asymptotic conditioned NM responding is reached in approximately 200 real trials (Gormezano et al., 1983), one simulated trial is approximately equivalent to 20 experimental trials.

Parameter values are $K_1 = 0.05$, $K_2 = 0.095$, $K_3 = 1.5$, $K_4 = 3$, $K_5 = 0.6$, $K_6 = 10$, $K_7 = 0.07$, $K_8 = 0.18$, $K_9 = 1$, $K_{10} = 1$, $K_{11} = 0.34$, $K_{12} = 1$, $K_{13} = 1$, $K_{14} = 0.0155$, $K_{15} = 0.04$, $K_{16} = 1$, $K_{17} = 0.0002$, $K_{18} = 0.001$, $k_{19} = 1.5$, $\beta_1 = 0.62$, $\beta_2 = 0.03$, $n = 70$, $m = 1.5$. The initial value of V_i was 0. The initial value of Z_i was 0.1. These values were kept constant for all simulations.

5 Storage and retrieval processes

As mentioned in Chapter 4, many theories have been proposed to account for latent inhibition (LI). Table 4.1 presents a taxonomy of the different models, summarizing and comparing their mechanisms. Models are first divided into those that describe LI as (1) a disruption of storage of CS–US associations and (2) disruption of retrieval of the CS–US associations. The storage group is further divided into (1) a group that assumes that the associability of the CS decreases during preexposure, thereby retarding the acquisition of CS–US associations, and (2) a group that proposes that the formation of CS–no consequence associations during preexposure interferes with the subsequent establishment of CS–US associations. Models that assume that CS associability decreases during preexposure are assigned to one of two groups depending on whether CS_i associability is modulated by (1) comparisons between observed and predicted events, or (2) the quality of the CS_i predictions of future events.

Chapter 5 describes a neural network model of classical conditioning, presented by Schmajuk, Lam, and Gray (1996), that describes LI. The model is an extension of the cognitive map model presented in Chapter 3. The model assumes variations in the effectiveness of both the CS and the US. According to the model, LI is the consequence of a decreased CS associability at the time of the storage of the CS–US association. Therefore, the network shares properties with other attentional models (such as that presented in Chapter 4) that describe the process of storing associations in LTM. However, the network introduced here is unique in that it also describes the process of retrieval of stored associations. Importantly, this retrieval property is not an additional assumption in the network but a simple and direct consequence (an emergent property) of the way this neural network stores and retrieves information. Therefore, according to the model, experimental manipulations before, during, or after conditioning can modify the retrieval of established associations, thereby modifying CR strength even when already stored associations remain constant.

According to Spear, Miller, and Jagielo (1990), current theories of conditioning tend to underscore the process of storage of associations but neglect the process of retrieval. In contrast, the model offered in this chapter combines both storage and retrieval processes and offers a quantitative description of both operations and their interactions. Importantly, storage and retrieval processes are two different aspects of the single neural mechanism.

A neural network theory of latent inhibition

The present chapter introduces a neural network theory of LI that formalizes some aspects of Sokolov's (1960) model and Gray's (1971) behavioral inhibition system. As mentioned, Sokolov (1960) suggested that animals build an

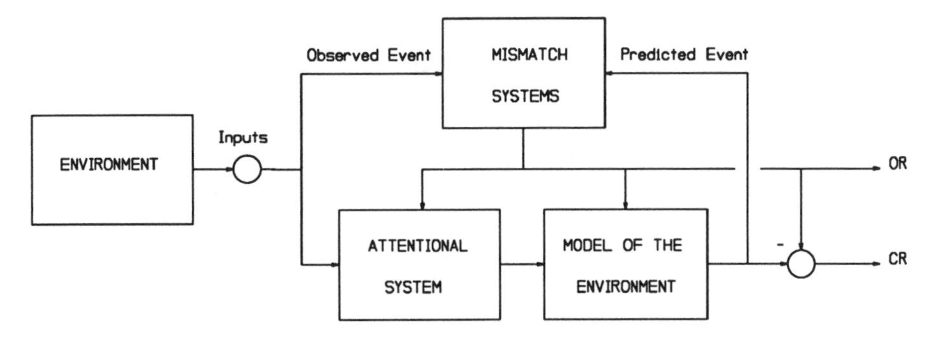

Figure 5.1. A block diagram of the SLG model. The model of the environment generates predictions of future events and controls the conditioned response (CR). The mismatch systems compare observed and predicted events to compute novelty. Novelty is then used to (1) control attention to environmental stimuli in the attentional system, (2) modify the model of the environment, and (3) inhibit ongoing conditioned behavior. CR: conditioned response; OR; orienting response.

internal model of their environment. Whenever novelty is detected (there is a mismatch between predicted and actual environmental events), (1) an orienting response (OR) is emitted and (2) the internal model is modified. When there is coincidence between the observed and the predicted stimulus, the animal may respond without changing its neural model of the world. Gray (1971) suggested that a behavioral inhibition system responds to signals of punishment, signals of nonreward, novel stimuli, and innate fear stimuli, by inhibiting ongoing behavior, increasing readiness for action (arousal level), and increasing attention to environmental stimuli.

The network described here provides a mechanistic account of Sokolov's (1960) notions of an internal model of the world and a mismatch system, as well as of Gray's concept of a behavioral inhibition system. As the model of the world generates predictions of future events, the mismatch system compares observed and predicted events to compute novelty. Novelty is then used to (1) control attention to environmental stimuli, (2) modify the model of the environment, and (3) inhibit ongoing behavior.

The Schmajuk–Lam–Gray (SLG) model

Figure 5.1 shows a block diagram of the model, henceforth designated the SLG model. In simple terms, the SLG model assumes that CSs activate internal representations. The attentional system enhances internal representations of CSs active at a time when the total environmental novelty is large (increased attention), and decreases internal representations of those CSs active at a time when the total novelty is small (decreased attention or inattention). The magnitude of the internal representations controls the storage of information into the model of the environment (associability) and the retrieval of information from the model (retrievability). Therefore, events with large internal representations show an increased capability to form new associations with other environmental events, as well as an augmented efficacy to retrieve old associations. Events with small internal representations show a decreased capability to form new associations with other environmental events, as well as a decreased efficacy to retrieve old associations.

Environmental CSs retrieve associations stored in the internal model of the world, thereby generating predictions of future events. The model of the world stores CS–CS and CS–US associations, built according to classical conditioning principles, that allow it to predict future CSs and USs based on the present CSs. Importantly, the model has cognitive mapping capabilities, that is, it can combine multiple CS–CS and CS–US associations to generate novel predictions. In the model, associations between environmental events change whenever there is a mismatch between predicted and observed events. When the predicted event is a US, the amplitude of the CR is proportional to the magnitude of the prediction of the US.

Novelty of a given event is defined as the absolute value of the difference between its predicted and observed magnitude. Novelty is high either when an unpredicted event is present or a predicted event is absent. Total novelty is defined as the sum of all individual novelties. As the internal model of the world becomes a more accurate replica of the actual world, total novelty and the attentional enhancement of internal representations decrease. Even when decreased, attention to some environmental events never becomes zero because some active internal representations are always needed to generate predictions and, therefore, to maintain a low level of total novelty. In addition to modulating attention to environmental events, total novelty also controls the generation of the OR. CR amplitude is inhibited in proportion to the magnitude of the OR.

In the context of the SLG model, LI is the consequence of the depressed internal representation of a CS after preexposure and of the time needed to restore this internal representation during conditioning.

The neural network

Figure 5.2 shows a detailed diagram of the SLG network. The SLG network is a *real-time model* (see Box 1.4). A formal description of the model is presented in Appendix 5.A.

The SLG theory is an extension of a neural network of classical conditioning described by Schmajuk and Moore (1988) and Schmajuk (1986, 1989) and presented in (Figure 3.11). As do Frey and Sears' (1978) and Wagner's (1978) models, the SLG network assumes that both the CS and the US vary in their effectiveness during classical conditioning. Whereas variations in the US representation are used to explain paradigms such as conditioned inhibition, blocking, and overshadowing, variations in the representation of the CS are used to describe paradigms such as LI and sensory preconditioning.

The SLG network incorporates different types of memory: (1) a trace short-term memory (STM) of CS_i, (2) an attentional long-term memory (LTM) for CS_i, (3) associative CS_i–CS_k and CS_i–US LTMs, and (4) intermediate-term memories of CS_i, US, and their predictions.

Trace STM

Trace STM $\tau_i(t)$ increases over time to a maximum following CS_i presentation and then decays back to zero (see Grossberg, 1975; Hull, 1943). In addition, we assume that $\tau_i(t)$ increases also when CS_i is predicted by other CSs (see Equation 5.1 in Appendix 5.A). The internal representation of CS_i, X_i, is proportional to τ_i and τ_i multiplied by its attentional LTM z_i.

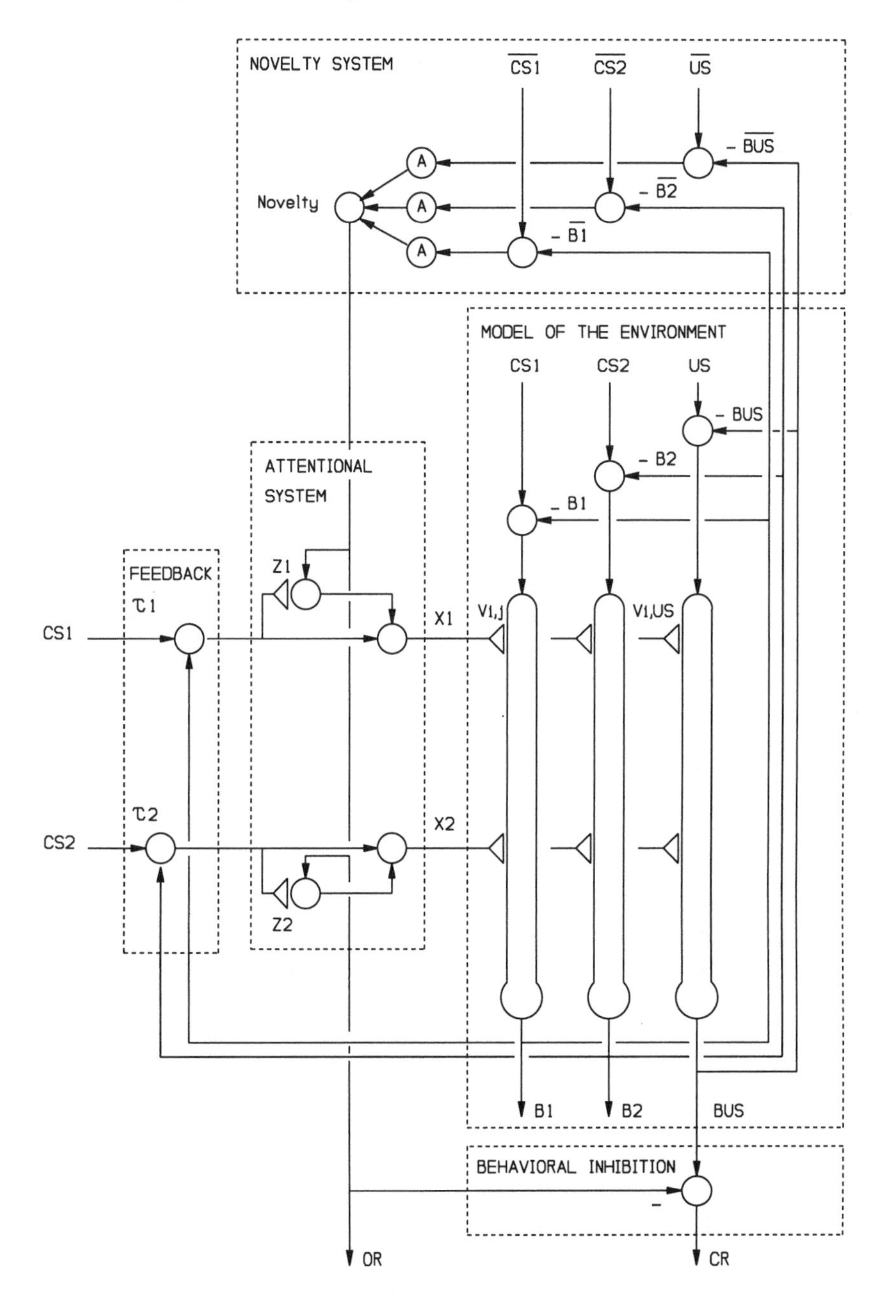

Figure 5.2. Diagram of the network. CS_i: conditioned stimulus; τ_i: trace STM of CS_i; X_i: internal representation of CS_i; z_i: CS_i–novelty association; $V_{i,US}$: CS_i–US association; $V_{i,j}$: CS_i–CS_j association; B_k: aggregate prediction of event k; B_{US}: aggregate prediction of the US; $\bar{\lambda}_k$: average observed value of CS_k; $\bar{B}_k$: average prediction of CS_k; CR: conditioned response; OR: orienting response. Arrows represent fixed synapses, open triangles variable synapses.

By simultaneously reaching a critical locus of learning, $X_i(t)$ allows CS_i to build associations with other CSs and the US even when they are separated by a temporal gap, such as in the case of trace conditioning.

Internal representations X_i determine both the read-in and read-out of associative LTM (see the discussion that follows). Simultaneous control of memory read-in and memory read-out by X_i, a property that characterizes memory storage in some neural networks, is a most important feature of the model. The magnitude of the internal representation of CS_i, X_i, modulates both (1) the rate of change of CS_i–CS_j and CS_i–US associations (see Equations 5.4 and 5.4') and (2) the magnitude of the CR (see Equations 5.3' and 5.10). That is, X_i controls both storage and retrieval of CS_i–CS_j and CS_i–US associations.

Attentional LTM

Attentional LTM z_i reflects the association between τ_i and the total environmental novelty, Novelty, $\Sigma_k |\bar{\lambda}_k - \bar{B}_k|$ (defined in the section that follows). Attentional LTM z_i represents the past experience that CS_i is accompanied by a certain amount of Novelty. This attentional LTM is established when Novelty temporally overlaps with the STM of CS_i, τ.

We assume that z_i can adopt both positive and negative values (see Equation 5.5). Whereas positive values of z_i can be interpreted as a measure of the attention directed to CS_i, negative values of z_i can be interpreted as a measure of the *inattention* to CS_i (see Lubow, 1989, p. 192). When Novelty is large, z_i becomes positive (attention to CS_i increases). When Novelty is small, z_i becomes negative (inattention to CS_i increases).

LI is the consequence of the depressed X_i that results from the decreased attention (small positive z_i) or the inattention (negative z_i) to the preexposed CS. The magnitude of the LI effect depends on the time needed to increase X_i by reversing inattention (negative z_i) or low attention into attention (positive z_i) during conditioning.

Associative LTM

As in the Schmajuk and Moore (1988) and Schmajuk (1989) models presented in Chapter 3, associative LTM $V_{i,k}$ represents the past experience that CS_i is followed by event *k* (CS_k or the US). This associative LTM is established when event *k* temporally overlaps with the internal representation of CS_i, X_i. As noted in Chapter 3, in a real-time network, periods of acquisition and extinction occur within the same "acquisition" trial. Acquisition occurs in those periods when X_i and the US temporally overlap and extinction occurs in those periods when X_i is active in the absence of the US. Asymptotic learning is reached when the amounts of acquisition and extinction are similar within an acquisition trial.

By controlling the magnitude of X_i (see Equation 5.2), attentional LTM z_i controls the rate of change of $V_{i,US}$ associations (see Equation 5.4). In addition, by controlling the magnitude of X_i, z_i also controls the size of B_{US} (see Equation 5.3') and the strength of the CR (see Equation 5.10).

Notice that whereas most models of classical conditioning incorporate CS–US associations, $V_{i,US}$, only Grossberg's (1975) and Schmajuk and DiCarlo's (1991a,b) models include an attentional LTM, z_i, to modulate the magnitude of an internal representation X_i.

Cognitive mapping

The network storing CS–CS and CS–US associations in Figure 5.2 is identical to the cognitive map network presented in Chapter 3 (Figure 3.11).

Aggregate predictions and learning

The internal representation of CS_i, X_i, becomes associated with the US in proportion to the difference $(\lambda_{US} - B_{US})$, where B_{US} is the aggregate prediction of the US by all CSs with representations active at a given time, and λ_{US} represents the intensity of the US. More generally, the aggregate prediction of CS_k by all CSs with representations active at a given time B_k is given by $B_k = \Sigma_i V_{i,k} X_i$. $V_{i,k}$ increases when $X_i(t)$ is active and event k is underpredicted, that is, $(\lambda_k - B_k)$ is greater than zero. $V_{i,k}$ decreases when $X_i(t)$ is active and event k is overpredicted, that is, $(\lambda_k - B_k)$ is smaller than zero. $V_{i,k}$ remains unchanged whenever $X_i(t)$ is zero or event k is perfectly predicted, that is, $(\lambda_k - B_k)$ equals zero.

Novelty

Novelty of an event k is defined as the absolute value of the difference between its predicted and observed magnitude at a given time t, $|\lambda_k - B_k|$. Novelty is high either when an unpredicted (unexpected) stimulus k is presented, or when a predicted (expected) stimulus k is absent. As mentioned, in real-time networks, periods of acquisition and extinction occur within the same reinforced acquisition trial. Although novelty tends to decrease in those periods when X_i and the US temporally overlap, novelty is present in those periods when X_i is active in the absence of the US. Therefore, in a real-time system, although novelty decreases as learning progresses, it is never totally eliminated.

Because in a real-time system novelty is not completely abolished during acquisition trials, the transition from acquisition to extinction trials might not result in a significant increase in novelty. In other words, whereas real-time systems can easily detect the unexpected presentation of a CS, they might not be able readily to detect the unexpected absence of a CS. In order to solve this problem, the neural network shown in Figure 5.2 computes novelty at time t as $|\bar{\lambda}_k - \bar{B}_k|$, that is, the absolute value of the difference between $\bar{\lambda}_k$, the running average of the observed value of event k, and $\bar{B}$, the running average of the aggregate prediction of event k. $\bar{\lambda}_k$ and $\bar{B}$ are computed by neural elements whose outputs are proportional to present and past values of λ_k and B_k (see Equations 5.6 and 5.7).

The total amount of novelty at a given time is indicated by the sum of the novelty of all stimuli present or predicted at a given time, Novelty, $\Sigma_k |\bar{\lambda}_k - \bar{B}_k|$. We assume that CS_k can be predicted by other CSs, the context, or itself. Therefore, either repeated presentations of CS_k in a given context or simply repeated presentations of CS_k lead to a decrease in CS_k novelty. Whereas CS_k–CS_k associations decrease CS_k novelty in a context-independent manner, CS_j–CS_k or CX–CS_k associations decrease CS_k novelty in a context-dependent way. Because, according to the SLG model, decrements in Novelty are responsible for LI, CS_k–CS_k associations are responsible for context-nonspecific LI, whereas CS_j–CS_k or CX–CS_k associations are responsible for context-specific LI (see Good and Honey, 1993).

Figure 5.2 shows that Novelty becomes associated with trace τ_i. This association is represented by the attentional memory z_i that modulates the amplitude of internal representation X_i, thereby modulating the rate of learning. Therefore, the SLG model suggests that attentional memory z_i is CS_i-specific but regulated by the nonspecific, total amount of environmental Novelty (see Equation 5.5).

BOX 5.1 Schmajuk, Lam, and Gray (1996) storage–retrieval model

The most relevant aspects of the SLG model are:

1. Presentation of CS_i or the prediction of CS_i, B_i, activates STM τ_i.
2. Attentional LTM z_i represents the association between τ_i or B_i and total novelty (see 10).
3. The internal representation of CS_i, X_i, is proportional to τ_i, B_i, and z_i.
4. Associative LTM $V_{i,US}$ represents the association of X_i with the US.
5. Changes in $V_{i,US}$ are proportional to X_i.
6. The prediction of the US, B_{US}, is proportional to X_i and $V_{i,US}$.
7. Attentional LTM z_i controls the rate of change of $V_{i,US}$ by controlling X_i (see 3 and 5).
8. Attentional LTM z_i controls the size of B_{US} by controlling X_i (see 3 and 6).
9. Novelty is the absolute value of the difference between the predicted and observed intensity of an event.
10. Total Novelty, the sum of the novelty of all stimuli present or predicted at a given time, decreases during conditioning and habituation.
11. Orienting response OR is proportional to total novelty.
12. CR amplitude increases with B_{US} and decreases with OR.

Also, we assume that z_i is nonspecific for different USs, that is, different USs affect the same value of z_i.

Orienting response

Sokolov (1960) proposed that the strength of the OR might be an index of the amount of processing afforded to a given stimulus, and that this amount of processing is proportional to the novelty of the stimulus. In agreement with Sokolov's view, Kaye and Pearce (1984) suggested that the strength of the OR elicited by CS_i would be proportional to its associability α_i, with $\alpha_i = |\lambda - \Sigma_j V_j|$, where $\Sigma_j V_j$ is the sum of the associative values of all CSs present on the preceding trial, and λ is the US intensity on the previous trial. In the Pearce and Hall (1980) model, associability of a given CS decreases as its association with the US increases. In line with this approach, we assume that the total environmental Novelty, $\Sigma_k |\bar{\lambda}_k - \bar{B}_k|$, determines the magnitude of the OR.

Behavioral inhibition

Gray (1971) suggested that a behavioral inhibition system, activated by signals of punishment or nonreward, innate fear stimuli, or novel stimuli, gives rise to behavioral inhibition, increased arousal, and increased attention to environmental stimuli. In line with this approach, we suggest that CR amplitude (1) increases proportionally to the magnitude of the prediction of the US and (2) decreases in proportion to the magnitude of the OR, $CR = \Sigma_i V_{i,k} X_i - \Sigma_k |\bar{\lambda}_k - \bar{B}_k|$. Similarly, Gabriel and Schmajuk (1990) suggested that the mismatch between

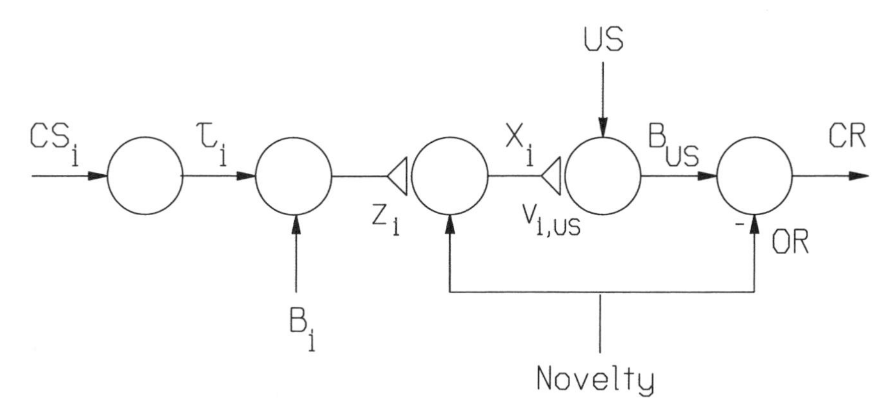

Figure 5.3. Basic mechanisms of the network. CS_i: conditioned stimulus; τ_i: trace STM of CS_i; X_i: internal representation of CS_i; z_i: CS_i–novelty association; $V_{i,US}$: CS_i–US association; US: unconditioned stimulus; B_i: aggregate prediction of CS_i; B_{US}: aggregate prediction of the US; CR: conditioned response; OR: orienting response. Arrows represent fixed synapses, open triangles variable synapses.

the actual and predicted intensity of the US might also control behavioral inhibition during avoidance: When a mismatch is detected, the avoidance response is inhibited.

As observed by McLaren, Kaye, and Mackintosh (1989), if the CR was simply described by $CR = \Sigma_i V_{i,k} X_i$, the model would predict increased responding after conditioning in a given context, followed by a change to another context where reinforcement continues (as the result of an increased Novelty, z_i, and an increased internal representation of CS_i, X_i, following the context change), a prediction not consistent with experimental evidence. However, in the SLG model, this potential increment in the amplitude of the CR is counterbalanced by an increase in the magnitude of the OR. Notice that this property of the SLG model is an essential part of Gray's (1971) behavioral inhibition system.

Box 5.1 summarizes the basic mechanisms of the network. According to the SLG model, LI is manifested because:

1. CS_i preexposure reduces Novelty (see item 10), therefore reducing z_i (see 2) and X_i (see 3).
2. A reduced X_i decreases (1) the rate of change (associability) of $V_{i,US}$ (see 5) and (2) the value of $V_{i,US}X_i$ (retrievability) that controls CR magnitude (see 6 and 12).
3. The combination of a small $V_{i,US}$ activated by a small X_i results in a reduced CR.

Interestingly, although CS preexposure results in a reduced $V_{i,US}$, the size of the CR can be increased by increasing z_i and X_i (see Figure 5.3). Therefore, experimental manipulations capable of activating X_i in the presence of large values of total Novelty will increase z_i (see 2) and X_i (see 3), thereby incrementing the magnitude of the CR without increasing $V_{i,US}$ (see 6). That is, even if CS preexposure results in a decreased $V_{i,US}$ value and CR magnitude during conditioning (LI), the magnitude of the CR can be increased by increasing z_i and X_i after conditioning (attenuation of LI).

The next section demonstrates with computer simulations how the SLG model deals with experimental data regarding LI.

CS PREEXPOSURE

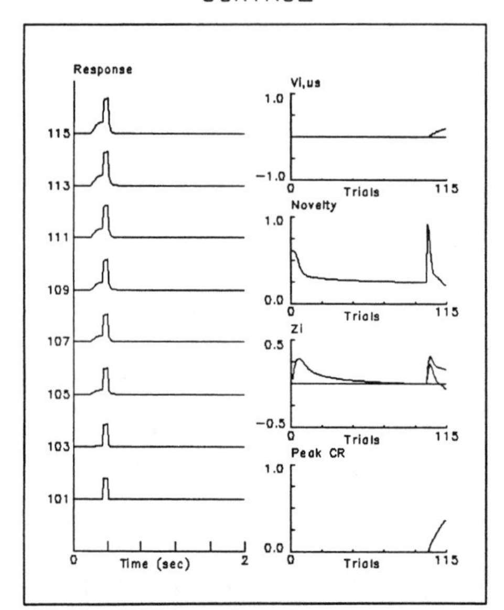

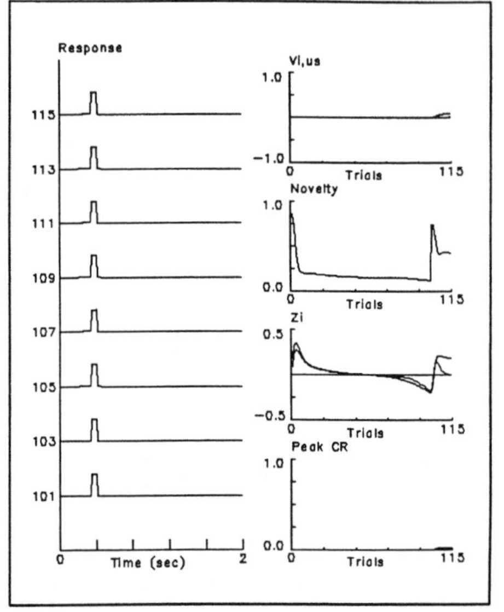

Figure 5.4. Real-time simulations of classical conditioning following CS preexposures or context-alone preexposure (control). *Left side of each panel.* Real-time simulated conditioned and unconditioned response in 15 successive trials. The first reinforced trial, trial 101, is represented at the bottom of the panel. *Right side of each panel.* Peak CR: peak CR as a function of trials; z_i: average z_i's as a function of trials; Novelty as a function of trials; $V_{i,\mathrm{US}}$ associations as a function of trials.

Computer simulations

By applying the neural network described in Figure 5.2, this section compares a large number of experimental results regarding LI, summarized in the section on latent inhibition in Chapter 2, with computer simulations. Figure 5.3 might be helpful to understand the basic mechanisms that explain the effects of different experimental designs on LI. All simulations were carried out with identical parameter values. Simulation results are very robust for a large range of parameter values. Parameter values used in the simulations are presented in Appendix 5.B.

Paradigms that yield retardation of excitatory or inhibitory conditioning

This section analyzes several paradigms that yield retardation of conditioning: (1) CS(A), CS(B), or context-alone preexposure preceding CS(A) excitatory conditioning (LI of excitatory conditioning), (2) CS(A) preexposure preceding CS(A) inhibitory conditioning (LI of inhibitory conditioning), and (3) CS(A)–weak US pairings preceding excitatory conditioning.

Latent inhibition of excitatory conditioning

In order to illustrate the operation of the SLG model, we describe a detailed simulation of LI. Figure 5.4 shows real-time simulations of 15 conditioning trials following 100 CS preexposure (CS preexposure) or context preexposure (control) trials. In the control case, Novelty decreases during preexposure to the context (trials 1–100). In trial 101, Novelty and z_i dramatically increase because neither the US nor the CS had been presented before. Therefore, conditioning proceeds at

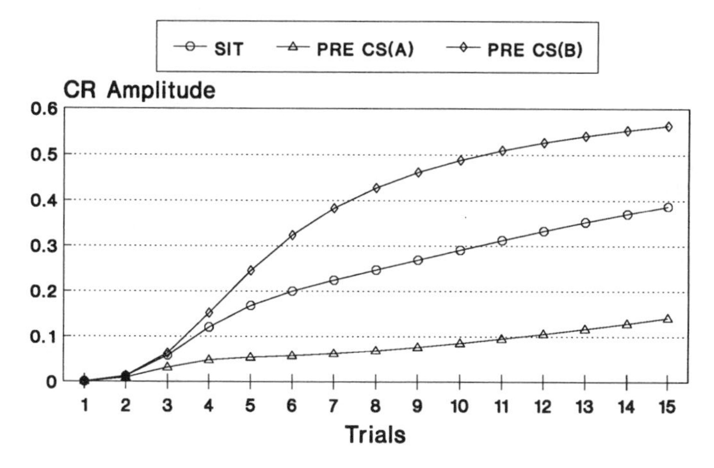

Figure 5.5a. Latent inhibition of excitatory conditioning. Peak CR amplitude during 15 CS(A)–US conditioning trials with US = 1 following (1) 50 context-alone preexposure trials (SIT), (2) 50 CS(A) preexposure trials [PRE CS(A)], or (3) 50 CS(B) preexposure trials [PRE CS(B)].

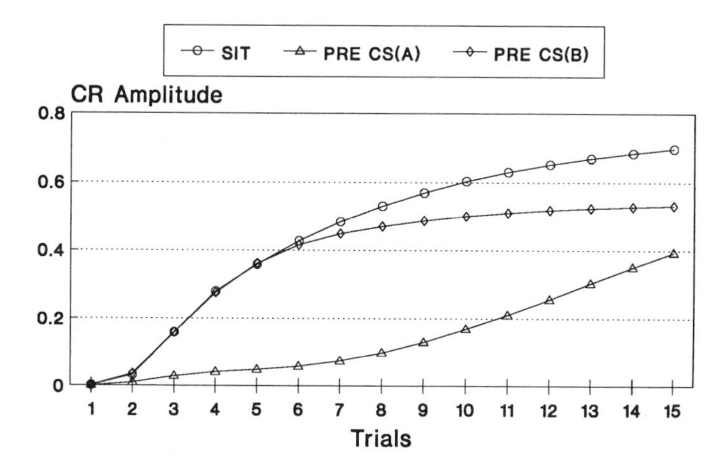

Figure 5.5b. Latent inhibition of excitatory conditioning. Peak CR amplitude during 15 CS(A)–US conditioning trials with US = 1.8 following (1) 50 context-alone preexposure trials (SIT), (2) 50 CS(A) preexposure trials [PRE CS(A)], or (3) 50 CS(B) preexposure trials [PRE CS(B)].

a fast rate. Although behavioral inhibition also increases, its effect on the CR is not large enough to compensate for the increased z_i. In the CS preexposure case, Novelty (which is initially larger than in the control case) decreases during CS preexposure (trials 1–100). In trial 101, Novelty and z_i show a moderate increment because only the US is novel. Therefore, conditioning proceeds at a slower rate than in the control case. Although behavioral inhibition also decreases, its effect on the CR is not large enough to compensate for the decreased z_i. Thus, according to the SLG model, LI results from decreasing Novelty and z_i (increased inattention) at the time of conditioning by preexposing the animal to the CS.

Stimulus specificity

Figure 5.5a shows peak CR amplitude during 15 CS(A)–US conditioning trials with US of intensity 1, following either 50 CS(A), 50 CS(B), or 50 context-alone preexposure (control or "situation," SIT) trials. Preexposure to CS(A), but not to CS(B) or the context alone, retards the magnitude of CS(A)–US asso-

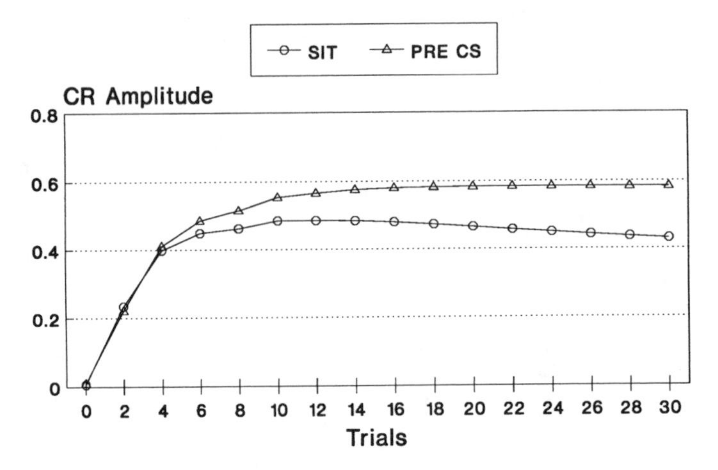

Figure 5.6. Latent inhibition of inhibitory conditioning. Peak CR amplitude during 30 CS(A)–CS(B) conditioning trials when 30 CS(A)–US trials are alternated with 30 CS(A)–CS(B) compound trials, following (1) 200 context-alone preexposure trials (SIT), or (2) 200 CS(B) preexposure trials (PRE CS).

ciations. According to the SLG model, because LI depends on Novelty at the time of the CS(A) presentation, CS(B) or context-alone presentations do not produce LI of CS(A), that is, LI is stimulus specific (e.g., Lubow and Moore, 1959, experiment 1). Like Figure 5.5a, Figure 5.19 also displays peak CR amplitude during CS(A)–US conditioning trials following preexposure to CS(B) (group SnEo, new CS in the old context).

Figure 5.5b shows that, according to the SLG model, when a US of intensity 1.8 is used (the value used in the following simulations), although preexposure to CS(B) yields a higher asymptotic level of responding than preexposure to CS(A), thereby demonstrating CS specificity, it yields a lower asymptotic level of responding than preexposure to the context alone. This lower asymptotic value is caused by the vigorous OR induced by the large Novelty originated by a combined presentation of the novel CS(A) and the strong US.

Latent inhibition of inhibitory conditioning

Figure 5.6 shows the effect of (1) preexposure to the context and (2) CS preexposure on the acquisition of conditioned inhibition. In the simulations, 30 reinforced trials with CS(A) alone alternated with 30 nonreinforced trials with a CS(A)–CS(B) compound, following 200 CS(B) or 200 context-alone (SIT) nonreinforced trials. Figure 5.6 shows that because the CR to CS(A)–CS(B) is stronger in the preexposed than in the control case, the SLG model correctly describes that CS preexposure retards the acquisition of conditioned inhibition (e.g., Rescorla, 1971a). As in the case of excitatory conditioning, acquisition of inhibitory conditioning is delayed because z_i is smaller in the preexposed than in the SIT case.

Preconditioning to CS–weak US presentations

Figure 5.7 shows peak CR amplitude during 15 conditioning trials with a strong US following 50 conditioning trials with a weak US. In agreement with Hall and Pearce's (1979) data, conditioning to a weak US retards the acquisition of

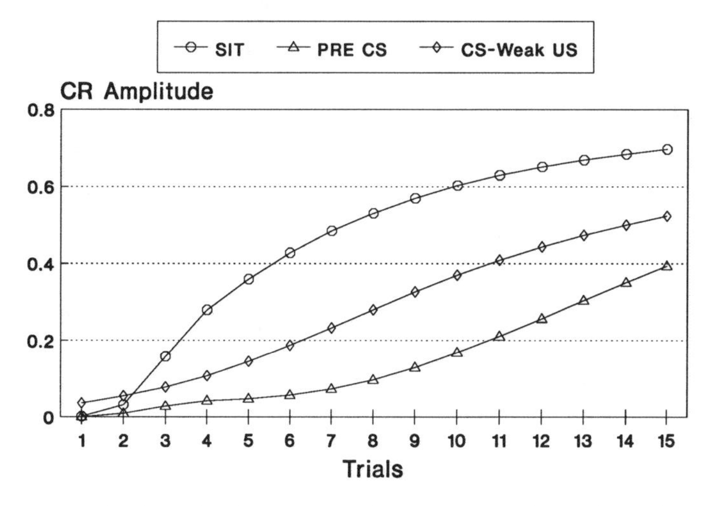

Figure 5.7. Conditioning following training with a weak US. Peak CR amplitude during CS(A)–US conditioning trials with a strong US, after (1) 50 context-alone preexposure trials (SIT), (2) 50 CS(A) preexposure trials (PRE CS), and (3) 50 reinforced CS(A)–weak US trials.

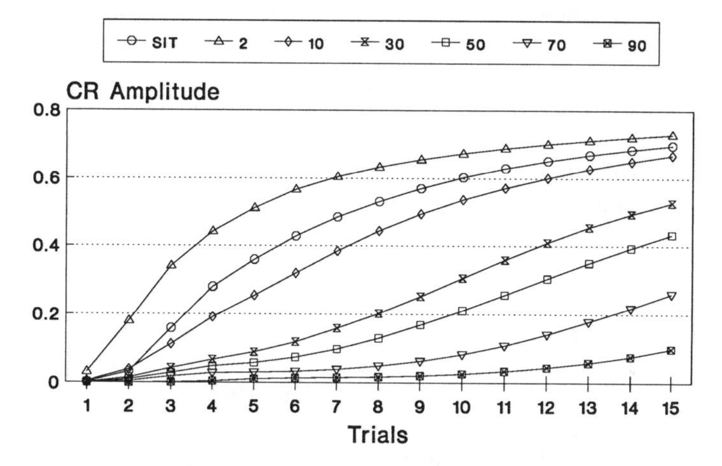

Figure 5.8. Latent inhibition as a function of the number of CS preexposures. Peak CR amplitude during 15 CS(A)–US conditioning trials following (1) 50 context-alone preexposure trials (SIT), (2) 2, (3) 10, (4) 30, (5) 50, (6) 70, and (7) 90 CS(A) preexposure trials.

conditioning to a strong US because Novelty and z_i decrease during CS–weak US conditioning, and consequently, conditioning is retarded during the CS–strong US conditioning phase. Although some CS–weak US association is accrued, it does not compensate for the decreased Novelty during conditioning.

Parameters of latent inhibition

This section examines the effects of different parameters of preexposure on the strength of LI: (1) number of CS preexposures, (2) CS duration, (3) total duration of CS preexposure, (4) CS intensity, and (5) ITI durations.

Number of CS preexposures

Figure 5.8 shows peak CR amplitude during 15 conditioning trials following 2, 10, 30, 50, 70, and 90 CS preexposure trials. In agreement with experimental

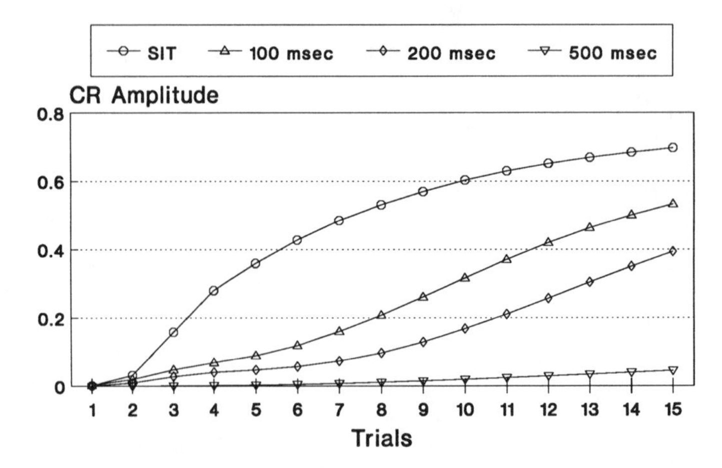

Figure 5.9. Latent inhibition as a function of CS duration. Peak CR amplitude during 15 CS(A)–US conditioning trials following (1) 50 context-alone preexposure trials (SIT), or 50 CS(A) preexposure trials with (2) 100-, (3) 200-, or (4) 500-msec CS(A).

data, LI is a nonmonotonic function of the number of CS preexposures: The CR increases for a small number of CS preexposures and later decreases with the number of CS preexposures. According to the SLG model, both Novelty and z_i increase during the first trials of CS preexposure (see Figure 5.4), thereby facilitating conditioning. With an increasing number of CS preexposures, CS–CS and CS–CX associations increase, thereby decreasing Novelty and z_i and consequently producing LI.

CS duration

Figure 5.9 shows peak CR amplitude during 15 conditioning trials following 50 CS preexposure trials with CS of 100-, 200-, and 500-msec duration. In agreement with experimental data (see Lubow, 1989, p. 63), simulation results indicate that LI increases with increasing CS durations. According to the SLG model, CX–CS and CS–CX associations are larger, and consequently, Novelty is smaller, as the duration of the CS increases. As Novelty decreases, so does z_i, and LI increases.

Total duration of CS preexposure

In agreement with Ayres et al.'s (1992) results, Figure 5.10 shows that 80 CS preexposure trials with a 100-msec CS, 40 CS preexposure trials with a 200-msec CS, and 20 CS preexposure trials with a 400-msec CS (total CS-preexposure time of 8,000 msec) yield similar levels of LI. As in Ayres et al.'s (1992) data, increasing the number of preexposure trials, from 20–80, has a small beneficial effect.

CS intensity

Figure 5.11 shows peak CR amplitude during 15 conditioning trials following 50 CS preexposure trials with CS of amplitude 0.5 and 1 during both preexposure and training. In agreement with Crowell and Anderson (1972), LI increases with the intensity of the preexposed CS. According to the SLG model, CS–CS

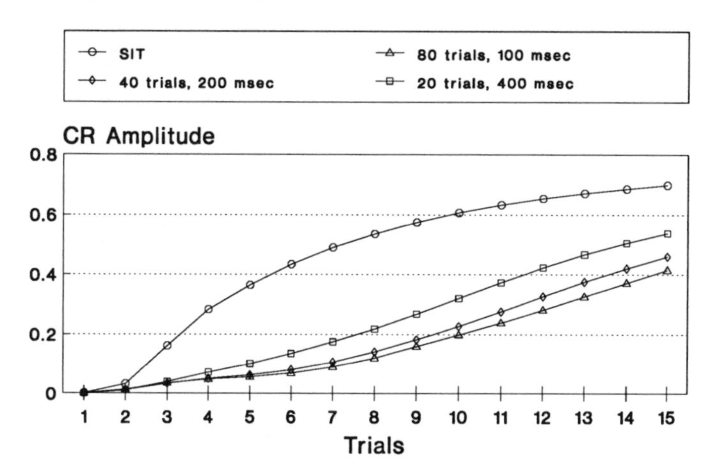

Figure 5.10. Latent inhibition and total CS–preexposure duration. Peak CR amplitude during 15 CS(A)–US conditioning trials following (1) 80 context-alone preexposure trials (SIT), (2) 80 preexposure trials with a 100-msec CS(A), (3) 40 preexposure trials with a 200-msec CS(A), or (4) 20 CS(A) preexposure trials with a 400-msec CS(A).

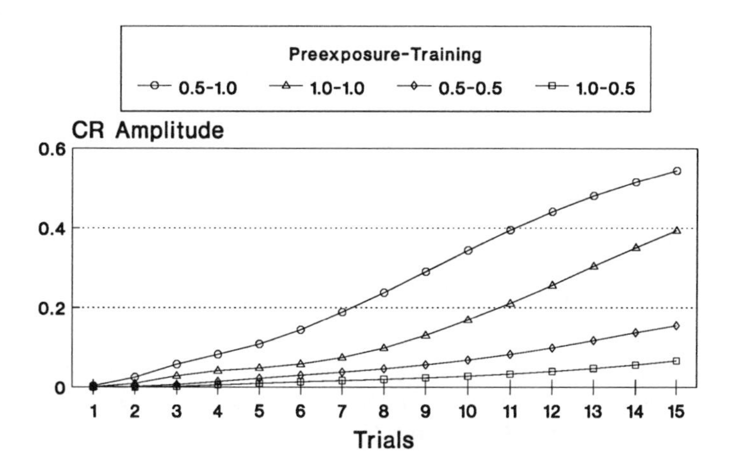

Figure 5.11. Latent inhibition as a function of CS intensity. Peak CR amplitude during 15 CS(A)–US conditioning trials following 50 CS(A) trials with different combinations of CS(A) intensities (0.5 vs. 1) during preexposure and training.

associations proceed at a faster rate with larger CS intensities (see Equations 5.4 and 5.4'). Faster increments in CS–CS associations entail smaller Novelty and z_i values and, therefore, larger LI effects.

Intertrial interval duration

Figure 5.12 shows peak CR amplitude during 15 conditioning trials with 2,000-msec ITI following 50 CS preexposure trials with 1,000-, 2,000-, 3,000-, and 5,000-msec ITI. In agreement with Lantz's (1973) and Schnur and Lubow's (1976) experimental results, LI increases with increasing ITIs. According to the SLG model, CX–CX associations increase with increasing ITIs, thereby decreasing Novelty and z_i, and facilitating LI. In addition, the model suggests that, because LI increases with increasing CS durations, the beneficial effects of in-

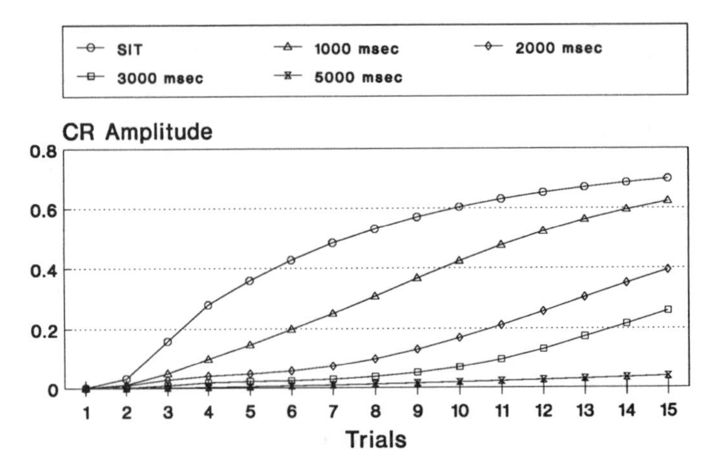

Figure 5.12. Latent inhibition as a function of intertrial interval duration. Peak CR amplitude during 15 CS(A)–US conditioning trials following (1) 50 context-alone preexposure trials (SIT), 50 CS(A) preexposure trials with (2) 1,000-, (3) 2,000-, (4) 3,000-, or (5) 5,000-msec ITI.

creasing ITI durations may become apparent only when a relatively short CS is used [5–10 sec as in Lantz's (1973) and Schnur and Lubow's (1976) experiments], but not when a relatively long CS is used [15–30 sec as in DeVietti and Barrett's (1986) and Crowell and Anderson's (1972) experiments].

Preexposing to different combinations of CSs

This section explores the consequences of preexposing to different combinations of CSs: (1) CS(A)–CS(B) preexposure preceding CS(A) conditioning, (2) CS(B) preexposure followed by CS(A)–CS(B) preexposure preceding CS(A) conditioning, (3) CS(B) preexposure followed by CS(A) preexposure preceding CS(A) conditioning, (4) preexposure to a compound or to the elements preceding compound conditioning, and (5) CS(A) preexposure followed by the introduction of a surprising event previous to CS(A) conditioning.

Preexposure to a compound CS

Figure 5.13 shows peak CR amplitude during 15 CS(A)–US conditioning trials following 50 CS(A)–CS(B) preexposure trials ["overshadowing" of LI, group CS(B)–CS(A)]. In agreement with Mackintosh (1973), Rudy, Kranter, and Gaffuri (1976), Holland and Forbes (1980), and Honey and Hall (1988, 1989), simultaneous presentation of the target CS(A) with a second stimulus, CS(B), during preexposure results in a reduced LI to CS(A). Lubow, Schnur, and Rifkin (1976) observed that the effect of adding CS(B) during preexposure of CS(A) could be the result of the absence of CS(B) during conditioning. In the same vein, according to the SLG model, simultaneous preexposure of CS(A) and CS(B) reduces LI because the absence of CS(B) during conditioning increases Novelty and z_A. In disagreement with Honey and Hall (1988, experiment 1), however, computer simulations show that LI is attenuated in the model by both simultaneous [200-msec CS(A) and CS(B)] and sequential [200-msec CS(A) and CS(B) with 400-msec interstimulus interval] presentations of CS(A) and CS(B).

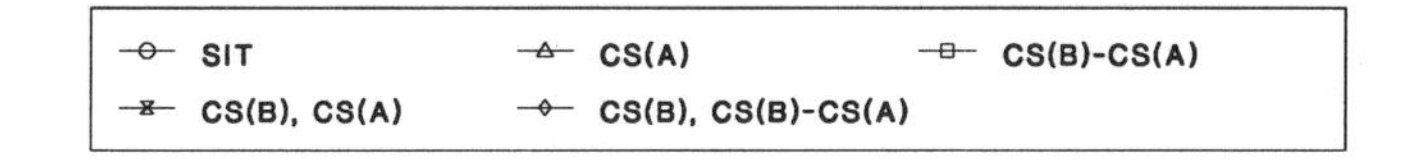

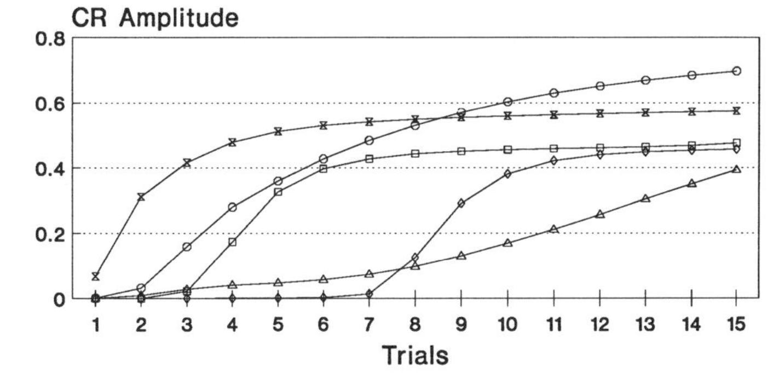

Figure 5.13. "Blocking" and "overshadowing" of latent inhibition (LI). Peak CR amplitude during 15 CS(A)–US trials preceded by (1) 50 context-alone preexposure trials (SIT), (2) 50 CS(A) preexposure trials [CS(A)], (3) 50 CS(B)–CS(A) preexposure trials [CS(B)–CS(A); overshadowing], (4) 50 CS(B) preexposure trials followed by 50 CS(A) preexposure trials [CS(B), CS(A)], and (5) 200 CS(B) preexposure trials followed by 50 CS(A)–CS(B) preexposure trials [CS(B), CS(B)–CS(A); blocking].

Figure 5.13 also shows peak CR amplitude during 15 CS(A)–US conditioning trials following 50 CS(A)–CS(B) preceded by 200 CS(B) preexposure trials ["blocking" of LI, group CS(B), CS(B)–CS(A)]. In agreement with Honey and Hall's (1988) data, whereas with relatively few (50) CS(B) preexposures blocking and overshadowing of LI reduce LI by similar amounts, blocking with more extensive (200) CS(B) preexposure tends to preserve LI [group CS(B), CS(A)]. In the framework of the model, the blocking of LI tends to preserve LI because CS(B) preexposure decreases Novelty at the time of CS(A)–CS(B) preexposure, thereby decreasing z_A at the time of conditioning and sparing LI.

Finally, Figure 5.13 shows that, in agreement with Rudy et al.'s (1976) results, LI is reduced by presentations of CS(B) prior to preexposing CS(A). According to the SLG model, CS(B) presentations result in increased Novelty when CS(B) is omitted during CS(A) preexposure, thereby increasing z_A and impairing LI.

Preexposure to a CS1–CS2 compound or its elements

In agreement with Baker, Hawkins, and Hall (1990), Figure 5.14 shows that 30 alternated preexposure trials to CS1–CS2 and the context (group CS1–CS2, CX/CS1–CS2+) are more effective than 30 alternated preexposure trials to CS1 and CS2, separately (group CS1, CS2/CS1–CS2+), in yielding LI to the CS1–CS2 compound. According to the SLG model, preexposure to the elements results in an increased Novelty when the elements are presented in a compound during conditioning, thereby producing a smaller LI effect than preexposure to the compound. Interestingly, although Baker et al.'s (1990) and Holland and Forbes's (1980) experiments are very similar, the model suggests that preexposure to the context might play an important role. In agreement with Holland and Forbes's (1980) data, preexposure to CS1–CS2 is *less* effective than alternated preexposure to separate presentations of CS1 and CS2 in yielding LI to the CS1–CS2 compound, when preexposure to the CS1–CS2 compound is not alternated with preexposure to the context.

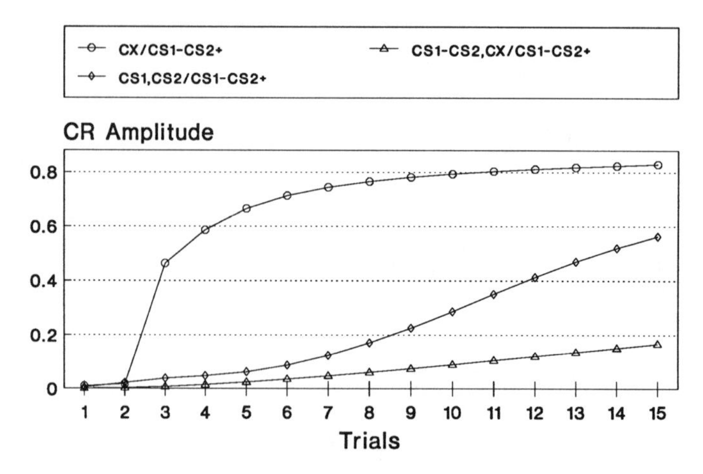

Figure 5.14. Compound or element preexposure. Peak CR amplitude during 15 CS1-CS2+ trials preceded by (1) 60 CX-alone preexposure trials, (2) 30 CS1–CS2 trials alternated with 30 CX-alone trials, and (3) 30 CS1 trials alternated with 30 CS2 trials.

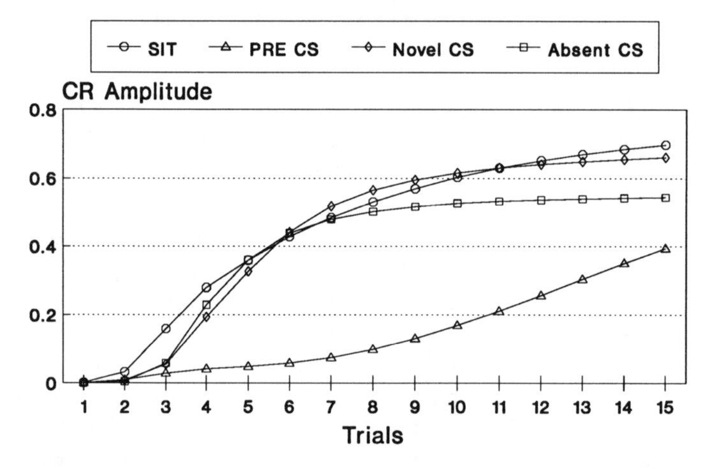

Figure 5.15. Surprising events prior to conditioning. Peak CR amplitude during 15 conditioning CS(A)–US trials after (1) 50 context-alone preexposure trials (SIT), (2) 50 CS(A) preexposure trials (PRE CS), (3) 50 CS preexposure trials followed by 5 trials in which a novel CS is introduced (novel CS), or (4) 50 CS preexposure trials in which a second CS (later omitted during conditioning) is presented (absent CS).

Preexposure to CS(A) followed by the introduction of a surprising event

Figure 5.15 shows peak CR amplitude during 15 conditioning CS(A)–US trials after (1) 50 CS preexposure trials followed by 5 trials in which a novel CS is introduced, or (2) 50 CS preexposure trials in which a second CS (later omitted during conditioning) is presented. In agreement with experimental data (e.g., Lantz, 1973, experiment 3; Hall and Pearce, 1982), both procedures attenuate LI. In the framework of the SLG model, both the presentation of a novel CS and the omission of a familiar CS increase Novelty during conditioning and, therefore, z_i, decreasing LI.

Contextual manipulations and the strength of latent inhibition

This section evaluates several contextual manipulations on the strength of LI: (1) preexposure to the apparatus prior to CS preexposure, (2) context presenta-

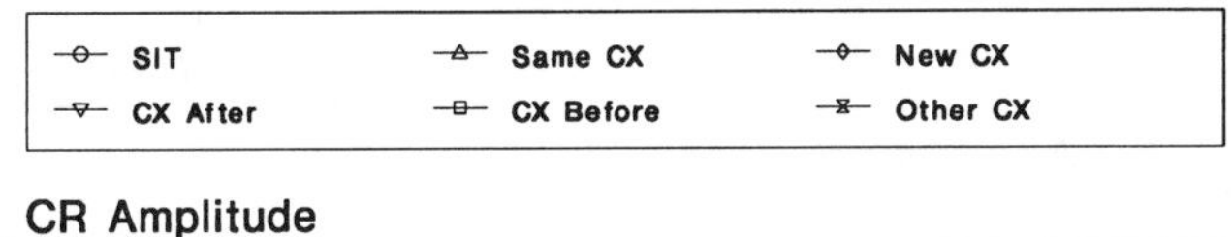

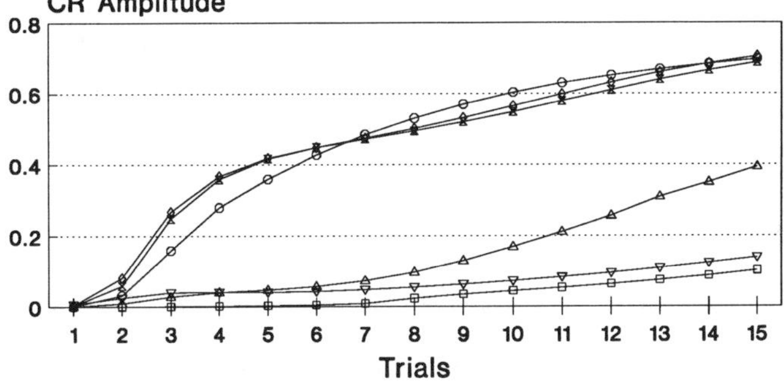

Figure 5.16. Contextual effects. Peak CR amplitude during 15 CS(A)–US trials preceded by (1) 50 context-alone preexposure trials (SIT), (2) 50 CS(A) preexposure trials in the same context used during conditioning (same CX), (3) 50 CS(A) preexposure trials in a context different from that of conditioning (new CX), (4) 50 CS(A) preexposure trials followed by 300 preexposure trials in the same context (CX after), (5) 500 context-alone preexposure trials followed by 50 CS(A) preexposure trials in the same context (CX before), and (6) 50 CS(A) preexposure trials in a context different from that of conditioning followed by 50 preexposure trials in the conditioning context (other CX).

tions following CS preexposure, (3) context changes following CS preexposure, (4) preexposure to the apparatus prior to CS preexposure combined with context changes following CS preexposure, (5) context changes and cuing treatments following CS preexposure, and (6) combinations of context and CS changes from preexposure to conditioning.

Preexposure to the apparatus prior to CS preexposure

Figure 5.16 (CX before) shows peak CR amplitude during 15 conditioning trials in context 1 following 500 context-alone preexposure trials in context 1 and 50 CS preexposure trials in context 1. In accord with Hall and Channel (1985a), LI is facilitated by context preexposure prior to CS preexposure. According to the SLG model, preexposure to the context decreases Novelty and therefore decreases z_i, facilitating LI.

Context presentations following CS preexposure

Figure 5.16 (CX after) shows peak CR amplitude during 15 conditioning trials in context 1 following 50 CS preexposure trials and 300 context-alone preexposure trials in context 1. In agreement with Hall and Minor (1984, experiment 5), exposure to the context alone, interposed between CS preexposure and conditioning in the same context, somewhat facilitates LI. In the framework of the SLG model, Novelty increases during exposure to the context after CS preexposure, but additional context presentations further reduce Novelty. Because Novelty is lower after context presentations following CS preexposure, z_i is reduced and LI enhanced.

Context changes following CS preexposure

Figure 5.16 shows peak CR amplitude during 15 CS(A)–US trials preceded by 50 CS(A) preexposure trials in the same context used during conditioning

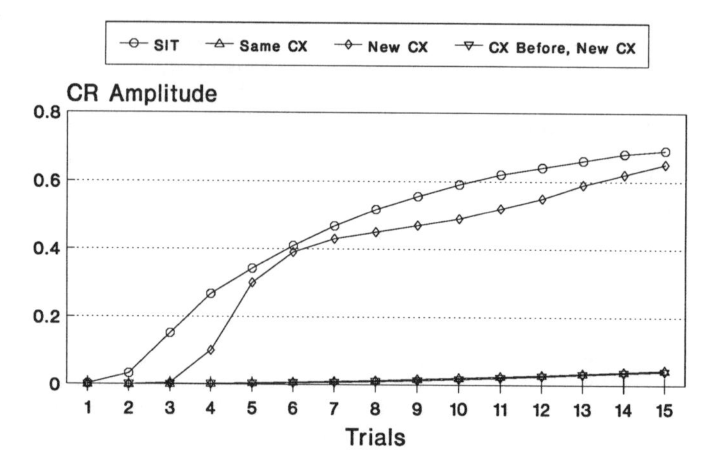

Figure 5.17. Restoration of LI in a novel context by context preexposure. Peak CR amplitude during 15 CS(A)–US trials preceded by (1) 100 context-alone preexposure trials in the same context used during conditioning (SIT), (2) 100 CS(A) preexposure trials in the same context used during conditioning (same CX), (3) 100 CS(A) preexposure trials in a context different from that of conditioning (new CX), (4) 500 context-alone preexposure trials followed by 100 CS(A) preexposure trials in a context different from that of conditioning (CX before, new context).

(same CX), 50 CS(A) preexposure trials in a context different from that of conditioning (new CX), and 50 CS(A) preexposure trials in a context different from that of conditioning followed by 50 preexposure trials in the conditioning context (other CX). In agreement with Wickens, Tuber, and Wicken's (1983) and Hall and Channel's (1985a) data, Figure 5.16 shows that LI is attenuated by a change in context from the preexposure to the conditioning stage (new CX). Also, in agreement with Hall and Minor (1984) and Lovibond, Preston, and Mackintosh (1984), LI attenuation occurs even when animals have been exposed to the conditioning context in the absence of the CS after CS preexposure (other CX).

In the framework of the SLG model, LI is attenuated by a change in context from preexposure to conditioning because Novelty increases due to the combined unconfirmed expectation and unexpected presentation when the context is changed and the CS remains the same. Because exposure to the conditioning context between CS preexposure and conditioning in the new context only decreases Novelty slightly at the time of conditioning, it still results in LI attenuation.

Preexposure to the apparatus prior to CS preexposure combined with context changes following CS preexposure

Figure 5.17 shows peak CR amplitude during 15 CS(A)–US trials preceded by (1) 100 context-alone preexposure trials in the same context used during conditioning (SIT), (2) 100 CS(A) preexposure trials in the same context used during conditioning (same CX), (3) 100 CS(A) preexposure trials in a context different from that of conditioning (new CX), (4) 500 context-alone preexposure trials followed by 100 CS(A) preexposure trials in a context different from that of conditioning (CX before, new context). In agreement with McLaren et al. (1994), the SLG model predicts that LI is preserved if animals are exposed to

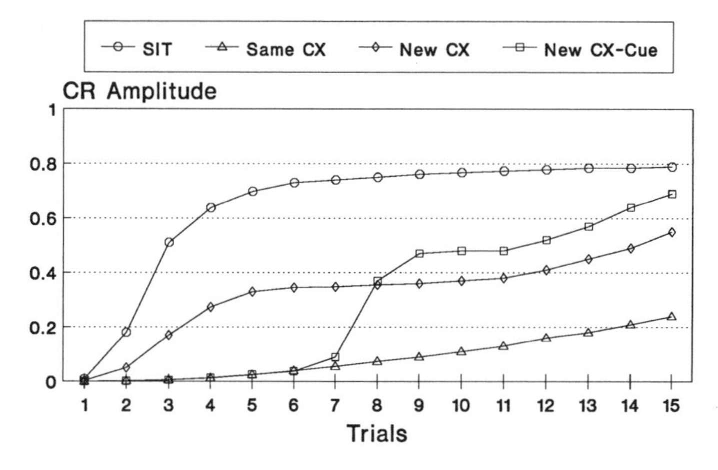

Figure 5.18. Restoration of LI in a novel context by a cuing treatment. Peak CR amplitude during (1) 15 CS(A)–CS(B) conditioning trials in context 1 after 50 context 1 preexposure trials (SIT), (2) 15 CS(A)–CS(B) reinforced trials in context 1 after 50 CS(A)–CS(B) preexposure trials in context 1 (same CS), (3) 15 CS(A) reinforced trials in context 2 after 50 CS(A)–CS(B) preexposure trials in context 1 (new CX), and (4) 15 CS(A)–CS(B) reinforced trials in context 2 after 50 CS(A)–CS(B) preexposure trials in context 1 (new CX-Cue).

the context of CS preexposure before the CS preexposure phase even when conditioning occurs in a different context.

The results can be easily explained in terms of Schmajuk and Moore's (1988) and McLaren et al.'s (1989) models. According to both models, exposure to the context generates CX–CX associations that slow down the formation of CX–CS associations during the CS preexposure phase. To the extent that LI is now the consequence of the CS–CS associations formed during CS preexposure, LI becomes context-independent (see the section on Novelty). According to the SLG model, context exposure decreases Novelty during CS preexposure, thereby decreasing the value of z_i to a greater extent than CS preexposure alone. During conditioning, the smaller initial value of z_i compensates for the increased Novelty introduced by the new context.

Restoration of LI in a novel context by a cuing treatment

Figure 5.18 shows peak CR amplitude during (1) 15 CS(A)–CS(B) conditioning trials in context 1 after 50 context 1 preexposure trials (SIT), (2) 15 CS(A)–CS(B) reinforced trials in context 1 after 50 CS(A)–CS(B) preexposure trials in context 1 (same CX), (3) 15 CS(A) reinforced trials in context 2 after 50 CS(A)–CS(B) preexposure trials in context 1 (new CX), and (4) 15 CS(A)–CS(B) reinforced trials in context 2 after 50 CS(A)–CS(B) preexposure trials in context 1 (new CX-cue). The intensity of contexts 1 and 2 was assumed to be 0.2. In agreement with Gordon and Weaver (1989), simulations show that the addition of a reminder CS(B) during conditioning reinstates LI otherwise attenuated by the change from context 1 during preexposure to context 2 during the first trials of conditioning.

In terms of the model, reinstatement of LI by a stimulus of the preexposure context is due to the decrease in Novelty that follows the prediction of the CS by the contextual CS(B). It should be noted, however, that because the model assumes relatively short CS traces (see Equation 5.1), it cannot describe the re-

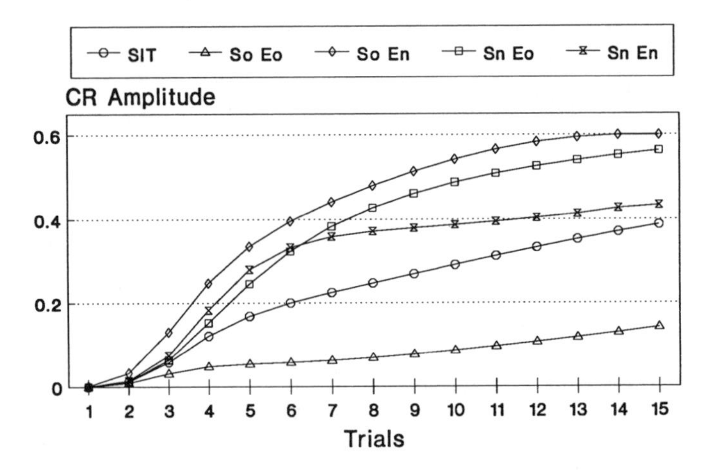

Figure 5.19. Stimulus–context interactions. Peak CR amplitude over 15 conditioning trials following 20 preexposure trials for SoEo (LI), SnEo, SoEn, and SnEn groups. Animals received CS preexposure in a given environment, followed by conditioning (1) with the old CS in the old environment (SoEo group), (2) with a new CS in the old environment (SnEo group), (3) with the old CS in a new environment (SoEn group), and (4) with a new CS in a new environment (SnEn group). The SIT group received context-alone preexposure in a given environment, followed by conditioning in the old environment.

instatement of LI by a short presentation of CS(B) in the novel context 2 and assumes simultaneous CS(A) and CS(B) presentations. Although not identical to the experimental design, the simulated results seem to capture well the essence of the paradigm.

Stimulus and context changes

Figure 5.19 shows peak CR amplitude over 15 conditioning trials following 20 preexposure trials for SoEo (old CS, old CX, LI), SnEo (new CS, old CX), SoEn (old CS, new CX), and SnEn (new CS, new CX) groups. In general, simulation results displayed in Figure 5.19 are very similar to Lubow, Rifkin, and Alek's (1976) data as displayed in Lubow, Rifkin, and Alek (1989, p. 77, Figure 5.5). In agreement with Lubow et al. (1976), the training of an old CS in a new environment (see the section on perceptual learning that follows) yields faster conditioning than training an old CS in an old environment (LI). Also in agreement with Balaz et al. (1982), conditioning to a new CS occurs faster in a familiar (SnEo) than in a novel environment (SnEn). In agreement with Lubow et al.'s (1976), Wickens et al.'s (1983), and Hall and Channel's (1985a) data, LI is attenuated by a change in context from the preexposure to the conditioning stage (SoEn group).

In the framework of the SLG model, LI is attenuated in the SnEo and SoEn cases because Novelty increases due to the combined unconfirmed expectation and unexpected presentation when either the context or the CS is changed and the other remains the same. The SnEn case shows a slower rate of learning than the SnEo and SoEn cases because Novelty is smaller in this case. Notice that in the SnEn case, Novelty results only from unexpected presentation of the new CX and CS, and not from the unconfirmed expectations of the now absent CX and CS.

Perceptual learning and latent inhibition

Figure 5.20a shows simulated results addressing Channel and Hall's (1981, experiment 1) data. It indicates peak CR amplitude during 10 discrimination trials for control, LI (same context), and perceptual learning (different context) cases. The control case received 50 context 1 preexposure trials, followed by 10 CS(A) reinforced trials alternated with 10 CS(B) nonreinforced trials in context 2. The same context case received 50 CS(A)–CS(B) compound preexposure trials in context 1, followed by 10 CS(A) reinforced trials alternated with 10 CS(B) nonreinforced trials in context 1. Finally, the different context case received 50 CS(A)–CS(B) preexposure trials in context 1, followed by 10 CS(A) reinforced trials alternated with 10 CS(B) nonreinforced trials in context 2. In agreement with Channel and Hall's (1981) and Lubow et al.'s (1976) data, the same context group shows LI [weaker conditioning to CS(A) than control animals], whereas the different context group shows perceptual learning [stronger conditioning to CS(A) than control animals]. Furthermore, computer simulations show that, consistent with experimental data (Chantrey, 1974), perceptual learning increases with increasing temporal separation of the CSs during preexposure.

Figure 5.20b shows simulated results addressing Trobalon's, Chamizo, and Mackintosh's (1992) discrimination data. It illustrates peak CR amplitude during 15 discrimination trials for control, same context, and different context cases. The control case received 10 context 1 preexposure trials, followed by 15 CS(A) reinforced trials alternated with 15 CS(B) nonreinforced trials in context 1. The same context case received 10 CS(A)–CS(B) compound preexposure trials in context 1, followed by 15 CS(A) reinforced trials alternated with 15 CS(B) nonreinforced trials in context 1. Finally, the different context case received 10 CS(A)–CS(B) compound preexposure trials in context 2, followed by 15 CS(A) reinforced trials alternated with 15 CS(B) nonreinforced trials in context 1. Simulations assumed that contextual cues are more salient in Trobalon et al.'s (1992) maze task (0.5) than in Channel and Hall's (1981) discrimination paradigm (0.1), an assumption supported by Diez-Chamizo, Sterio, and Mackintosh's (1985) results demonstrating that extramaze landmarks overshadow intramaze cues, but not vice versa. In agreement with Trobalon et al.'s data (1992), preexposure to intramaze cues resulted in perceptual learning only when these cues were presented in the same context during preexposure and discrimination training.

Apparently, the SLG model can explain Channel and Hall's (1981) and Trobalon et al.'s (1992) conflicting results. Although the simulations differ in the number of preexposure trials, the basis for the different simulated results is the intensity of the context, assumed to be stronger in the maze discrimination case. In the Channel and Hall (1981) experiment, preexposure to CS(A) and CS(B) in context 1 results in the formation of multiple CX1–CS and CS–CS associations (including self-associations) that (1) increase the internal representation of each CS (thereby fostering perceptual learning) and (2) decrease total Novelty, z_A, and z_B (thereby fostering LI). When conditioning occurs in the same context, context 1, the small initial z_A and z_B yield LI. When conditioning occurs in a different context, context 2, old CX1–CS associations are removed; total Novelty increases, thereby increasing z_A and z_B; LI is attenuated;

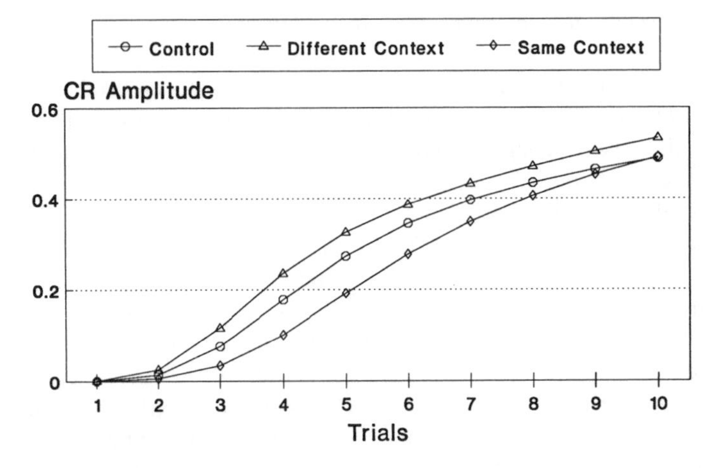

Figure 5.20a. Perceptual learning. Peak CR amplitude during 10 CS(A)–US trials, during discrimination acquisition [CS(A)+, CS(B)-] in context 1, preceded by (1) 50 context 1 preexposure trials (control), (2) 50 CS(A)–CS(B) preexposure trials in context 1 (same context), and (3) 50 CS(A)–CS(B) preexposure trials in context 2 (different context). Amplitude of contexts 1 and 2 is 0.1.

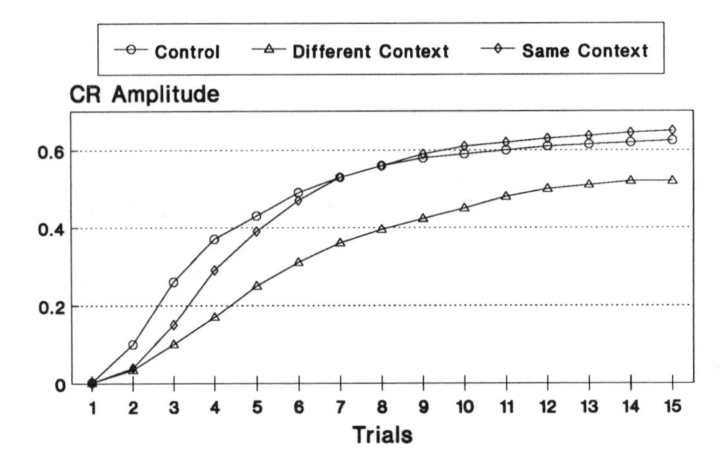

Figure 5.20b. Perceptual learning. Peak CR amplitude during 15 CS(A)–US trials, during discrimination acquisition [CS(A)+, CS(B)-] in context 1, preceded by (1) 10 context 1 preexposure trials (control), (2) 10 CS(A)–CS(B) preexposure trials in context 1 (same context), and (3) 10 CS(A)–CS(B) preexposure trials in context 2 (different context). Amplitude of contexts 1 and 2 is 0.5.

and perceptual learning (based on the remaining CS–CS associations) becomes apparent.

In the Trobalon et al. (1992) experiment, preexposure to CS(A) and CS(B) in the salient context 1 also results in the formation of multiple CX1–CS and CS–CS associations (including self-associations) that (1) increase the internal representation of each CS (thereby fostering perceptual learning) and (2) strongly decrease total novelty z_A and z_B (thereby fostering LI). Because context 1 is very salient, Novelty, z_A and z_B are very small at the end of preexposure. Either in the same context or in a different context, z_A and z_B grow to similar values and, therefore, the magnitude of the CR is determined by the strength of the CX1–CS associations. Therefore, when conditioning occurs in the same context, context 1, old CX1–CS associations are still present and perceptual learning is enhanced. When conditioning occurs in a different context, context 2, old CX1–CS associations are removed, Novelty increases, the OR increases and inhibits the CR, and perceptual learning (based on the remaining CS–CS associations) is attenuated.

As mentioned, McLaren et al. (1989) proposed a model that describes perceptual learning. According to the model, perceptual learning is the consequence of the greater decline in the associability (LI) of the common elements than that of the unique elements of two stimuli that share multiple features during preexposure. Since the context is part of the common elements of the stimuli, the model probably is able to predict different results according to the salience of the context.

Orienting response and latent inhibition

As mentioned, Pearce, Kaye, and Hall (1983) and Kaye and Pearce (1984) suggested that both (1) the strength of the OR elicited by CS_i and (2) CS_i associability were proportional to the novelty of the US, $|\lambda - \Sigma_j V_j|$. Furthermore, they proposed that the strength of the OR provides a direct index of the associability of CS_i. However, contrary to their prediction that manipulations affecting the OR should affect LI in an identical manner, Hall and Channel (1985b) and Hall and Schachtman (1987) found clear dissociations between the OR and LI.

In the same vein of Pearce and Hall's (1980) model, the SLG model assumes that OR amplitude is proportional to Novelty, $\Sigma_k |\bar{\lambda}_k - \bar{B}_k|$ (see Equation 5.8). However, in contrast to their model, the SLG model assumes that the CS_i associability is modulated by, but not identical to, Novelty or the OR. Therefore, this section applies the SLG model to illustrate the differential effects that contextual and temporal manipulations might have on the OR and LI.

Figure 5.21 shows simulated results addressing Hall and Channel's (1985b) data. Figure 5.21a displays the value of OR amplitude during 10 trials in which (1) the CS is presented in context 1 following 10 preexposure trials in context 1 (group CS–CX1/CS–CX1), (2) the CS is presented in context 2 following 10 preexposure trials in context 1 (group CS–CX1/CS–CX2), (3) the CS is presented in context 1 following 10 preexposure trials in context 1 alternated with 10 preexposure trials in context 2 (group CS–CX1, CX2/CS–CX1), (4) the CS is presented in context 2 following 10 preexposure trials in context 1 alternated with 10 preexposure trials in context 2 (group CS–CX1, CX2–/CS–CX2), (5) the CS is presented with the US in context 1 following 10 preexposure trials in context 1 alternated with 10 preexposure trials in context 2 (group CS–CX1, CX2/CS–CX1+), and (6) the CS is presented with the US in context 2 following 10 preexposure trials in context 1 alternated with 10 preexposure trials in context 2 (group CS–CX1, CX2/CS-CX2+). In accord with Hall and Channel (1985b, experiment 1), ORs strongly dishabituate when the CS is presented in the novel context 2, group (CS–CX1/CS–CX2) versus group (CS–CX1/CS–CX1). Also, in agreement with Hall and Channel (1985b, experiment 2), CS presentations in context 1 alternated with exposure to context 2 produce a relatively small dishabituation of the OR when the CS is presented in context 2, group (CS–CX1,

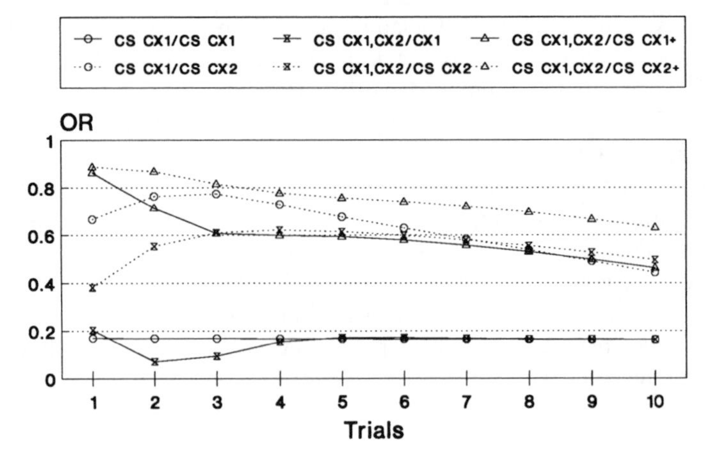

Figure 5.21a. Orienting response and latent inhibition: effects of changing context. Orienting response (OR) amplitude during 10 trials in which (1) the CS is presented in context 1 following 10 preexposure trials in context 1 (CS CX1/CS CX1), (2) the CS is presented in context 2 following 10 preexposure trials in context 1 (CS CX1/CS CX2), (3) the CS is presented in context 1 following 10 preexposure trials in context 1 alternated with 10 preexposure trials in context 2 (CS CX1, CX2/CS CX1), (4) the CS is presented in context 2 following 10 preexposure trials in context 1 alternated with 10 preexposure trials in context 2 (CS CX1, CX2/CX CX2), (5) the CS is presented with the US in context 1 following 10 preexposure trials in context 1 alternated with 10 preexposure trials in context 2 (CS CX1, CX2/CS CX1+), and (6) the CS is presented with the US in context 2 following 10 preexposure trials in context 1 alternated with 10 preexposure trials in context 2 (CS CX1, CX2/CS CX2+).

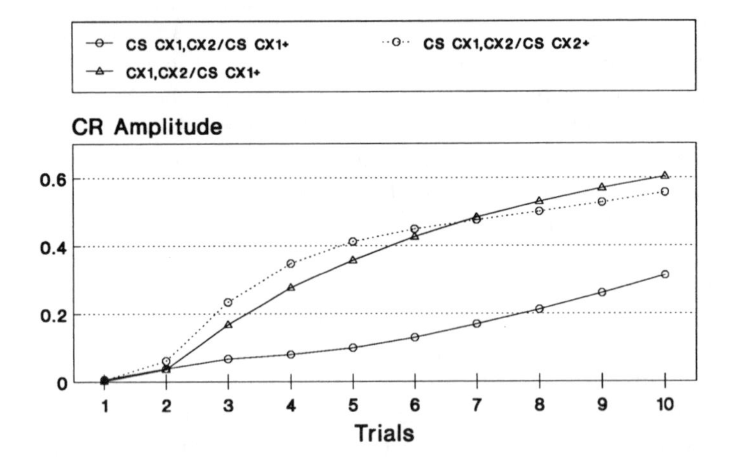

Figure 5.21b. Orienting response and latent inhibition: effects of changing context. Peak CR amplitude during (1) 10 CS–US trials in context 1 following 10 trials of CS preexposure in context 1 alternated with 10 trials of preexposure to context 2 (CS CX1, CX2/CS CX1+), (2) 10 CS–US trials in context 2 following 10 trials of CS preexposure in context 1 alternated with 10 trials of preexposure to context 2 (CS CX1, CX2/CS CX2+), and (3) 10 CS–US trials in context 2 following 10 trials of preexposure in context 1 alternated with 10 trials of preexposure to context 2 (CX1, CX2/CS CX1+).

CX2/CS–CX2) versus group (CS–CX1, CX2/CS–CX1), because in the first trials increases in Novelty are relatively small.

Figure 5.21b shows the peak CR amplitude during (1) 10 CS–US trials in context 1 following 10 trials of CS preexposure in context 1 alternated with 10 trials of preexposure in context 2 (CS–CX1, CX2/CS-CX1+), (2) 10 CS–US trials in context 1 following 10 trials of CS preexposure in context 1 alternated with 10 trials of preexposure in context 2 (CS–CX1, CX2/CS-CX1+), and (3) 10 CS–US trials in context 2 following 10 trials of preexposure in context 1 alternated with 10 trials of preexposure to context 2 (CX1, CX2/CS-CX2+).

As shown in Figure 5.21a, CS presentations in context 1 alternated with exposure in context 2 produce a relatively small dishabituation of the OR in the first reinforced trials, when the CS and the US are presented in context 2. However, whereas the OR rapidly decreases in the (CS–CX1, CX2/CS-CX1+) group, it slowly decreases in the (CS–CX1, CX2/CS-CX2+) group. As shown in Figure 5.21b, this slow rehabituation decreases LI in the (CS–CX1, CX2/CS-CX2+) group.

In summary, in agreement with Hall and Channel (1985b, experiments 2 and 3), the SLG model predicts that a context change that produces a relatively modest dishabituation, group (CS–CX1, CX2/CS–CX2) versus group (CS–CX1, CX2/CS–CX1) in Figure 5.21a, clearly impairs LI, group (CS–CX1, CX2/CS-CX2+) versus group (CS–CX1, CX2/CS-CX1+) in Figure 5.21b. According to the SLG model, CS presentations in context 1 alternated with exposure to context 2, followed by CS presentations in context 2 in the absence of the US, produce a relatively small dishabituation of the OR. However, the same preexposure procedure followed by CS presentations in context 2 in the presence of the US produces a large dishabituation of the OR. This large, slowly rehabituating OR impairs LI.

Figure 5.22 shows simulated results addressing Hall and Schachtman's (1987) data. Figure 5.22a displays the value of OR amplitude during 10 trials in which (1) the CS is presented in context 1 following 20 CS preexposure trials in context 1 and 20 preexposure trials in CX1 alone [CS–CX1/(T)/CS–CX1] (the exposure to CX1 alone is intended to represent the effect of exposure to contexts similar to the preexposure context during the period of time between CS preexposure and testing), (2) the CS is presented in context 1 following 20 preexposure trials in context 1 alone and 20 CS preexposure trials in context 1 (CX1/CS–CX1/CS–CX1), and (3) the CS is presented in context 1 following 40 preexposure trials in context 1 alone (CX1/CX1/CS–CX1). In the simulations, context amplitude is assumed to be 0.5. In agreement with Hall and Schachtman (1987), the OR dishabituates after a period of time between CS preexposure and testing, group (CS–CX1/(T)/CS–CX1) versus group (CX1/CS–CX1/CS–CX1) in Figure 5.22a. Also, in agreement with Hall and Schachtman (1987), preexposure to context 1 alone is followed by a large dishabituation when the CS is introduced in context 1, group (CX1/CX1/CS–CX1).

Figure 5.22b shows peak CR amplitude during 10 trials in which (1) the CS is presented with the US in context 1 following 20 preexposure trials in context 1 and 20 preexposure trials in CX1 alone [CS–CX1/(T)/CS-CX1+] (the exposure to CX1 alone is intended to represent the effect of exposure to contexts similar to the preexposure context during the period of time between CS preexposure and testing), (2) the CS is presented with the US in context 1 following 20 preexposure trials in context 1 alone and 20 CS preexposure trials in context 1 (CX1/CS–CX1/CS-CX1+), and (3) the CS is presented with the US in context 1 following 40 preexposure trials in context 1 alone (CX1/CX1/CS-CX1+). As shown in Figure 5.22b, CS presentations in context 1 followed by exposure to context 1 (representing the period of time between CS preexposure

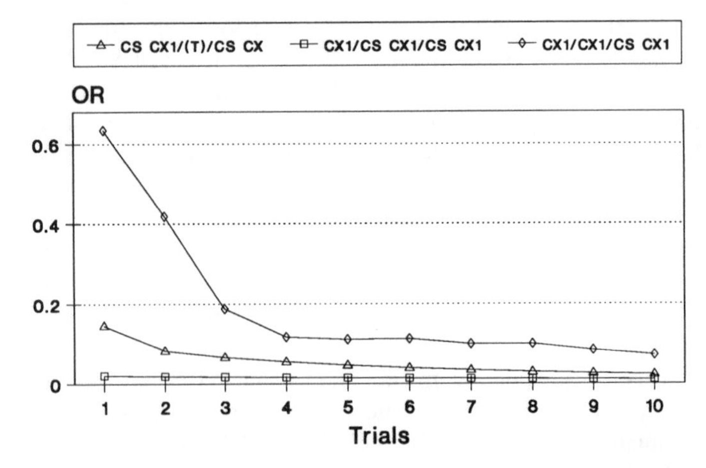

Figure 5.22a. Orienting response and latent inhibition: effects of a time interval. OR amplitude during 10 trials in which (1) the CS is presented in context 1 following 20 CS preexposure trials in context 1 and 20 preexposure trials in CX1 alone [CS CX1/CX1(T)/CS CX1] [exposure to the CX1 represents the effect of a period of time (T) between CS preexposure and testing], (2) the CS is presented in context 1 following 20 preexposure trials in context 1 alone and 20 CS preexposure trials in context 1 (CX1/CS CX1/CS CX1), and (3) the CS is presented in context 1 following 40 preexposure trials in context 1 alone (CX1/CX1/CS CX1).

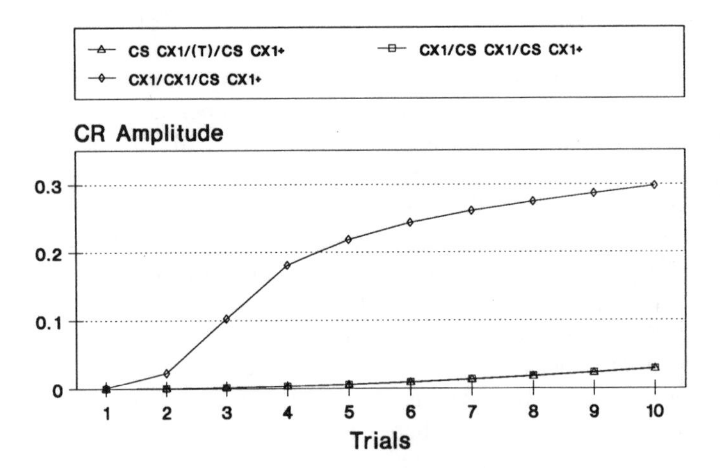

Figure 5.22b. Orienting response and latent inhibition: effects of a time interval. Peak CR amplitude during 10 trials in which (1) the CS is presented with the US in context 1 following 20 preexposure trials in context 1 and 20 preexposure trials in CX1 alone [CS CX1/CX1(T)/CS CX1+] [exposure to the CX1 represents the effect of a period of time (T) between CS preexposure and testing], (2) the CS is presented with the US in context 1 following 20 preexposure trials in context 1 alone and 20 CS preexposure trials in context 1 (CX1/CS CX1/CS CX1+), and (3) the CS is presented with the US in context 1 following 40 preexposure trials in context 1 alone (CX1/CX1/CS CX1+).

and testing) do not decrease LI, group [CS–CX1/(T)/CS-CX1+] versus group (CX1/CS–CX1/CS-CX1+) in Figure 5.22b.

In summary, in agreement with Hall and Schachtman (1987), the SLG model predicts that a period of time producing dishabituation of the OR (presumably because the animal is exposed to contexts similar to the preexposure context during this interval) does not decrease LI. According to the SLG model, LI is not attenuated even when the OR is dishabituated because z_i (which controls the

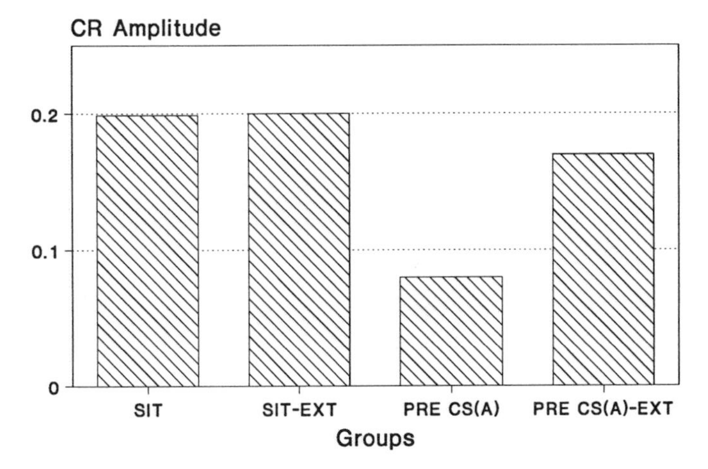

Figure 5.23. Attenuation of LI by extinction of the training context. Peak CR to CS(A) in the testing context 3 after (1) 50 context 1 preexposure trials, 5 CS(A) reinforced trials in context 1, 50 nonreinforced (extinction) trials in context 2, and 1 habituation trial in context 3 (SIT); (2) 50 context 1 preexposure trials, 5 CS(A) reinforced trials in context 1, 50 nonreinforced (extinction) trials in context 1, and 1 habituation trial in context 3 (SIT–EXT); (3) 50 CS(A) preexposure trials in context 1, 5 CS(A) reinforced trials in context 1, 50 nonreinforced (extinction) trials in context 2, and 1 habituation trial in context 3 (PRE CS(A)); and (4) 50 CS(A) preexposure trials in context 1, 5 CS(A) reinforced trials in context 1, 50 nonreinforced (extinction) trials in context 1, and 1 habituation trial in context 3 (PRE CS(A)–EXT).

rate of conditioning) increases at a slower rate than the OR (which reflects the magnitude of Novelty at a given time).

Postconditioning manipulations

Experimental data showing that some postconditioning manipulations might attenuate LI have been used to support the view that LI is a retrieval rather than a storage phenomenon. Because in the SLG model attentional memory z_i controls not only the storage, but also the retrieval of CS_i–US associations (see Equations 5.3' and 5.10), the model is able to describe the effects of postconditioning manipulations.

Attenuation of LI by extinction of the training context

Figure 5.23 shows peak CR to CS(A) in the testing context 3 after (1) 50 context 1 preexposure trials, 5 CS(A) reinforced trials in context 1, 50 nonreinforced (extinction) trials in context 2, and 1 habituation trial in context 3 (SIT); (2) 50 context 1 preexposure trials, 5 CS(A) reinforced trials in context 1, 50 nonreinforced (extinction) trials in context 1, and one habituation trial in context 3 (SIT–EXT); (3) 50 CS(A) preexposure trials in context 1, 5 CS(A) reinforced trials in context 1, 50 nonreinforced (extinction) trials in context 2, and 1 habituation trial in context 3 (PRE CS(A)); and (4) 50 CS(A) preexposure trials in context 1, 5 CS(A) reinforced trials in context 1, 50 nonreinforced (extinction) trials in context 1, and 1 habituation trial in context 3 (PRE CS(A)–EXT). In agreement with Grahame et al.'s (1992) results, the extinction of context 1 results in the attenuation of LI.

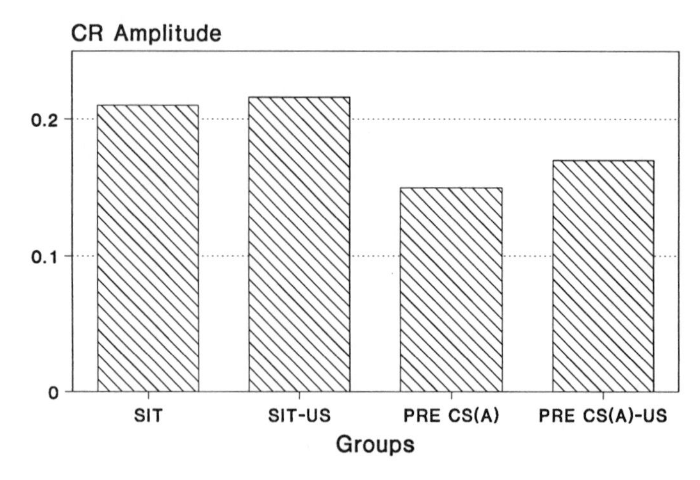

Figure 5.24. Attenuation of LI by US presentations in another context. Peak CR to CS(A) in testing context 3 after (1) 50 context 1 preexposure trials, 5 CS(A) reinforced trials in context 1, and 10 nonreinforced trials in context 2 (SIT); (2) 50 context 1 preexposure trials, 5 CS(A) reinforced trials in context 1, and 10 reinforced trials in context 2 (SIT–US); (3) 50 CS(A) preexposure trials in context 1, 5 CS(A) reinforced trials in context 1, and 10 nonreinforced trial in context 2 [PRE CS(A)]; and (4) 50 CS(A) preexposure trials in context 1 preexposure trials, 5 CS(A) reinforced trials in context 1, and 10 reinforced trials in context 2 [PRE CS(A)–US].

According to the SLG model, Novelty increases during extinction of the training context 1, thereby increasing the z_i of CS_i whose representation is activated in context 1 through CX1–CS associations. In the testing context 3, the increased z_i value of CS_i results in larger X_i and larger CRs (attenuation of LI).

Attenuation of LI by US presentations in another context

Figure 5.24 shows peak CR to CR to CS(A) in testing context 3 after (1) 50 context 1 preexposure trials, 5 CS(A) reinforced trials in context 1, 10 reinforced trials in the reminder context 2 (SIT); (2) 50 context 1 preexposure trials, 5 CS(A) reinforced trials in context 1, and 10 reinforced trial in the reminder context 2 (SIT–US); (3) 50 CS(A) preexposure trials in context 1, 5 CS(A) reinforced trials in context 1, and 10 nonreinforced trial in the reminder context 2 (PRE CS(A)); and (4) 50 CS(A) preexposure trials in context 1 preexposure trials, 5 CS(A) reinforced trials in context 1, and 10 reinforced trial in the reminder context 2 (PRE CS(A)–US). Because the reminder context was similar in some aspects to the training context 1, context 2 was represented by context 1, but its intensity was changed from 0.1 to 0.3. The testing context 3 was radically different from the training and reminder contexts. In agreement with, but not as dramatic as, Kasprow et al.'s (1984) results, Figure 5.24 shows that US presentations in another context increase responding to CS(A), thereby decreasing LI.

According to the SLG model, Novelty increases during US presentations in the reminder context 2, thereby increasing the z_i of CS_i whose representation is activated in context 2 through CX1–CS associations activated by the common elements in contexts 1 and 2. In the testing context 3, the increased z_i value of CS_i results in larger X_i and larger CRs (attenuation of LI). According to the SLG model, the similarity among preexposure, conditioning, and reminder contexts is essential to produce the observed effects. Therefore, the model predicts that the effect should be smaller as the similarities between the contexts decrease.

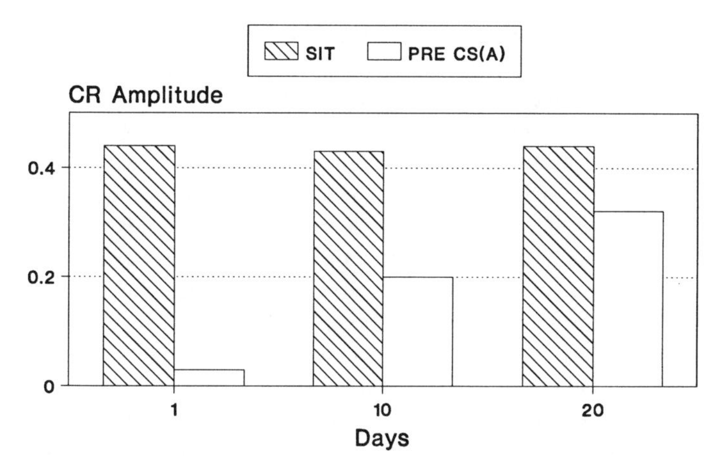

Figure 5.25. Attenuation of LI by delayed testing. Peak CR to CS(A) in context 1 after 50 context 1 preexposure trials in the presence of a continuous CS(R) representing the other rats (SIT); 15 CS(A) reinforced trials in the presence of CS(R); and 1, 1, 10, or 20 trials in the presence of CS(R) representing the other rats in the colony room during the different retention intervals (SIT). The LI cases received similar treatment with the addition of CS(A) presentations in the first phase.

Attenuation of LI by delayed testing

Figure 5.25 shows peak CR to CS(A) in context 1 after 50 context 1 preexposure trials in the presence of a contextual CS(R) representing other rats in the cage (SIT); 15 CS(A) reinforced trials in the presence of CS(R); and 1, 1, 10, or 20 trials in the presence of CS(R) representing other rats in the colony room during the different retention intervals (SIT). The LI cases received similar treatment with the addition of CS(A) presentations in the first phase. In agreement with Kraemer, Randall, and Carbary (1991), LI decreases with increasing retention intervals. Because similar simulated results are obtained in the absence of CS(R), LI attenuation seems to be independent of the experimental procedure used by Kraemer et al. (1991) consisting of collective preexposure and conditioning but individual testing.

According to the SLG model, during the retention interval Novelty increases because the CS(R) is presented in the absence of the US, thereby increasing the Novelty and z_i of CS_i whose representation is activated by CS(R) through CS(R)–CS associations. When tested, the increased z_i value of CS_i results in larger CRs (attenuation of LI). As the retention interval increases in duration, the CS representation increases its association z_i with Novelty, thereby decreasing LI in proportion to the duration of the retention interval. Importantly, CS–US associations remain constant during the retention interval.

Discussion

Comparison with other models of classical conditioning and latent inhibition

As mentioned in Chapter 4, many theories have been proposed to account for LI. Table 4.1 presents a taxonomy of the different models, summarizing and comparing their mechanisms. Models were first divided into those that describe

LI as a (1) disruption of storage of CS–US associations and (2) disruption of retrieval of the CS–US associations. The storage group was further divided into (1) a group that assumes the associability of the CS decreases during preexposure, thereby retarding the acquisition of CS–US associations, and (2) a group that proposes the formation of CS–no consequence associations during preexposure interferes with the subsequent establishment of CS–US associations. Models that assume CS associability decreases during preexposure were assigned to one of two groups, depending on whether CS_i associability is modulated by (1) comparisons between observed and predicted events, or (2) the quality of the CS_i predictions of future events.

Among other models that assume CS associability decreases during preexposure, the SLG model proposes that CS_i associability is controlled by the Novelty of US and CSs, computed as the sum of absolute values of the differences between observed values and the aggregate predictions of active US and CSs. As an associability theory, the SLG model shares some aspects with other learning theories. For example, the SLG model is similar to Schmajuk and Moore's (1988), Grossberg's (1975), and Frey and Sears's (1978) in the use of an attentional memory to modulate CS representation. Also, the SLG model shares with Lubow et al.'s (1981) model the idea that CS_i associability ranges from full attention to full inattention, in contrast to most models in which CS_i associability varies between no attention and full attention. The model is also similar to Rescorla and Wagner's (1972) in the use of a simple delta rule to regulate CS–US associations.

Although the SLG model shares properties with other associability models, it is unique in that the retrievability of CS_i–CS_j and CS_i–US associations is also controlled by the Novelty at the time of CS_i presentations, which modulates the magnitude of the internal representation of CS_i, X_i (see Figure 5.3). This property is not an additional assumption, but a simple and direct consequence (an emergent property) of the way the neural network stores and retrieves information: X_i controls both the rate of change in CS_i–CS_j and CS_i–US associations [$dV/dt = f(X)$, see Equations 5.4 and 5.4') and the output of the neuron storing the association, thereby controlling the CR [CR $= f(X)$, see Equations 5.3', and 5.10]. Because X_i is controlled by Novelty [$X = f$(Novelty), see Equations 5.2 and 5.5], both storage [$dV/dt = f$(Novelty)] and retrieval [CR $= f$(Novelty)] are functions of Novelty. Therefore, according to the SLG model, experimental manipulations of Novelty after conditioning can modify the retrieval of already established associations, thereby modifying CR strength even when CS_i–US associations remain constant.

As a retrieval theory (see Table 4.1), the SLG model shares some aspects with theories that emphasize the significance of mechanisms operating during memory retrieval (Spear et al., 1990). According to Spear et al. (1990), current theories of conditioning tend to underscore the process of acquisition of CS–US associations, but neglect the process of expression of the association during performance. In contrast, the SLG model combines both acquisition and expression processes and offers a quantitative description of both operations and their interactions. As pointed out, however, the model is not a pure retrieval model either, that is, storage and retrieval processes are two different aspects of the single neural mechanism illustrated in Figure 5.3. In a sense, the model reconciles Spear's (1981) suggestion that many phenomena typically attributed to storage processes are the consequence of retrieval processes with more traditional views that stress storage operations.

Simulation results

Table 5.1 lists the results of different types of experimental manipulations, and can be regarded as a benchmark for comparing the power of competing theories. Table 5.1 shows that the SLG model correctly describes many of the features that characterize LI.

1. The SLG model describes several paradigms that yield retardation of excitatory or inhibitory conditioning: (1) Specific preexposure to CS(A), but not to CS(B) or the context alone, produces LI to excitatory conditioning of CS(A), (2) CS(A) preexposure yields LI for the inhibitory conditioning of CS(A), and (3) CS(A)–weak US pairings generate LI to the excitatory conditioning of CS(A). In general, these results are explained in terms of the decreased Novelty, and the consequently decreased z_i during conditioning, that follows CS preexposure.

Whereas Pearce and Hall's (1980) model can describe LI after conditioning with a weak US, this result is a problem for Mackintosh's (1975), Schmajuk and DiCarlo's (1981), Frey and Sears's (1978), and Wagner's (1978) models, all of which predict a positive transfer following conditioning with a weak US. Like the Pearce and Hall (1980) model, the SLG model correctly describes LI (negative transfer) following conditioning with a weak US.

2. The SLG model describes the effects of different parameters of preexposure on the strength of LI: (1) LI is a nonmonotonic function of the number of CS preexposures, (2) LI increases with increasing CS durations, (3) LI increases with increasing total duration of CS preexposure, (4) LI increases with increasing CS intensities, and (5) LI increases with increasing ITI durations. In general, these results are explained in terms of the increasingly smaller Novelty, and the consequently increasingly smaller z_i, with increasing CS durations, intensities, and ITI durations.

Because the SLG model combines a low initial value of associability with increased Novelty during the first preexposure trials, it is able to describe LI as a nonmonotonic function of the number of CS preexposures. Importantly, although most models can describe LI augmentation with an increasing number of CS preexposures and increasing CS intensities, only real-time models able to capture the temporal arrangement of a trial can portray LI enhancement with increasing CS or increasing ITI durations.

3. The SLG model correctly portrays the consequences of preexposing to different combinations of CSs before conditioning: (1) Simultaneous CS(A)–CS(B) preexposure impairs LI to CS(A) (overshadowing of LI), (2) CS(B) preexposure followed by CS(A)–CS(B) preexposure tends to preserve LI to CS(A) (blocking of LI), (3) CS(B) preexposure followed by CS(A) preexposure impairs LI to CS(A), (4) preexposure to a compound is more effective than preexposure to the elements in yielding LI to the compound, and (5) CS(A) preexposure followed by the introduction of a surprising event impairs LI to CS(A). In general, these results are explained in terms of the increased Novelty and the consequently increased z_i that accompanies the presentation of CS(A) during conditioning. However, the model incorrectly predicts that sequential CS(A)–CS(B) preexposure also impairs LI to CS(A). Further investigation of this issue is warranted.

Wagner's (1978) model describes attenuation of LI to CS(A) by simultaneous CS(A)–CS(B) preexposure (overshadowing of LI) in terms of a decreased CX–CS(A) association in the presence of CS(B)–CS(A) associations. Wagner's

Table 5.1. *Paradigms described with the Schmajuk, Lam, and Gray (1996) storage–retrieval model.*

Paradigms	Phase 1	Phase 2	Phase 3	Results	Figure
1. LI of excitatory or inhibitory conditioning					
CS(A) preexposure and excitatory conditioning	CS(A) CX1–	CS(A) CX1+		LI	5.5ab
CS(B) preexposure and excitatory conditioning	CS(B) CX1–	CS(A) CX1+		No LI	5.5ab
CS(A) preexposure and inhibitory conditioning	CS(A) CX1–	CS(B) CX1+	CS(A) CS(B) CX1–	LI	5.6b
CS(A)–weak US pairings	CS(A) CX1+	CS(A) CX1++		Similar to LI	5.7b
2. Parameters of LI	CS(A) CX1–	CS(A) CX1+			
Number of CS preexposures	CS(A) CX1–	CS(A) CX1+		Increased LI	5.8
CS duration	CS(A) CX1–	CS(A) CX1+		Increased LI	5.9
Total duration	CS(A) CX1–	CS(A) CX1+		Same LI	5.10
CS intensity	CS(A) CX1–	CS(A) CX1+		Increased LI	5.11
ITI duration	CS(A) CX1–	CS(A) CX1+		Increased LI	5.12
3. Combinations of CSs during preexposure					
Simultaneous CS(A)–CS(B) preexposure	CS(A)–CS(B) CX1–	CS(A) CX1 +		Decreased LI	5.13
Sequential CS(A)–CS(B) preexposure	CS(A) → CS(B) CX1–	CS(A) CX1 +		Decreased LI[a]	—
Few CS(B) preexposures, CS(A)–CS(B) preexposure	CS(B) CX1–	CS(A)–CS(B) CX1–	CS(A) CX1+	Decreased LI	—
Many CS(B) preexposures, CS(A)–CS(B) preexposure	CS(B) CX1–	CS(A)–CS(B) CX1–	CS(A) CX1+	Increased LI	5.13
CS(B), CS(A) preexposure	CS(B) CX1–	CS(A) CX1–	CS(A) CX1+	Decreased LI	5.13
CS(A)–CS(B) preexposure, CS(A)–CS(B) conditioning	CS(A)–CS(B) CX1–	CS(A)–CS(B) CX1+		LI	5.14
CS(A), CS(B) preexposure, CS(A)–CS(B) conditioning	CS(A) CX1–	CS(B) CX1–	CS(A) CS(B) CX1+	Decreased LI	5.14
CS(A) preexposure, surprising event	CS(A) CX1–	CS(A)–CS(B) CX1–	CS(A) CX1+	Decreased LI	5.15
	CS(A)–CS(B) CX1–	CS(A) CX1+		Decreased LI	5.15

4. Contextual manipulations					
Context, CS(A) preexposure	CX1–	CS(A) CX1–	CS(A) CX1+	Increased LI	5.16
CS(A) preexposure, context	CS(A) CX1–	CX1–	CS(A) CX1+	Increased LI	5.16
Change context after CS(A) preexposure	CS(A) CX1–	CS(A) CX2+		Decreased LI	5.16
Context, CS(A) preexposure, change context	CX1–	CS(A) CX1+	CS(A) CX2+	Restored LI	5.17
Change context and cuing	CS(A)–CS(B) CX1–	CS(A) CS(B) CX2+		Restored LI	5.18
Changes in CS and context (SoEo)	CS(A) CX1–	CS(A) CX1+		LI	5.19
Changes in CS and context (SnEo)	CS(A) CX1–	CS(B) CX1+		No LI	5.19
Changes in CS and context (SoEn)	CS(A) CX1–	CS(A) CX2+		PL	5.19
Changes in CS and context (SnEn)	CS(A) CX1–	CS(B) CX2+		No LI	5.19
5. Perceptual learning					
Discrimination, same context	CS(A)–CS(B) CX1–	CS(A) CX1+	CS(B) CX1–	LI	5.20a
Discrimination, different context	CS(A)–CS(B) CX1–	CS(A) CX2+	CS(B) CX2–	PL	5.20a
Maze discrimination, same context	CS(A)–CS(B) CX1–	CS(A) CX1+	CS(B) CX1–	Increased PL	5.20b
Maze discrimination, different context	CS(A)–CS(B) CX1–	CS(A) CX1+	CS(B) CX2–	Decreased PL	5.20b
6. Orienting response and latent inhibition					
Context change	CS(A) CX1–	CX2–	CS(A) CX1–	OR	5.21a
	CS(A) CX1–	CX2–	CS(A) CX2–	Similar OR	5.21a
	CS(A) CX1–	CX2–	CS(A) CX1+	LI	5.21b
	CS(A) CX1–	CX2–	CS(A) CX2+	No LI	5.21b
Time period	CX1–	CS(A) CX1–	CS(A) CX1–	OR	5.22a
	CS(A) CX1–	CX1–(T)	CS(A) CX1–	Increased OR	5.22a
	CX1–	CS(A) CX1–	CS(A) CX1+	LI	5.22b
	CS(A) CX1–	CX1–(T)	CS(A) CX1+	Identical LI	5.22b
7. Postconditioning manipulations					
Extinction of the training context	CS(A) CX1–	CS(A) CX1+	CX1–	Decreased LI	5.23
US presentations in a different context	CS(A) CX1–	CS(A) CX1+	CX2+	Decreased LI	5.24
Delayed testing	CS(A) CX1–	CS(A) CX1+	CX1 (T)	Decreased LI	5.25

Note. LI: latent inhibition; PL: perceptual learning; T: Time. [a]: the model fails to describe the experimental result.

model suggests that CS(B) presentations preceding CS(A)–CS(B) presentations decrease the formation of CS(B)–CS(A) associations (blocking of LI) and are, therefore, less efficient in disrupting LI, a result in agreement with experimental results.

Whereas the SLG model explains reduced LI after the unexpected presence or absence of a stimulus previous to conditioning in terms of the resulting increased Novelty, other formal models cannot describe the absence of a predicted CS and, therefore, fail to predict the ensuing LI attenuation.

4. The SLG model explains the effects of several contextual manipulations on the strength of LI: (1) Context changes following CS preexposure impair LI, (2) preexposure to the apparatus prior to CS preexposure facilitates LI, and (3) context presentations following CS preexposure facilitate LI. Although context changes after CS preexposure impair LI, the model can also explain why (1) cuing treatments following CS preexposure in a novel context and (2) exposure to the context of CS preexposure before the CS preexposure phase preserve LI. Context changes following CS preexposure impair LI because they increase Novelty during conditioning. In different ways, the other manipulations decrease Novelty at the time of CS presentation.

In contrast to Schmajuk and DiCarlo's (1991a, b), Mackintosh's (1975), Frey and Sears's (1978), and Pearce and Hall's (1980) theories, the SLG model is able to describe correctly contextual effects. According to Wagner's (1978) model, context presentations following CS preexposure decrease CX–CS associations, thereby attenuating LI. As mentioned, although Wagner (1979) and Baker and Mercier (1982) reported that sometimes this is the case, in general, LI is robust and the extinction effect small or nonexistent, facts that disprove Wagner's model. In agreement with experimental results showing that LI is immune to contextual extinction (Hall and Minor, 1984), the SLG model shows that context presentations following CS preexposure do not attenuate LI. In addition, in contrast to Wagner's (1978) suggestion that preexposure to the apparatus prior to CS preexposure interferes with the formation of CX–CS associations, thereby attenuating LI, the SLG model correctly predicts that context preexposure facilitates LI.

Gordon and Weaver (1989) showed that cuing treatments following CS preexposure in a novel context can restore LI. As one possible explanation, Gordon and Weaver (1989) suggested that CS–no consequence associations are encoded along with the preexposure context. When animals are conditioned in a novel context, they fail to retrieve the CS–no consequence association and, therefore, conditioning proceeds normally. When the background noise present during reinforcement is presented again during conditioning, retrieval of the CS–no consequence association is facilitated and, therefore, LI is manifested. In contrast to the retrieval explanation, the SLG model suggests that reinstatement of LI by a cue present during preexposure is due to the decrease in Novelty that follows the prediction of the CS by that cue. This explanation is similar to that offered by Wagner (1978) in terms of decreased associability (see Gordon and Weaver, 1989).

5. According to the SLG model, perceptual learning is the result of the enhancement of the internal representations of CS(A) through its associations with itself, other CSs, and the context. When changes in context produce large changes in z_A, perceptual learning is stronger in the new context due to the increased z_A. When changes in context produce smaller changes in z_A, perceptual learning is stronger in the old context due to CX–CS associations.

Although Wagner's (1978) theory easily accounts for the attenuation of LI subsequent to context changes, that theory cannot account for the facilitatory effect of preexposure after a change in context (perceptual learning). Channel and Hall (1981) and Lubow et al. (1976) proposed that context changes following CS preexposure attenuate LI, thereby allowing the expression of perceptual learning. In contrast, for McLaren et al. (1989) perceptual learning is the consequence of the greater decline in the associability of the common elements than the unique elements of two stimuli that share multiple features when preexposure occurs in the same context.

6. Although in the SLG model Novelty controls both the strength of the OR and attention (thereby controlling LI), the model can describe experimental data that suggest a dissociation between changes in the OR and LI effects. According to the SLG model, a procedure that produces a relatively small dishabituation of the OR in the absence of the US can still produce a large dishabituation of the OR (thereby impairing LI) in the presence of the US during conditioning (Hall and Channel, 1985b). Furthermore, the SLG model suggests that a period of time producing dishabituation of the OR might not result in the attenuation of LI (Hall and Schachtman, 1987) because z_i (which controls the rate of conditioning) increases at a slower rate than the OR (which reflects the magnitude of Novelty at a given time).

Although sharing similar principles with the Pearce and Hall (1980) model, the SLG model seems more competent in describing the differential effects that contextual and temporal manipulations have on the OR and LI. A major difference in the models is that, whereas in the Pearce and Hall model the strength of the OR and CS_i associability are identical, in the SLG model they are only functionally related.

7. The SLG model can describe data showing that postconditioning manipulations might atenuate LI. Although these experimental data have been used to support the view that LI is a purely retrieval rather than storage phenomenon, the SLG model suggests that extinction of the training context, US presentations in a context different from the training context, and delayed testing yield increased z_i, thereby increasing X_i and the magnitude of the CR. In other words, even when the SLG model describes LI partly as a storage phenomenon, its retrieval attributes can explain increases in CR responding following different postconditioning manipulations. Probably most intriguing are the simulated results describing the effect of delayed testing: Because the CS–US association stays constant but z_i and X_i change, this paradigm clearly illustrates how changes in z_i modulate retrieval. It is important to notice that, although the SLG model combines both storage and retrieval processes, both operations are described in terms of one single neural mechanism.

In contrast to the SLG model, most current theories of conditioning are associability theories and, therefore, not capable of explaining the effect of postconditioning manipulations. However, Miller and Schachtman's (1985) comparator hypothesis explains recovery from LI after extinction of the training context or a retention interval following conditioning, by suggesting that both procedures decrease the indirect activation of the US representation, thereby increasing excitatory responding.

In summary, the SLG model can provide explanations for many attributes of LI: (1) Retardation of excitatory or inhibitory conditioning is explained in terms of the decreased Novelty, and the consequently decreased z_i during conditioning,

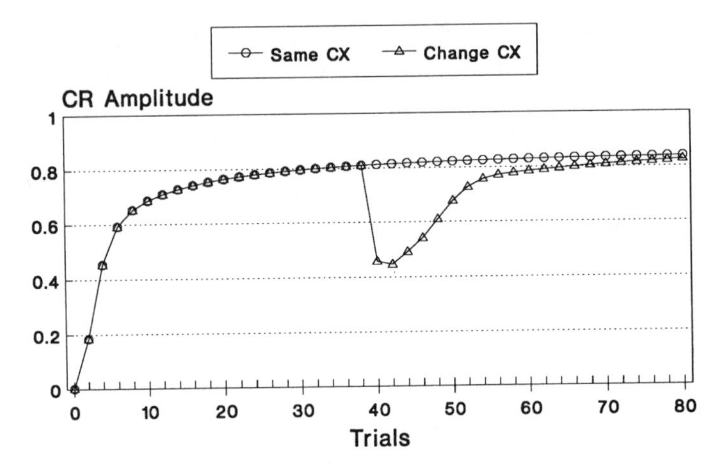

Figure 5.26. Effect of changing context during acquisition. Peak CR amplitude during 80 conditioning CS(A)–US trials (1) in the same context, or (2) changing context on trial 40.

that follows CS preexposure; (2) the effects of different parameters of preexposure on the strength of LI are explained in terms of the increasingly smaller Novelty and the consequently increasingly smaller z_i during conditioning, with increasing CS durations, intensities, and ITI durations; (3) the consequences of preexposing to different combinations of CSs before conditioning are explained in terms of the increased Novelty, and the consequently increased z_i, that accompanies the presentation of CS(A) during conditioning; (4) the effects of several contextual manipulations on the strength of LI are explained in terms of the increased Novelty that follows the change in context and results in an increased z_i during conditioning; (5) perceptual learning is the result of the enhancement of the internal representations of CS(A) through its associations with itself, other CSs, and the context; (6) the dissociation between changes in the OR and LI effects is explained in terms of the different rates of change of z_i and the OR during conditioning; and (7) the effects of postconditioning manipulations, such as extinction of the training context, US presentations in a context different from the training context, and delayed testing, are explained in terms of increments in z_i and X_i that increase the magnitude of the CR during testing (retrieval).

Other classical conditioning paradigms

In order to explore the general power of the SLG model beyond its descriptions of LI, we carried out simulations for other classical conditioning paradigms. The neural network correctly describes (1) acquisition of delay and trace conditioning; (2) conditioned reinforcement; (3) decreased CR responding after extensive training in one environment followed by testing in a second environment (e.g., Penick and Solomon, 1991; see also McLaren et al., 1989, p. 118), as illustrated in Figure 5.26; (4) the disruption of the CR by presenting a novel stimulus shortly before the CS (external inhibition; Pavlov, 1927), as illustrated in Figure 5.27; (5) extinction, (6) partial reinforcement with different percentages of reinforced trials (see Gormezano and Moore, 1969, for a review); (7)

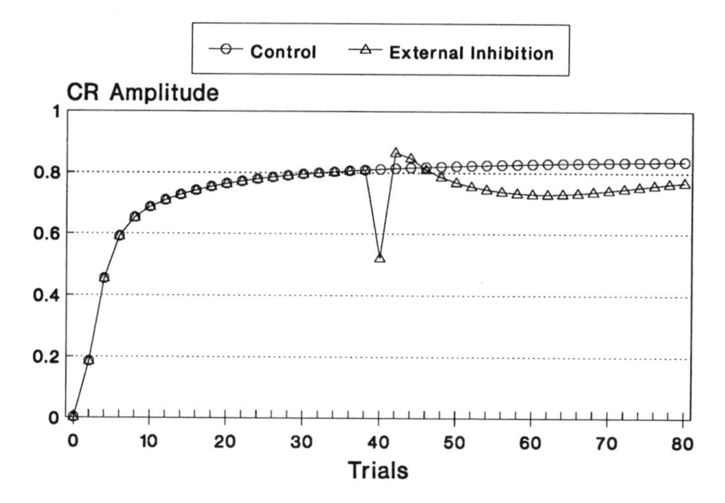

Figure 5.27. External inhibition. Peak CR amplitude during 80 conditioning CS(A)–US trials with and without introducing an external inhibitor CS(B) on trial 40.

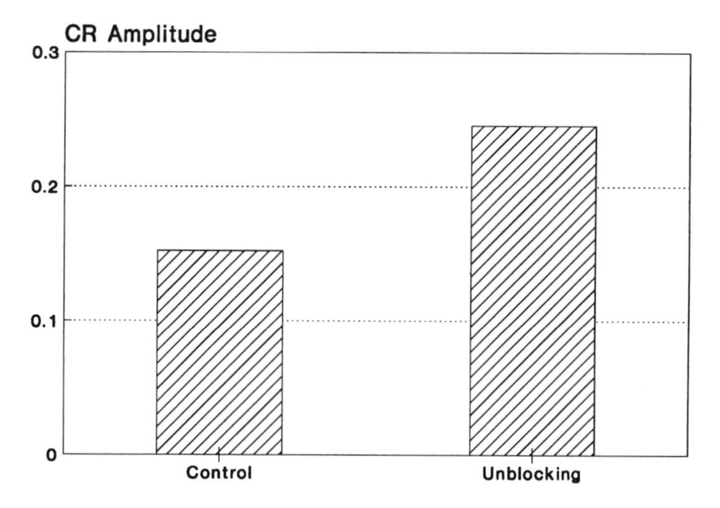

Figure 5.28. Unblocking. Peak CR amplitude evoked by CS_2 after (1) 10 CS_1 trials reinforced with 2 USs and 10 CS_1–CS_2 trials reinforced with 2 USs in the case of blocking, and (2) 10 CS_1 trials reinforced with 2 USs and 10 CS_1–CS_2 trials reinforced with 1 US in the case of unblocking. The CSs are 200 msec in duration; both USs are 50 msec in duration and presented with 150-msec and 1700-msec ISI.

conditioned inhibition; (8) blocking (Kamin, 1968); (9) the "reinforcer nonspecific" aspects of unblocking (Dickinson and Mackintosh, 1979), as illustrated in Figure 5.28; (10) overshadowing (Wagner et al., 1968); (11) discrimination acquisition and reversal; and (12) sensory preconditioning (Brogden, 1939).

In addition, preliminary simulated results suggest that the model can describe the effect of CS extinction in alleviating blocking (Balaz et al., 1982) and overshadowing (Kasprow et al., 1982).

Extending the power of the model

Notwithstanding its numerous correct predictions, the SLG model cannot explain saving effects in extinction–acquisition series (Frey and Ross, 1968;

Table 5.2. *Simulations obtained with different versions of the Schmajuk, Lam, and Gray (1996) storage–retrieval model*

	Properties of the Model	
Paradigms	Novelty	Novelty Predictive Value
Delay conditioning	Y	Y
Trace conditioning	Y	Y
Partial reinforcement	Y	Y
Extinction	Y	Y
Acquisition series		Y
Extinction series		Y
Latent inhibition	Y	Y
Blocking	Y	Y
Unblocking	Y	Y
Overshadowing	Y	Y
Discrimination	Y	Y
Conditioned inhibition	Y	Y
Inhibition by differential conditioning	Y	Y
Inhibition by decrease in the US	Y	Y
Counterconditioning	Y	Y
Superconditioning	Y	Y
Overprediction effects	Y	Y
Generalization	Y	Y
Compound conditioning		
Simultaneous feature-positive discrimination		
Simultaneous feature-negative discrimination		
Serial feature-positive discrimination		
Serial feature-negative discrimination		
Negative patterning		
Positive patterning		
Sensory preconditioning	Y	Y
Second-order conditioning	Y	Y
Context switching	Y	Y
External inhibition	Y	Y

Note. Y: The model can describe the experimental data.

Smith and Gormezano, 1965). This deficiency is based on the fact that z_i reflects the association of CS_i with Novelty, which decreases as learning progresses, and not with its own associations with other CSs or the US (predictive value of CS_i), which increases with increasing learning. Computer simulations show that the addition of an attentional variable y_i that reflects the predictive quality of CS_i allows the SLG model to successfully describe savings in extinction–acquisition series and overtraining extinction (North and Stimmel, 1960) effects. Therefore, a more complete description of attentional phenomena in classical conditioning seems to require mechanisms not only based on comparisons between observed and predicted events, but also on the quality of predictions of future events. Table 5.2 shows the paradigms that the SLG model correctly describes when an attentional variable y_i reflecting the predictive quality of CS_i is incorporated into the model.

Moreover, a complete description of additional complex classical conditioning paradigms (e.g., negative patterning) seems to dictate the adoption of a generalized delta rule (see Schmajuk and DiCarlo, 1992). As explained in Chapter 6, when combined with the Schmajuk and DiCarlo (1992) model, the SLG model can describe an even wider variety of classical conditioning data.

Appendix 5.A

A formal description of the SLG model

This section formally describes the SLG model as depicted in Figure 5.2.

Short-term memory trace of the CS

CS_i generates a STM trace τ_i according to

$$d(\tau_i)/dt = K_1(CS_i - \tau_i) \tag{5.1}$$

where K_1 represents the rate of increase and decay of τ_i.

Internal representation of the CS

The STM trace of CS_i, τ_i, can be modified by the attentional value z_i to yield the internal representation of CS_i, X_i, according to

$$X_i = K_2(\tau_i + K_3B_i)(K_4 + z_i) \tag{5.2}$$

where K_3 represents a reinjection coefficient for the prediction of CS_i (B_i), and K_4 an unmodifiable connection between input ($\tau_i + K_3B_i$) and X_i. By Equation 5.2, X_i is active either when (1) CS_i is present, or (2) when CS_i is predicted by other CSs. Increasing CS_i–CS_i associations also increases the magnitude of X_i.

Aggregate predictions

The *aggregate prediction* of event k by all CSs with representations active at a given time B_k is given by

$$B_k = \Sigma_i V_{i,k} X_i \tag{5.3}$$

where $V_{i,k}$ represents the association of X_i with CS_k.

The *aggregate prediction* of the US by all CSs with representations active at a given time B_{US} is given by

$$B_{US} = \Sigma_i V_{i,US} X_i \tag{5.3'}$$

where $V_{i,US}$ represents the association of X_i with the US.

Long-term memory CS–US associations

Changes $V_{i,US}$ are given by

$$d(V_{i,US})/dt = K_5 X_i(\lambda_{US} - B_{US})(1 - |V_{i,US}|) \tag{5.4}$$

where X_i represents the internal representation of CS_i, λ_{US} the intensity of the US, B_{US} the aggregate prediction of the US by all X's active at a given time. By Equation 5.4, $V_{i,US}$ increases whenever X_i is active and $\lambda_{US} > B_{US}$ (with $K_5 = K_5'$) and decreases when $\lambda_{US} < B_{US}$ (with $K_5 = K_5'$). In order to prevent the extinction of conditioned inhibition or the generation of an excitatory CS by presenting a neutral CS with an inhibitory CS, we assume that when $B_{US} < 0$, then $B_{US} = 0$.

By Equation 5.4, $V_{i,\mathrm{US}}$ varies between 1 and –1. Because real synapses do not change from excitatory to inhibitory or vice versa (but see Alkon et al., 1992), connectionist models have been criticized for allowing weights to change from positive and negative and, conversely, from negative to positive. Appendix 3.B describes equations that, employing positive excitatory and positive inhibitory associations, still yield similar results to those obtained with Equation 5.4. However, for simplicity and speed in the simulations, the present study makes use of Equation 5.4.

Long-term memory CS–CS associations

Changes $V_{i,j}$ are given by

$$d(V_{i,j})/dt = K_6 X_i(\lambda_j - B_j)(1 - |V_{i,j}) \qquad (5.4')$$

where X_i represents the internal representation of CS_i, B_j the aggregate prediction of event j by all X's active at a given time, and λ_j the intensity of CS_j. By Equation 5.4', $V_{i,j}$ increases whenever X_i is active and $\lambda_j > B_j$ (with $K_6 = K_6'$) and decreases when $\lambda_j < B_j$ (with $K_6 = K_6''$). When $B_j < 0$, then $B_j = 0$. $V_{i,i}$ represents the association of CS_i with itself.

Like $V_{i,\mathrm{US}}$, $V_{i,j}$ also varies between 1 and -1. Although $V_{i,j}$ can be described by equations that employ positive excitatory and positive inhibitory associations (see Appendix 3.B), for simplicity and speed in the simulations, the present study makes use of Equation 5.4'.

Attentional memory

Changes in the association of X_i with Novelty, z_i, are given by

$$d(z_i)/dt = (\tau_i + K_3 B_i)\,[K_7 \text{ Novelty } |1- z_i| - K_8 (1 + z_i)] \qquad (5.5)$$

where K_7 represents the rate of increase and K_8 the rate of decay of z_i. In the absence of Novelty, z_i decreases until $z_i = -1$.

By Equation 5.5, z_i varies between 1 and –1. Equation 5.5 can be rewritten as two separate equations, one for the excitatory values of z_i, ze_i:

$$d(ze_i)/dt = (\tau_i + K_3 B_i)\,[K_7 \text{ Novelty}(1 - ze_i) - K_8 ze_i] \qquad (5.5')$$

and another for the inhibitory values of z_i, zi_i:

$$d(zi_i)/dt = (\tau_i + K_3 B_i)\,[-K_7 \text{ Novelty } zi_i + K_8(1 - zi_i)] \qquad (5.5'')$$

that are combined to yield $z_i = ze_i - zi_i$. By Equations 5.5' and 5.5", ze_i and zi_i have always positive values. For simplicity and speed in the simulations, the present study makes use of Equation 5.5.

We assume that when $z_i \leq 0$, then $X_i = K_2(\tau_i + K_3 B_i)\,K_4$ (see Equation 5.2). This means that when z_i becomes negative, input $(\tau_i + K_3 B_i)$ activates X_i only through the unmodifiable connnection K_4. By Equations 5.4 and 5.4', whereas a CS_i with negative z_i has a relatively small X_i and, therefore, its associations change slowly, a CS_i with positive z_i has a relatively large X_i and, therefore, its associations change rapidly. LI is the consequence of the small positive or the negative value assumed to be acquired by z_i during CS preexposure (as Novelty decreases) and, consequently, of the time needed to restore z_i to its positive value (as Novelty increases) during conditioning. Interestingly, in line with Lubow et al.'s (1981) suggestion, positive values of z_i ($ze_i > 0$) can be interpreted as a measure of the attention directed to CS_i, whereas negative values of z_i ($zi_i > 0$) can be interpreted as a measure of the *inattention* to CS_i.

Total novelty

The average observed value of event k is given by

$$\bar{\lambda}_k = K_9(1 - \bar{\lambda}_k)\,\lambda_k - K_{10}\,\bar{\lambda}_k, \tag{5.6}$$

where K_9 and K_{10}, respectively, represent the rates of increase and decay of $\bar{\lambda}_k$.

The average prediction of event k is given by

$$\bar{B}_k = K_{11}(1 - \bar{B}_k)\,B_k - K_{12}\,\bar{B}_k \tag{5.7}$$

where K_{11} and K_{12}, respectively, represent the rates of increase and decay of $\bar{B}_k$

Total Novelty is given by

$$\text{Novelty} = \Sigma_k\,|\bar{\lambda}_k - \bar{B}_k| \tag{5.8}$$

where k includes all CSs and the US.

Novelty is normalized between 0 and 1 by applying

$$\text{Novelty} = \text{Novelty}^2/(K_{13}^{\,2} + \text{Novelty}^2) \tag{5.9}$$

The magnitude of the OR is given by OR = Novelty.

Performance rules

The amplitude of the CR is given by

$$\text{CR} = K_{14}B'_{\text{US}}(1 - K_{15}\,\text{OR}) \tag{5.10}$$

where B'_{US} is given by $B'_{\text{US}} = B_{\text{US}}^{\,2}/(K_{16}^{\,2} + B_{\text{US}}^{\,2})$. According to Equation 5.10, the magnitude of the CR increases with increasing predictions of the US and decreases with increasing ORs elicited by Novelty.

The amplitude of the total response is given by

$$\text{R} = \text{CR} + K_{17}\text{US} \tag{5.11}$$

Appendix 5.B

Simulation parameters

In our computer simulations, each trial is divided into 200 time units. Each time unit represents approximately 10 msec. CSs of amplitude 1 are presented in general between 200 and 400 msec. In the simulations of the introduction of a surprising event, the amplitude of the novel CS is 2. A 50-msec US of intensity 1.8 is applied at 350 msec for all simulations, with the exception of those for perceptual learning and context changes after CS preexposure in which a US of intensity 1.0 is used. Unless otherwise indicated, context amplitude is 0.1 in all simulations.

Parameters values used in all simulations are $K_1 = 0.2$, $K_2 = 2$, $K_3 = 0.4$, $K_4 = 0.1$, $K_5 = 0.005$, $K_6 = 0.005$, $K_7 = 0.02$, $K_8 = 0.005$, $K_9 = 1$, $K_{10} = 0.005$, $K_{11} = 1$, $K_{12} = 0.005$, $K_{13} = 0.75$, $K_{14} = 1$, $K_{15} = 0.7$, $K_{16} = 0.15$, and $K_{17} = 0.5$.

6 Configural processes

As mentioned in previous chapters, Sokolov (1960) suggested that the brain constructs a neural model of environmental events by storing classical conditioning associations. When there is coincidence between external events and the predictions generated by the brain model, animals respond without changing their internal model of the environment. Chapter 3 introduced a real-time neural model that incorporates a delta rule. This rule describes changes in the synaptic connections between the two neural populations by way of minimizing the squared value of the difference between the output of the population controlling the CR generation and the US. According to the "simple" delta rule, CS_i–US associations are changed until $f(US) = US - \Sigma_j V_{j,US} CS_j$ is zero. As shown in Chapter 3 (Table 3.1), this rule describes (1) acquisition and extinction of delay conditioning, (2) acquisition and extinction of trace conditioning, (3) ISI curves, (4) partial reinforcement, (5) overshadowing, (6) blocking, (7) conditioned inhibition, (8) counterconditioning, (9) simultaneous feature-positive and feature-negative discriminations, (10) superconditioning, (11) overprediction effects, (12) discrimination, and (13) generalization. However, the model fails to describe important paradigms such as positive and negative patterning.

Kehoe (1988) presented a network that incorporates a hidden-unit layer trained according to a delta rule. In addition to the paradigms described by the Rescorla–Wagner model, the network describes stimulus configuration, learning to learn, savings effects, and importantly, positive and negative patterning.

Schmajuk and DiCarlo (1992; see also Gluck and Myers, 1993; Tesauro, 1990) introduced a model that, by employing a "generalized" delta rule (also known as backpropagation; see Box 1.2) to train a layer of hidden units that "configure" simple CSs, is able to solve negative and positive patterning. The present chapter introduces the network and shows that it provides descriptions of acquisition of delay and trace conditioning, extinction, acquisition–extinction series, blocking, overshadowing, discrimination acquisition and reversal, compound conditioning, feature-positive discrimination, conditioned inhibition, negative patterning, positive patterning, generalization, and occasion setting.

Stimulus configuration

As described in Chapter 2, in negative patterning two CSs are introduced separately in the presence of the US, and presented together in the absence of the US. Combined presentation of both CSs generates a response smaller than the sum of the responses to each CS. In order to account for negative patterning, Spence (1952; see also Kehoe, 1988; Mishkin and Petri, 1984; Rudy and Sutherland, 1989) suggested that when both CSs are presented together, they generate a "patterned" or "configural" stimulus. As a result of this configuration, the strength of a CR is proportional to the summation of the excitatory effects of the CSs and the inhibitory force of the configural stimulus. Therefore, when presented separately, each CS elicits a CR larger than that elicited by both CSs presented together.

In positive patterning, two CSs are introduced separately in the absence of the US and together in the presence of the US. Combined presentation of both CSs generates a response larger than the sum of the responses to each CS. In order to account for positive patterning, it is assumed that when both CSs are presented together, they generate a configural stimulus. During training, the configural stimulus is always reinforced, whereas the individual CSs are reinforced only on half of the trials. If we assume that the individual CSs and configural stimulus compete to form associations with the US, the association of the configural stimulus with the US becomes increasingly stronger and the associations of the individual CSs with the US become increasingly weaker. After a certain number of trials, combined presentation of the CSs elicits a CR stronger than the sum of the CRs elicited by each CS when presented in isolation. In logical terms, whereas negative patterning is equivalent to learning an "exclusive-or" response rule, positive patterning is similar to learning an "and" response rule.

Rumelhart, Hinton, and Williams (1986) showed that exclusive-or problems can be solved by networks that incorporate a layer of "hidden" units positioned between input and output units. In such multilayer networks, the information coming from the input units is recoded by hidden units into an "internal representation." The exclusive-or problem can be solved if this internal representation, active only with the simultaneous presentation of both inputs, acquires an inhibitory association with the output. Because the output is proportional to the sum of the excitatory effects of the individual inputs and the inhibitory force of the internal representation, the output will be large when each input is presented separately and small when both inputs are presented together. Rumelhart et al. (1986) suggested that, rather than fixed internal representations, the system might be able to learn internal representations adequate for performing the task in question. It is easy to notice the similarity between Spence's (1952) account of negative patterning, suggesting that when both inputs are presented together, they generate a configural stimulus, and Rumelhart et al.'s (1986) solution for exclusive or problems in multilayer networks, proposing the learning of an internal representation active with the simultaneous presentation of both inputs. Based on such correspondence, this chapter introduces a multilayer neural network that describes stimulus configuration. In this network, the information coming from the input units is recoded by hidden units into *internal representations* assumed to be equivalent to *configural stimuli* (CN). The recoding of input-unit activity into hidden-unit activity represents the process of stimulus configuration.

Figure 6.1 shows a neural network that, like the model shown in Chapter 3 (Figure 3.1), modifies the internal model of the environment when predicted events differ from observed events. As before, more reliable CSs prevent less reliable or salient ones from becoming associated with the US. In addition, in the model depicted in Figure 6.1, when the difference between predicted and observed events cannot be reduced by modifying CS–US associations, elementary CSs are combined into internal configural representations that are associated with the US to solve the problem at hand.

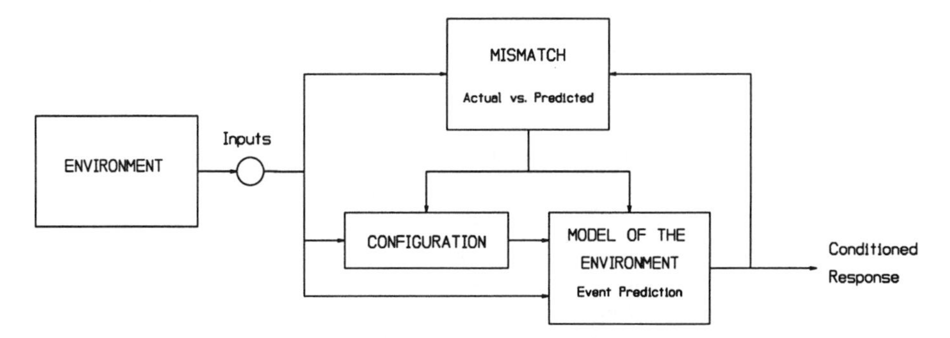

Figure 6.1. A block diagram of a configural model of the environment. The model of the environment generates predictions of future events and controls the conditioned response (CR). The mismatch system compares observed and predicted events to compute novelty. Novelty is then used to (1) generate compound stimuli in the configural system and (2) modify the model of the environment.

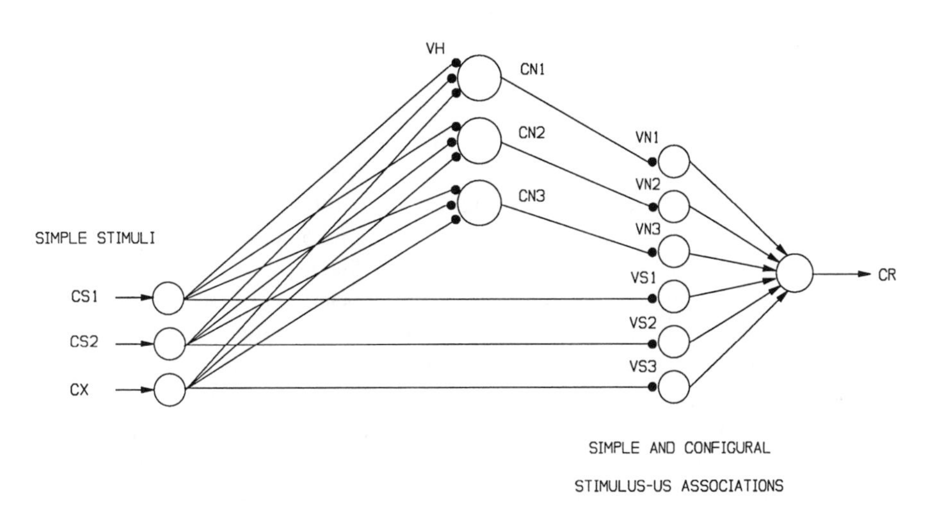

Figure 6.2. The SD model. Diagram of a network that incorporates a layer of hidden units capable of describing stimulus configuration in classical conditioning. CS: conditioned stimulus; CN: configural stimulus; CX: Context; VS: CS–US association; VN: CN–US associations; VH: CS–CN association; CR: conditioned response. Arrows represent fixed synapses, solid circles variable synapses.

The network

Figure 6.2 shows the Schmajuk and DiCarlo (SD) (1992) model, a multilayer network that accounts for stimulus configuration. The network (1) describes behavior in real time, (2) incorporates a layer of hidden units positioned between input and output units that internally codes *configural stimuli,* (3) includes inputs that are connected to the output directly and indirectly through the hidden-unit layer, and (4) employs a biologically plausible backpropagation procedure to train its hidden-unit layer. In addition, the network can be implemented using neurons whose adaptive weights assume only positive values. As shown in Chapter 13, all these features contribute to the biological relevance of the network: (1) Behavior and brain activity are portrayed as they unfold in real time,

(2) direct and indirect input–output connections are mapped over parallel brain circuits, and (3) error signals are propagated back to hidden units through an "error network" that extends over different regions of the brain, including the hippocampus.

Figure 6.2 shows a multilayer network with one input layer, one hidden-unit layer, and two output layers. Input units are activated by conditioned stimuli CS_1 and CS_2 and the context CX. Input units form direct associations, VS_1, VS_2, and VS_3, with the first output layer. In addition, input units form associations $VH_{i,j}$ with the hidden-unit layer. In turn, hidden units form associations, VN_1, VN_2, and VN_3, with the first output layer. The output activities of the hidden-unit layer are assumed to code configural stimuli denoted by CN_1, CN_2, CN_3.

As explained in Appendix 3.B in Chapter 3 , we assume that VS_1, VS_2, VS_3, VN_1, VN_2, and VN_3, whether excitatory or inhibitory, assume only positive values. Therefore, in order to obtain inhibitory effects, excitatory and inhibitory activities of the first output layer are summed with their corresponding signs in a single-unit second output layer. This unit determines the output of the system, the CR. As explained later, the assumption that associations of simple CS_i's with the output units VS_i and configural CN_j's with the output units VN_j are stored in a separate layer of neural elements, instead of in a single output unit as in a traditional backpropagation architecture, has important implications for the present implementation of backpropagation.

Figure 6.2 shows that a CS accrues a *direct* association with the output units and becomes *configured* with other CSs in a hidden unit that, in turn, acquires associations with the output units. Stimulus configuration is achieved by adjusting CS–hidden unit associations $VH_{i,j}$. Because some hidden units might be activated by different CSs, initial values of $VH_{i,j}$ define the amount of generalization between CSs. We assume that generalization is stronger between stimuli in the same modality than between stimuli in different modalities.

The SD model weights more heavily the output of the hidden units than the output of the direct inputs, a difference that gives an advantage to the hidden units to establish associations with the US and to control the CR output. The model also assumes that the rate of change in the CS–hidden unit associations $VH_{i,j}$ is faster than the rate of change in the associations of simple and configural CSs with the US, VS_i and VN_j. In short, configural learning is faster than CS_i–US and CN_j–US learning, and configural stimuli may establish faster and stronger associations with the US.

Figure 6.3 results from adding to Figure 6.2 an "error network" that regulates the associations formed by simple and configural stimulus with the US, VS_i and VN_j, and the associations of simple stimuli with the hidden units $VH_{i,j}$. VS_i and VN_j associations are controlled by a delta rule that reduces the output error EO between the aggregate prediction of the US and the actual value of the US. This aggregate prediction equals the sum of the activation by simple and configural stimuli of their associations with the US ($B = \Sigma_i VS_i CS_i + \Sigma_j VN_j CN_j$). Therefore, the output error is given by EO = US – B.

Biologically plausible version of backpropagation

The original backpropagation procedure (see Box 1.2) describes changes in the synaptic strength between input units and a given hidden unit $VH_{i,j}$ as proportional to (1) the value of the association of the hidden unit with the single output

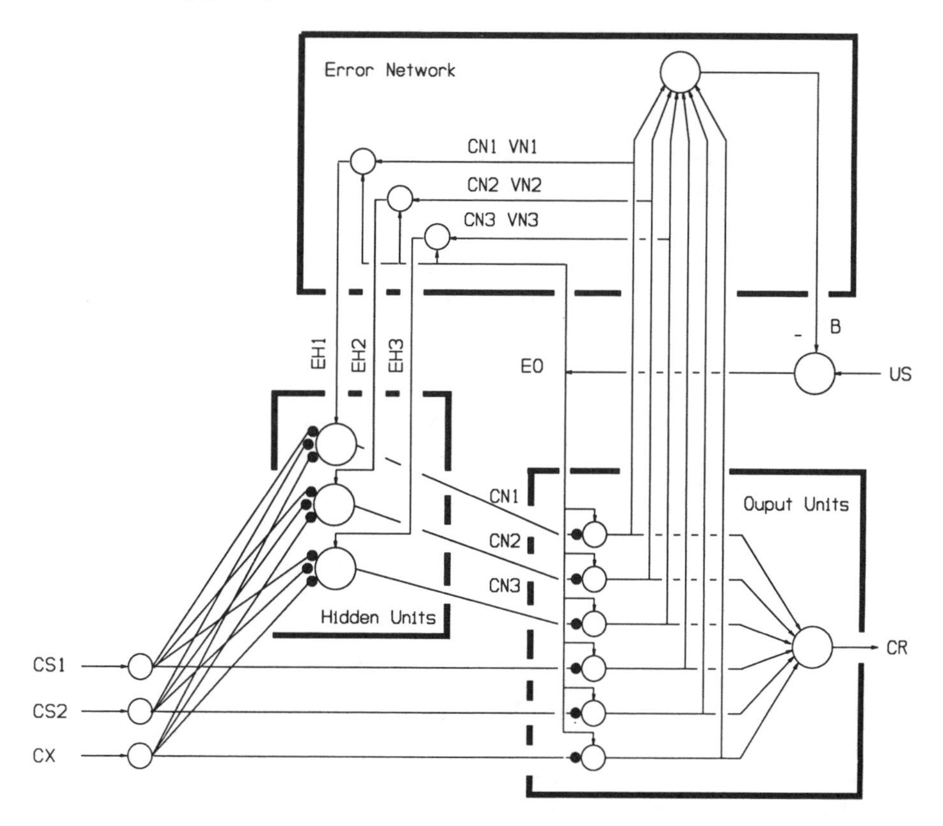

Figure 6.3. The SD model. Diagram of the network of Figure 6.2 showing error signals sent to the hidden units and output units. CS: conditioned stimulus; CN: configural stimulus; VS: CS–US association; VN: CN–US associations; VH: CS–CN association; US: unconditioned stimulus; *B:* aggregate prediction; CR: conditioned response; EH: error signal for hidden units; EO: error signal for output units. Arrows represent fixed synapses, solid circles variable synapses.

unit, (2) the output error, and (3) the derivative of the output of the hidden unit. In mathematical terms, the error signal that regulates the associations of hidden unit *j* with the input CSs is proportional to $d(\mathrm{CN}_j)\ \mathrm{VN}_j\mathrm{EO}$, where $d(\mathrm{CN}_j)$ is the derivative of the output of hidden unit *j*, VN_j the association of hidden unit *j* with the output, and EO the output error. It was not clear, however, how a neurophysiological mechanism can perform this essential computation.

Several implementations of backpropagation rules in biologically plausible circuits and consistent with accepted neurobiological principles have been proposed. Parker (1985) suggested that backpropagation of the error signals may be accomplished in a biologically plausible way by using an error network connected to the original network. Zipser and Rumelhart (1990) further developed this idea and proposed a neural architecture composed of two networks, "one for the forward propagation of activity and the other for the backward propagation of error." Briefly, errors in the output units are backpropagated to the hidden units by associating the output of the units computing output errors to the output of the hidden units. Because the changes in the weights of the error signals are similar to the changes in the weights of the hidden units, the output of a backpropagation unit to a given hidden unit is proportional to the sum of

Box 6.1 Schmajuk and DiCarlo's (1992) configural network

The most important aspects of the network are:

1. It includes one input layer, one hidden-unit layer, and two output layers.
2. Input units are activated by conditioned stimuli, CS_1 and CS_2 and the context CX.
3. Presentation of CS_i (CS_1 and CS_2, or the context CX) activates STM trace X_i.
4. Input units form direct associations, VS_1, VS_2, VS_3, with the first output layer. These associations are regulated by a simple delta rule (see Boxes 1.1 and 2.1).
5. Input units form associations $VH_{i,j}$ with the hidden-unit layer (stimulus configuration). These associations are regulated by a biologically plausible version of the generalized delta rule or backpropagation (see Box 1.2).
6. Hidden units form associations, VN_1, VN_2, VN_3, with the first output layer. These associations are regulated by a simple delta rule (see Box 1.1). The output activities of the hidden-unit layer are assumed to code configural stimuli denoted by CN_1, CN_2, CN_3.
7. B_{US} is the aggregate prediction of the US by all active CSs and CNs.
8. CR amplitude is proportional to the aggregate prediction B_{US}.

the output errors multiplied by weights connecting the corresponding hidden unit with the output unit. Recently, Tesauro (1990) proposed a similar scheme to describe backpropagation of errors. Both in Zipser and Rumelhart's (1990) and Tesauro's (1990) cases, the approach requires that the weights in both forward and backward networks be homologous and this homology be maintained during the learning process. Because exactly homologous weights are difficult to guarantee, Schmajuk and DiCarlo (1992) suggested an alternative procedure to backpropagate errors in a neurophysiological possible manner.

Schmajuk and DiCarlo (1992) proposed that simple stimulus–hidden-unit associations $VH_{i,j}$ are regulated by a biologically plausible backpropagation procedure consistent with accepted neurobiological principles. In the network shown in Figure 6.3, the associations of hidden units with the US, VN_j, are stored in a separate layer of neural elements and, therefore, the output activity of each of these neural elements is proportional to CN_jVN_j. Consequently, the error signal that regulates $VH_{i,j}$ associations in hidden units can be calculated by multiplying VN_jCN_j activities by the value of the output error EO. Accordingly, a given hidden unit j changes its associations with the input CSs, VH_{ij}, in proportion to the error signal $EH_j = f(VN_jCN_jEO)$. As in backpropagation, the synaptic strengths between input CSs and a given hidden unit are changed proportionally to (1) the value of the association of the hidden unit with the output unit and (2) the output error. However, in contrast to backpropagation, the derivative of the output of the hidden unit, $d(CN_j)$, is replaced by the output of the hidden unit CN_j. Because virtually identical results are obtained with both procedures (see Appendix C in Schmajuk and DiCarlo, 1992), the neural architecture shown in Figure 6.3 implements a biologically plausible version of backpropagation. Furthermore, Appendix 13.A in Chapter 13 shows that the error signal $EH_j = f(VN_jCN_jEO)$ might be computed by pyramidal cells in the hippocampus.

Box 6.1 summarizes the most important features of the S–D network. A formal description of the S–D model as well as a detailed comparison of the present and original versions of backpropagation are presented in Appendix

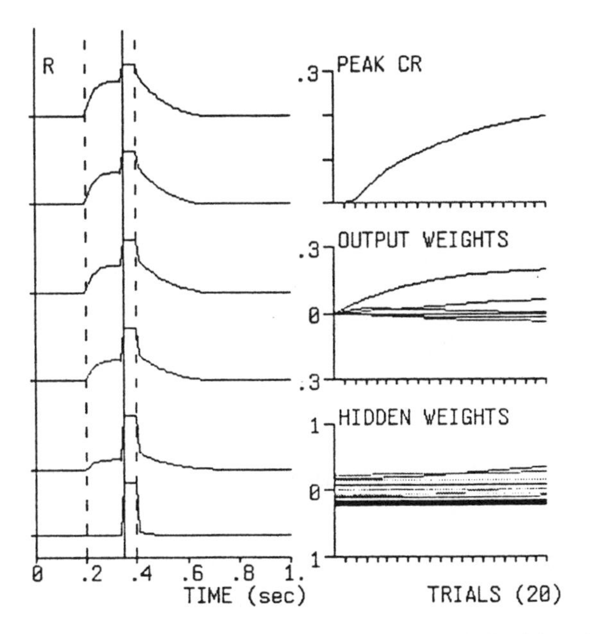

Figure 6.4. Acquisition of classical conditioning. *Left panels.* Real-time simulated conditioned and unconditioned response on trials 1, 4, 8, 12, 16, and 20. Vertical dashed lines indicate CS onset and offset, vertical solid line US onset. Trial 1 is represented at the bottom of the panel. *Right panels.* Peak CR: peak CR as a function of trials; output weights: average VSs and VNs as a function of trials; hidden weights: average VHs as a function of trials.

6.A. In addition, Appendix 3.B in Chapter 3 shows how the model can be physiologically implemented by employing adaptive weights that assume only positive values.

Computer simulations

By applying the S–D neural network described in Figure 6.2, this section contrasts the experimental results in classical conditioning with computer simulations of the following paradigms: (1) acquisition of delay conditioning, (2) extinction, (3) acquisition–extinction series, (4) compound conditioning, (5) positive-feature discrimination, (6) acquisition and retention of negative patterning, and (7) acquisition and retention of positive patterning. Parameter values, shown in Appendix 6.B, were kept constant for all simulations.

This book includes computer software that allows the simulation of the SD model on DOS computers. The software allows the simulation of the paradigms described in this chapter. In addition, it allows users to simulate their own experimental designs and examine the model's predictions.

Acquisition of delay conditioning

Figure 6.4 shows real-time simulations on trials 1, 4, 8, 12, 16, and 20 in a delay conditioning paradigm with a 200-msec CS, 50-msec US, and 150-msec ISI. As CR amplitude increases over trials, output weights VS_i and VN_j, and hidden weights $VH_{i,j}$, may increase or decrease over trials.

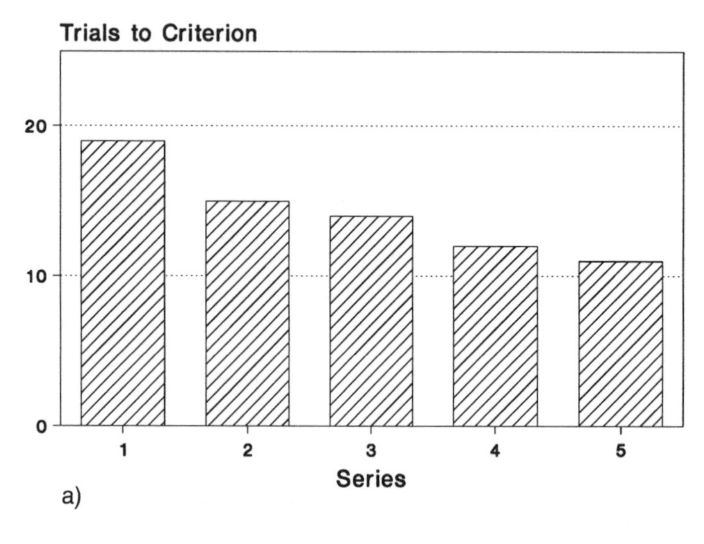

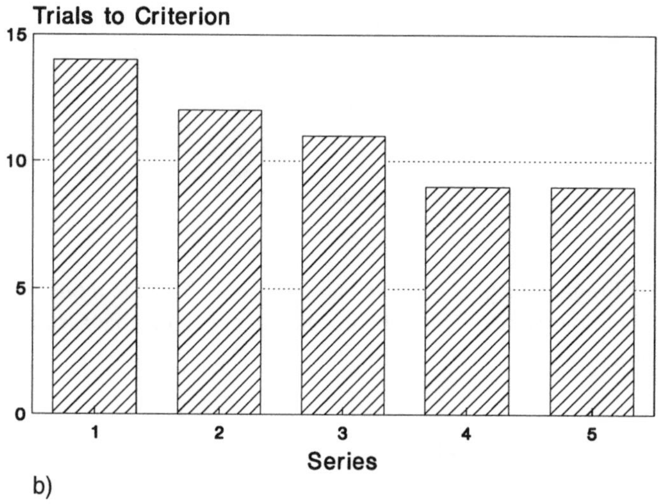

Figure 6.5. Acquisition–extinction series. (a) *Acquisition:* simulated trials to acquisition criterion (0.19) in five series of acquisition trials. (b) *Extinction:* simulated trials to extinction criterion (0.02) following acquisition.

Acquisition–extinction series: savings effects

Five acquisition–extinction series were simulated by alternating (1) delay conditioning until a maximum CR amplitude of 0.19 was reached with (2) extinction until a maximum CR amplitude of 0.02 was attained.

Figure 6.5a shows the number of trials to acquisition in acquisition series. These results are in agreement with the acquisition–extinction series of the Schmaltz and Theios (1972) study showing that normal animals decrease the number of trials to acquisition criterion. Reacquisition is faster than acquisition because the output of the hidden units increases over trials and causes VN_j associations to increase. The increasing output of the hidden units causes the system to rely more on VN_j associations than VS_i associations over successive series. Because the learning rate in hidden units is faster than that of VN_j associations, the system becomes increasingly faster (savings effect).

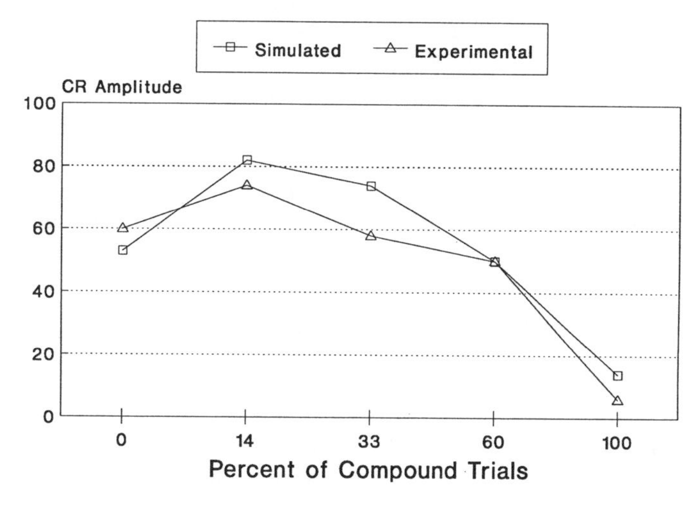

Figure 6.6. Compound conditioning. Relationship between simple stimulus and compound stimulus responding as a function of the proportion of simple and compound acquisition trials. Average of CR peak amplitude evoked by CS_1 and CS_2, after training with different proportions of CS_1, CS_2, and CS_1–CS_2 reinforced trials. Peak CR amplitude is expressed as a percentage of the peak CR to CS_1–CS_2. Percentage of compound trials is computed by the number of CS_1–CS_2 trials divided by the total number of CS_1, CS_2, and CS_1–CS_2 trials. Experimental data from Kehoe (1986).

Figure 6.5b shows the number of trials to extinction in an extinction series. The model predicts faster extinction over series. These results are in agreement with the acquisition–extinction series of the Schmaltz and Theios (1972) study showing that normal animals decrease the number of trials to extinction criterion. Stronger reliance on hidden units implies faster extinction over successive series.

Compound conditioning

In compound conditioning, two or more stimuli are presented together in the presence of the US. Kehoe (1986) explored the relation between configuration and summation by comparing the responding to a CS_1–CS_2 compound and its components, CS_1 and CS_2, under different proportions of reinforced presentations of CS_1–CS_2, CS_1, and CS_2. He found that responding to each component, CS_1 and CS_2, decreased with the increasing proportion of CS_1–CS_2 presentations. These results suggest that CS_1–CS_2, CS_1, and CS_2 have separate representations that compete to establish associations with the US.

Kehoe's (1986) results are readily captured by the parallel input–output architecture of the SD model shown in Figure 6.6. In the SD network, a CS (1) accrues a direct association with the output units and (2) becomes configured with other CSs in a hidden unit that, in turn, acquires associations with the output units. Direct connections, relaying stimulus components, and indirect connections, relaying configured stimuli from the hidden-unit layer, compete to establish associations with the US at the output units. This competition is regulated by a delta rule.

Figure 6.6 shows experimental and simulated relationships between simple stimulus and compound stimulus responding as a function of the proportion of simple and compound acquisition trials. It displays the average of CR peak amplitude evoked by either component, CS_1 or CS_2, after training with different proportions of CS_1, CS_2, and CS_1–CS_2 reinforced trials. Peak CR amplitude is expressed as a percentage of the peak CR relative to the CR evoked by the compound CS_1–CS_2. The percent of compound trials is proportional to the number of CS_1–CS_2 trials divided by the total number of CS_1, CS_2, and CS_1–CS_2 trials. During CS_1 and CS_2 reinforced trials, only direct CS_i–US associations are increased. During CS_1–CS_2 reinforced trials, simple (CS_i) and configural (CN_j) stimuli compete to gain association with the US. Therefore, small percentages of compound trials result in increased responding to the components and larger percentages of compound trials result in decreased responding to the components (because of the competition between simple and configural stimuli established by the delta rule). Figure 6.6 shows that the SD model describes well Kehoe's (1986) compound conditioning data. As shown in Chapter 13, the parallel input–output architecture of the SD model not only provides an adequate description of compound conditioning, but also contributes to the correct mapping of the model over parallel brain circuits.

Simultaneous and serial feature-positive discrimination (occasion setting)

Much of the research in occasion setting grew from investigations of the nature of learning in feature-positive discriminations. In a feature-positive discrimination, presentations of the "feature" cue *X* and the "target" cue *A* are accompanied by the US, and presentations of *A* alone are not reinforced. According to most conditioning theories (e.g., Pearce and Hall, 1980; Rescorla and Wagner, 1972), in feature-positive discriminations, *X* should acquire conditioned excitatory associations with a representation of the US. However, under some circumstances, notably when *X* precedes *A* on compound trials in a serial feature-positive discrimination, *X* would instead acquire the ability to depress or enhance the action of associations between representations of *A* and the US (see Holland, 1992). Thus, although a simultaneous feature-positive discrimination procedure encourages solution of the discrimination by simple conditioning of *X*, in a serial feature-positive discrimination procedure, *X* comes to "set the occasion" for responding on an association between *A* and the US.

Figure 6.7 shows simulated results for simultaneous and serial feature-positive discriminations. In the figure, feature *X* corresponds to CS_1 and target *A* to CS_2. Figure 6.7 depicts peak CR amplitude evoked by CS_1–CS_2 and CS_2 after 75 reinforced CS_1–CS_2 trials alternated with 75 nonreinforced CS_2 trials. Peak CR amplitude is expressed as a percentage of the peak CR to CS_1–CS_2. Simulations show feature-positive patterning because the response to CS_1–CS_2 is large and the response to CS_2 almost negligible. Notice that responding to the CS_1–CS_2 compound is due to CS_1–US associations.

Figure 6.7 also shows peak CR amplitude evoked by CS_1 followed by CS_2 after 100 msec, CS_1, and CS_2 after 75 reinforced CS_1-followed-by-CS_2 trials alternated with 75 nonreinforced CS_2 trials. Peak CR amplitude is expressed as a percentage of the peak CR to CS_1 followed by CS_2 after 100 msec. Although the response to CS_1–CS_2 is large and the response to CS_2 almost negligible, responding to the CS_1–CS_2 compound is not due to CS_1–US associations, but CS_2–US associations modulated by configural stimulus–US associations. These results are in agreement with those of Holland (1992).

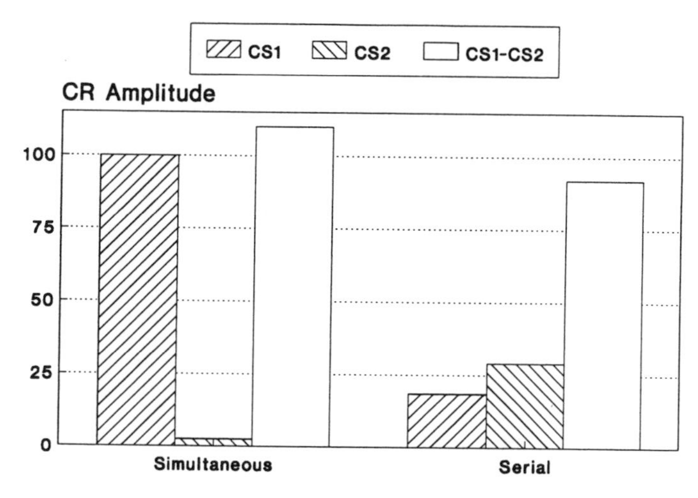

Figure 6.7. Simultaneous and serial feature-positive discrimination. *Simultaneous:* peak CR amplitude evoked by CS_1–CS_2, CS_1, and CS_2 after 75 reinforced CS_1–CS_2 trials alternated with 75 nonreinforced CS_2 trials. Peak CR amplitude is expressed as a percentage of the peak CR to CS_1–CS_2. *Serial (occasion setting):* peak CR amplitude evoked by CS_1 followed by CS_2 after 100 msec, CS_1, and CS_2 after 75 reinforced CS_1 followed by CS_2 trials alternated with 75 nonreinforced CS_2 trials. Peak CR amplitude is expressed as a percentage of the peak CR to CS_1 followed by CS_2 after 100 msec.

Positive patterning

Figure 6.8 exhibits simulated data regarding the temporal course of positive patterning acquisition. Figure 6.8 shows the percent of CRs evoked by CS_1, CS_2, and CS_1–CS_2 over 60 blocks of trials. Each block consisted of one nonreinforced CS_1 trial, one nonreinforced CS_2 trial, and one reinforced CS_1–CS_2 trial. CR percentage is expressed as a proportion of the peak CR after 20 acquisition trials. As in the Bellingham, Gillette-Bellingham, and Kehoe (1985) experiment, at the beginning animals respond to both compound and components, and this is followed by a gradual decline of the response to the components. The temporal course of acquisition of positive patterning is very similar in simulations and experimental data.

Negative patterning

Figure 6.9 displays simulated data regarding the temporal course of negative patterning acquisition. The figure shows the percent of CRs evoked by CS_1, CS_2, and CS_1–CS_2 over 60 simulated blocks of trials. Each block consisted of one reinforced CS_1 trial, one reinforced CS_2 trial, and one nonreinforced CS_1–CS_2 trial. CR percentage is expressed as a proportion of the peak CR after 20 acquisition trials. As in the Bellingham et al. (1985) experiment, at the beginning animals respond to both compound and components, and this is followed by a gradual decline of the response to the compound. However, the initial difference between responding to compound and components is larger in the simulations than in the experimental data.

Although it is usually accepted that negative patterning can be solved by creating a configural stimulus that is active when both CSs are active together and that this configural stimulus acquires inhibitory association with the US, com-

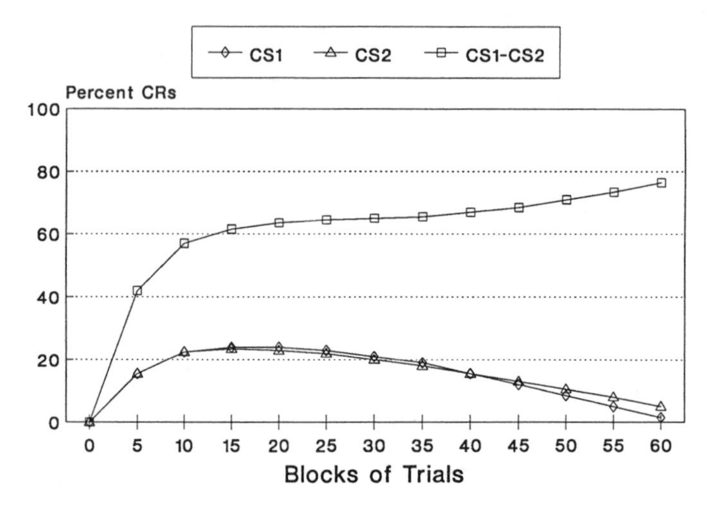

Figure 6.8. Temporal course of positive patterning acquisition. Percent of CRs evoked by CS_1, CS_2, and CS_1–CS_2 over 60 blocks of trials. Each block consisted of one nonreinforced CS_1 trial, one nonreinforced CS_2 trial, and one reinforced CS_1–CS_2 trial. CR percentage is expressed as a proportion of the peak CR after 20 acquisition trials.

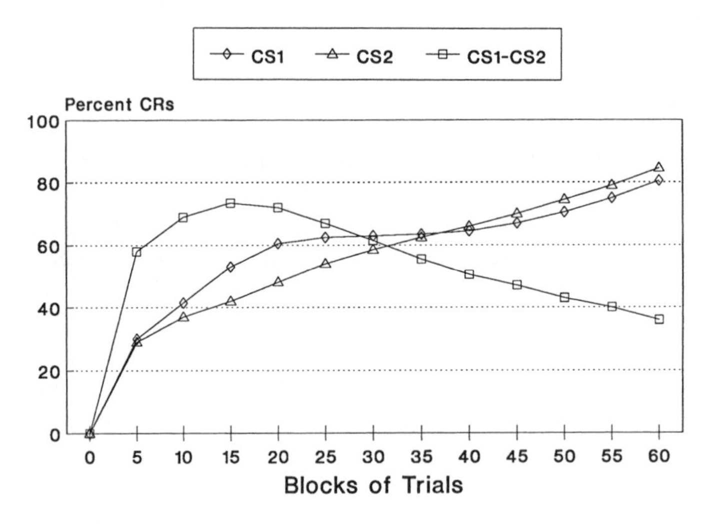

Figure 6.9. Temporal course of negative patterning acquisition. Percent of CRs evoked by CS_1, CS_2, and CS_1–CS_2 over 60 blocks of trials. Each block consisted of one reinforced CS_1 trial, one reinforced CS_2 trial, and one nonreinforced CS_1–CS_2 trial. CR percentage is expressed as a proportion of the peak CR after 20 acquisition trials.

puter simulations show that a variety of configural stimuli are built during negative patterning and that they acquire excitatory and inhibitory associations with the output unit.

Schmajuk and DiCarlo (1992) analyzed the effects that (1) varying the number of hidden units or (2) adding noise to the error used to train hidden units have on the acquisition of negative patterning. They showed that at least four hidden units are required to solve negative patterning 75 percent of the time, and that performance degrades rapidly with error noise levels greater than 5 percent.

Table 6.1. *Simulations obtained with different versions of the Schmajuk and DiCarlo (1992) configural model.*

Paradigm	Configural Model	Configural Novelty Predictive Value
Delay conditioning	Y	Y
Trace conditioning	Y	Y
Partial reinforcement	Y	Y
Extinction	Y	Y
Acquisition series	Y	Y
Extinction series	Y	Y
Latent inhibition		Y
Blocking	Y	Y
Unblocking		Y
Overshadowing	Y	Y
Discrimination	Y	Y
Conditioned inhibition	Y	Y
Inhibition by differential conditioning	Y	Y
Inhibition by decrease in the US	Y	Y
Counterconditioning	Y	Y
Superconditioning	Y	Y
Overprediction effects	Y	Y
Generalization	Y	Y
Compound conditioning	Y	Y
Simultaneous feature-positive discrimination	Y	Y
Simultaneous feature-negative discrimination	Y	Y
Serial feature-positive discrimination	Y	Y
Serial feature-negative discrimination	Y	Y
Negative patterning	Y	Y
Positive patterning	Y	Y
Sensory preconditioning		Y
Second-order conditioning		Y
Context switching	Y	Y
External inhibition	Y	Y

Note. Y: The model can simulate the experimental data.

Discussion

Computer-simulated results were contrasted with experimental data. By evaluating the model in the large variety of learning paradigms listed in Table 6.1 (using the same parameter values), we expect it to reflect properties of classical conditioning and to reveal more than a fortuitous selection of parameter values.

The biologically plausible, real-time rendition of backpropagation offered here differs from the original version in that the error signal used to train hidden units, instead of including the derivative of the activation function of the hidden units, simply contains their activation function.

Alternatives to backpropagation

The SD model accomplishes stimulus configuration by incorporating a hidden layer trained through a backpropagation procedure. This section examines alternative models that also provide internal representations that can be used to solve configural tasks.

One alternative to the backpropagation procedure used in the SD network is Kehoe's (1988) model. Kehoe (1988) presented a three-layer network model that describes configuration, learning to learn, and savings effects. As in Kehoe's (1988) model, the SD model adopts a unique-stimulus hypothesis that assumes that during learning organisms "maintain a fully active representation of the separate components as well as synthesizing a configural representation of the compound" (Kehoe and Graham, 1988). Also as in Kehoe's (1988) model, the connections of the hidden layer to the output units in the SD model change at a slower rate than those from the inputs to the hidden units, thereby describing savings.

In contrast to the backpropagation procedure employed in the SD model, in Kehoe's model all configural units are trained with the same US signal and, therefore, they all learn the same configuration, a procedure that might subtract power from the system. Unlike the SD model, Kehoe's model does not offer an adequate description of stimulus compounding (p. 422), the increased responding to the components in positive patterning (p. 420), or a good description of savings in successive extinctions (p. 416). The major distinction, however, between the SD and the Kehoe model resides in the way they configure stimuli. Whereas in the Kehoe model configuring is equivalent to creating a compound stimulus active when the component stimuli are presented together (e.g., CS_1 and CS_2), in the SD model configuring is accomplished by training hidden units that may respond to various combinations of stimuli (e.g., respond to CS_1 in the absence of CS_2).

As simple and elegant as the backpropagation procedure might be, it has been shown that the procedure suffers from catastrophic interference (McClosky and Cohen, 1989). Catastrophic interference refers to the fact that, under certain conditions, the storage of new information in the network interferes with previously stored information. However, Levandowsky (1991) showed that, when stimuli and responses are represented by uncorrelated vectors and a continuous retrieval measure is used, backpropagation networks do not exhibit catastrophic interference.

One alternative to backpropagation that does not show catastrophic interference is Carpenter and Grossberg's (1987) adaptive resonance theory (ART) model. ART learns stimulus patterns in real time. In ART, interactions between an attentional system and orienting subsystem enable the network to stabilize its learning, without an external teacher. The attentional system learns bottom–up codes and top–down expectancies. Mismatches between the learned top–down expectancies against input patterns activate an orienting subsystem, which resets incorrect codes and searches for new codes.

Extending the power of the model

Multiple response systems

The SD model (as well as all other extant conditioning models) assumes that all inputs activate essentially the same CR, that is, the form of the CR is determined by the choice of US. However, much investigation of occasion setting has exploited the fact that the form of the CR is often determined not only by the US, but also by the nature of the CS. For example, rats exhibit very different CR forms during visual (e.g., rearing on hind legs) and auditory (rapid head movements, or "head jerk" behavior) signals for food (see Holland, 1992). Consequently, both to describe more completely conditioned behavior in general and to address the outcomes of experiments in occasion setting in particular, Schmajuk, Lamoreaux, and Holland (1997) extended the SD model to describe

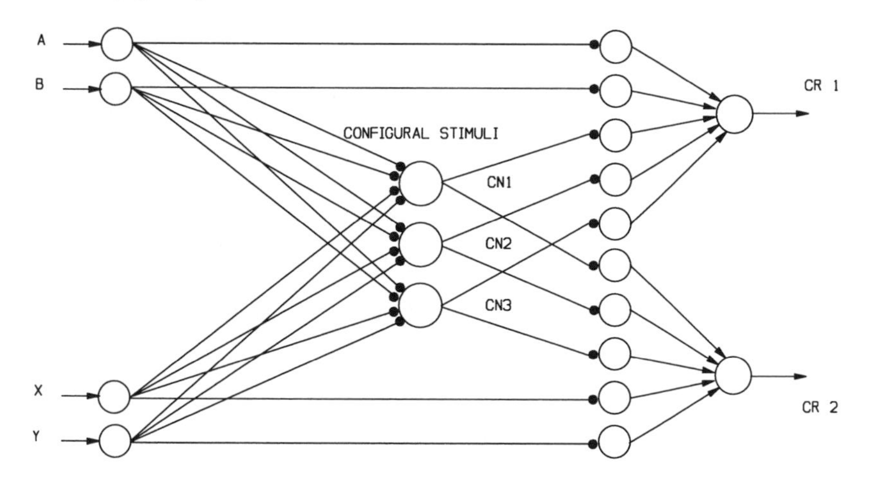

Figure 6.10. The SD model for multiple-response systems. Diagram of a network that incorporates a layer of hidden units capable of describing stimulus configuration in classical conditioning. *A* and *B*: inputs to response system 1; *X* and *Y*: inputs to response system 2; CN: configural stimulus. Arrows represent fixed synapses, solid circles variable synapses.

multiple-response systems. It is important to notice that this extension does not modify the model's learning rules for the CR, but only the computation of the responses for the different systems.

Figure 6.10 shows a diagram of the SD neural network for multiple-response systems. It illustrates how the associations computed by the network are organized to generate CR_1 and CR_2. The figure also shows a multilayer network with one input layer, one hidden-unit layer, and two sets of output layers, one controlling CR_1 and another one controlling CR_2. Input units are activated by targets *A* and *B*, assumed to be in one sensory modality (e.g., auditory), and by features *X* and *Y*, assumed to be in a different sensory modality (e.g., visual).

Figure 6.10 shows that inputs *A* and *B* (e.g., auditory) activate CR_1 (e.g., head jerk) through their direct *excitatory* associations with the US (VS_A, VS_B). Symmetrically, inputs *X* and *Y* (e.g., visual) activate CR_2 (e.g., rearing) through their direct *excitatory* associations with the US (VS_X, VS_Y). The output activities of the hidden-unit layer (assumed to code configural stimuli CN_1, CN_2, CN_3) excite or inhibit through their associations with the US (VN_1, VN_2, VN_3) both CR_1 and CR_2. Although not indicated in Figure 6.10, inputs *A*, *B*, *X*, and *Y* *inhibit* both CR_1 and CR_2 through their direct inhibitory associations with the US ($-VS_A$, $-VS_B$, $-VS_X$, $-VS_Y$).

The model can be applied to many of the paradigms that distinguish simple conditioning from occasion setting, for instance, response form as a function of whether serial or simultaneous compound stimuli were used in feature-positive discrimination training.

Schmajuk et al. (1997) applied the SD model for multiple-response systems to occasion setting. Computer simulations of serial discriminations show most of the properties reported in occasion setting (see Holland, 1992), including those related to response form, extinction and counterconditioning, transfer ef-

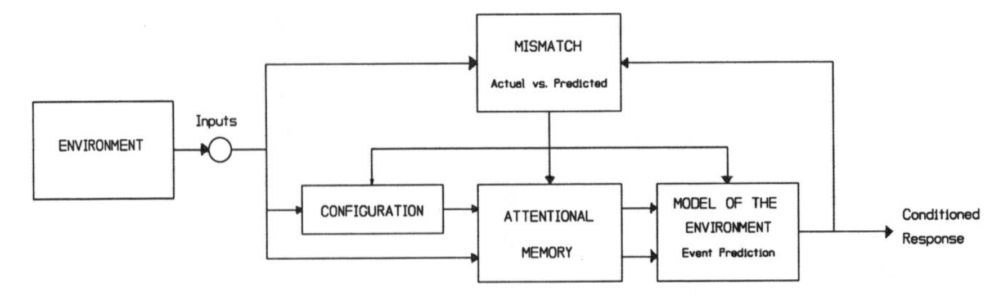

Figure 6.11. A configural-attentional model of the environment. The model of the environment generates predictions of future events and controls the conditioned response (CR) based on simple and configural associations. The mismatch systems compare observed and predicted events to compute novelty. Novelty is then used to (1) control attention to environmental stimuli in the attentional system, (2) modify the model of the environment, and (3) inhibit ongoing conditioned behavior. CR: conditioned response; OR: orienting response.

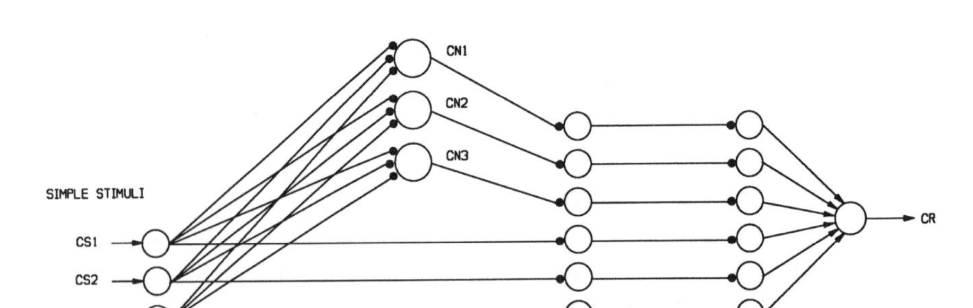

Figure 6.12. A configural-attentional network. Diagram of a network that incorporates a layer of hidden units capable of describing stimulus configuration in classical conditioning and a layer of attentional units capable of describing latent inhibition phenomena. CS: conditioned stimulus; CN: configural stimulus; VS: CS–US association; VN: CN–US associations; VH: CS–CN associations; CR: conditioned response. Arrows represent fixed synapses, solid circles variable synapses.

fects, CS_1–US, CS_1–CS_2, and CS_2–US temporal intervals, termination asynchrony, relative time intervals, context, CS_1–CS_2 similarity, CS_2 intensity, CS_2 pretraining, and CS_1 and CS_2 pretraining.

Latent inhibition and sensory preconditioning

Although the SD model is able to describe all the paradigms listed in Table 6.1, two important classical conditioning paradigms are missing, namely, latent inhibition and sensory preconditioning. However, when combined with the model presented in Chapter 5, the SD model can describe an even wider variety of classical conditioning data. Such integration is illustrated in Figures 6.11 and 6.12 showing that simple stimuli CS_i become *configured* with other CSs in a hidden unit that represents configural stimulus CN_j. As before, it is assumed that $f(CS_i)$ and $f(CN_i)$ are modulated by the association of the internal representations of CS_i and CN_j with the total environmental Novelty. Both $f(CS_i)$ and $f(CN_i)$ become associated with the US. Table 6.1 also shows the paradigms that the SLG model correctly describes when (1) an attentional variable y_i reflecting the predictive quality of CS_i and (2) stimulus configuration are incorporated into the model, as illustrated in Figure 6.12.

Appendix 6.A

A formal description of the SD model

The SD model is depicted in Figures 6.2 and 6.3.

1. Short-term memory (STM) traces

The STM trace of CS_i, X_i, is defined by

$$d(X_i)/dt = -K_1X_i + K_2(K_3 - X_i)\,CS_i \tag{6.1}$$

where $-K_1X_i$ represents the passive decay of the STM of CS_i, K_2 represents the rate of increase of X_i, and constant K_3 is the maximum possible value of X_i.

We assume that the input representing context has a constant STM trace, $X_{context} = 0.5$.

2. Neural activity

Hidden unit j is activated by the STM of different CS_i's in proportion to their connections with the hidden unit:

$$\text{sum}_j = \Sigma_i X_i \text{VH}_{i,j} \tag{6.2}$$

where $\text{VH}_{i,j}$ represents the association between the STM trace X_i and hidden unit j.

The output of the hidden units is a sigmoid given by

$$an_j = K_4(\text{sum}_j^n/\beta_1^n + \text{sum}_j^n). \tag{6.3}$$

Whereas Rumelhart et al. (1986) assumed that the output of neural units was active even in the absence of any input, we assume that the hidden units are active only when inputs are present, that is, $an_j = 0$ if $\text{sum}_j = 0$.

The output of the input units is given by

$$as_i = K_5X_i \tag{6.4}$$

We assume $K_4 > K_5$, which gives an advantage to the hidden units over the direct inputs to establish associations with the US.

3. Input–output associations

Changes in the associations between input i and the US, VS_i, are given by a modified delta rule:

$$d(\text{VS}_i)/dt = K_6 as_i(1 - |\text{VS}_i|)\,\text{EO}_i \tag{6.5}$$

By Equation 6.5, VS_i changes only when as_i is active and the output error EO_i is not zero. Equation 6.5 is a Hebbian rule (VS_i changes with concurrent presynaptic activity as_i and postsynaptic activity EO_i). The term $(1 - |VS_i|)$ bounds VS_i $(-1 \leq VS_i \leq 1)$.

Output error EO_i is given by

$$EO_i = US - VS_i as_i - B_i \qquad (6.5')$$

where $B_i = \Sigma_{l \neq i} VS_l as_l + \Sigma_j VN_j an_j$, that is, the aggregate prediction of the US based on every node associated with the US, different from as_i, including hidden-unit–output connections. By Equation 6.6, EO_i is linearly controlled by the individual *local* prediction of the US, $VS_i as_i$, and the *global* aggregate prediction of the US, B_i.

4. Hidden-unit–output associations

Changes in the association between hidden unit j and the US, VN_j, are given by a modified delta rule:

$$d(VN_j)/dt = K_6 an_j (1 - |VN_j|)\, EO_j \qquad (6.6)$$

By Equation 6.5', VN_j changes only when an_j is active and EO_j is not zero. As in Equation 6.5, Equation 6.5' is a Hebbian rule (VN_j changes with concurrent presynaptic activity an_j and postsynaptic activity EO_j). The term $(1 - |VN_j|)$ bounds VN_j $(-1 \leq V_j \leq 1)$.

Output error EO_j is given by

$$EO_j = US - VN_j an_j - B_j \qquad (6.6')$$

where $B_j = \Sigma_{h \neq j} VN_h an_h + \Sigma_i VS_i as_i$, that is, the aggregate prediction of the US based on every node associated with the US, different from an_j, including input–output connections. Notice that output errors EO_i and EO_j used in Equations 6.6 and 6.6' are identical.

5. Input–hidden-unit associations

Changes in the association between input i and hidden unit j, $VH_{i,j}$, are given by a modified delta rule:

$$d(VH_{i,j})/dt = K_7 as_i (1 - |VH_{i,j}|)\, EH_j \qquad (6.7)$$

where term $(1 - |VH_{ij}|)$ bounds VH_{ij} between -1 and 1. By Equation 6.7, $VH_{i,j}$ changes only when $VH_{i,j}$ is active and the hidden-unit error EH_j, is not zero.

Hidden-unit error EH_j is given by

$$EH_j = 1/[1 + e^{-k_8\, an_j\, VN_j (US - B)}] - 0.5 \qquad (6.8)$$

where $B = \Sigma_i VS_i as_i + \Sigma_j VN_j an_j$, that is, the aggregate prediction of the US based on every node associated with the US. As explained in Chapter 13, Equation 6.8 approximately describes the result of the interaction between medial septal and entorhinal cortex inputs to dentate gyrus, CA3, and CA1 hippocampal fields, as proportional to $\sqrt{|US - B| VN_i an_i}$.

The error signal used to train hidden units, described by Equation 6.8, is similar to the error signal for hidden units used by Rumelhart et al. (1986) in that it is proportional to (1) the value of the association of the hidden units with the output units, and (2) the output error (in this case, there is only one output error). Equation 6.8, however, differs from the original backpropagation hidden-unit error equation in that, instead of including the derivative of the activation function, it simply contains the activation function an_j. Although the original error expression is not guaranteed to avoid the fatal problem of local minima (Rumelhart et al., 1986, p. 324), it is possible that the change introduced in the SD model decreases the power of the network to circumvent local minima. In order to explore this possibility, we carried out several computer simulations. First, simulations using Equation 6.8 either with activation function an_j or its derivative yielded almost identical results. This similarity might be due to the fact that $d(an_j) = (1 - an_j)\, an_j$ becomes $d(an_j) \propto an_j$ for small values of an_j, a condition generally met in our simulations. Second, simulations carried out with the original backpropagation network, using either the hidden-unit activation function or its derivative in the error term, also yielded similar results for several tasks, including the exclusive or, parity, and encoder problems. Importantly, however, although similar results were obtained in these simulations, more hidden units were required to solve the tasks both with the SD model and the original backpropagation network. Although in both cases the need of more hidden units might result from the use of an_j instead of $d(an_j)$, an additional factor in the SD network might be the real-time characteristics of the system, that is, more combinations of inputs should be configured. As mentioned, because of the large number of neurons available in association cortices, the use of more hidden units is not a serious obstacle for a neurophysiologically viable system.

6. CR generation

The CR output of the system is given by

$$\mathrm{CR} = \mathrm{R}_1[\Sigma_i \mathrm{VS}_i as_i + \Sigma_j \mathrm{VN}_j an_j] \tag{6.9}$$

where $\mathrm{R}_1[x] = 0$ if $x < K_9$, and $\mathrm{R}_1[x] = x - K_9$ if $x > K_9$. K_9 is the behavioral threshold of the system.

The total output of the system, nictitating membrane (NM) response, is given by

$$\mathrm{NMR} = \mathrm{CR} + \mathrm{UR} \tag{6.10}$$

Appendix 6.B

Simulation parameters

Our simulations assume that one simulated time step is equivalent to 10 msec. Each trial consists of 500 steps, equivalent to 5 sec. Unless specified, the simulations assume 200-msec CSs, the last 50 msec of which overlaps the US. CS onset occurs at 200 msec. Parameters are selected so that simulated asymptotic values of CR are reached in around 10 acquisition trials. Since asymptotic conditioned NM responding is reached in approximately 200 real trials (Gormezano, Kehoe, and Marshall, 1983), one simulated trial is approximately equivalent to 20 experimental trials.

Parameter values are $K_1 = 0.07$, $K_2 = 0.18$, $K_3 = 1$, $K_4 = 2.5$, $K_5 = 1.5$, $K_6 = 0.005$, $K_7 = 0.33$, $K_8 = 5$, $K_9 = 0.03$, $\beta_1 = 0.5$, $n = 1.5$. The initial values of VS_i and VN_j were 0. Because five hidden units are needed in order to attain reliable negative patterning, this number of hidden units was used in all simulations. Input–hidden-unit association weights $\mathrm{VH}_{i,j}$ were randomly assigned using a uniform distribution ranging between $\pm$ 0.25 in all simulations.

7 Timing

The previous chapters introduced several neural networks that build internal models of the environment serving to predict what event follows what event. However, more comprehensive models of the environment need not only predict what follows what, but also when and where different events take place. Whereas "temporal mapping" properties of the environmental model reflect the relationships among temporal events, "spatial mapping" properties reflect the relationships among spatial locations. Temporal maps are used in "temporal navigation" to predict the timing of environmental events in classical and operant conditioning. Spatial maps, used in spatial navigation to determine the location where those events occur, are described in Chapter 10.

This chapter presents Grossberg and Schmajuk's (1989) spectral timing model, a neural network capable of learning the temporal relationships between two events. The model consists of three layers of neural elements. A step function, activated by the CS presentation, excites the first layer that contains many elements, each one having a different reaction time. The output of each element in the first layer is a sigmoid function that activates a second layer of habituating transmitter gates. In turn, the output of each transmitter gate activates a long-term memory element. Those memory elements active at the time of the US presentation become associated with the US in proportion to their activity. The outputs of all memory elements are added in order to determine the magnitude of the CR. During testing, the CR shows a peak at the time when the elements that have been active simultaneously with the US are active again. The model is able to describe ISI curves with single and multiple USs, and a Weber law for temporal generalization.

Theories of timing

Several theories have been proposed to account for timing: internal clock (Church, 1984), time encoding by associative values (Schmajuk, 1986; Schmajuk and Moore, 1986), delay lines (Desmond and Moore, 1988), a spectral timing neural network (Grossberg and Schmajuk, 1989), a tuned network (Schmajuk, 1990), and a discrete Fourier transform network (Gluck, Reifsnider, and Thompson, 1990.)

Church (1984) proposed a timing theory (scalar timing theory) that was later implemented as a connectionist model by Church and Broadbent (1991). The model comprises the following components: (1) a pacemaker that emits pulses, (2) a switch that is opened at the onset of the event to be timed and closed at its offset, (3) a counter that accumulates pulses, (4) a "reference memory" that accumulates pulses of reinforced times and a "working memory" that stores the total number of pulses accumulated in a particular trial, and (5) a comparator that compares the values stored in both memories. Values stored in working memory are compared to values stored in reference memory and, if they are similar, a response is produced. Notice that the number of stored pulses increases with the measured time. The model is able to describe timing with single but not multiple USs and describes a Weber law for temporal generalization.

Schmajuk (1986, Schmajuk and Moore, 1986) proposed a model in which the time interval between the CS onset and the US presentation is encoded by the CS–US associative value. Making use of a trace of the CS, the model indicates that CS–US associations decrease with increasing ISIs. By assuming that CR onset is inversely proportional to the magnitude of the CS–US association, the model accounts for timing with different ISIs. The model is able to describe ISI curves with single but not multiple USs, and it does not describe a Weber law for temporal generalization. Similarly, the real-time, delta rule model presented in Chapter 3 is able to generate simple ISI curves.

Desmond and Moore (1988) proposed a neural network model that simulates the temporal characteristics of a CR in classical conditioning. The model assumes that the CS onset activates an onset process in one set of neural elements, and the CS offset activates an offset process in another set of elements. Each process consists of sequential and overlapping activation of neural elements *X*. Once activated, each element *X* remains activated for a certain time. Each element *X* makes contact, through a modifiable synapse, with two neural elements *V* and *E*. Element *V* receives input from the US and from element *E*. Element *E* receives input only from the US. The output of the *V* element describes the CR. The model is able to describe ISI curves with single and multiple USs. The model does not describe, however, a Weber law for temporal generalization.

Schmajuk (1990) presented a real-time neural network capable of describing temporal discrimination and spatial learning in a unified fashion (see Chapter 10). The network incorporates detectors that can be tuned to the value of the temporal trace of CS_i at the time when the US is presented. After a detector has been tuned, its output generates an effective stimulus that peaks at the time when the US was presented before. The outputs of the tuned detectors become associated with the US. This timing theory can be regarded as a computational and neural version of Gormezano's (1972) trace amplitude recognition hypothesis. Computer simulations show that the network correctly describes temporal discrimination in classical conditioning and instrumental learning, as well as classical conditioning under different ISIs and with mixed ISIs.

Gluck et al. (1990) proposed that the onset of the CS generates a collection of input signals that resemble sine waves of varying phase and frequency (like those used in a traditional discrete Fourier transform) and that those signals become associated with the US using a delta rule. Although Gluck et al. (1990) pointed out that the inputs need not be so regular or periodic as the sine waves used in their simulations, but need only be sufficiently rich, varied, and complex, they do not offer any biological justification for the existence of such waves. The model is able to capture the CR peak timing, discrete CR shifts with ISI changes, and a double-peak CR with mixed ISI training. The model fails, however, to describe the anticipatory character of the CR, the decrease in CR onset latency during training, and the dependence of CR acquisition on ISI.

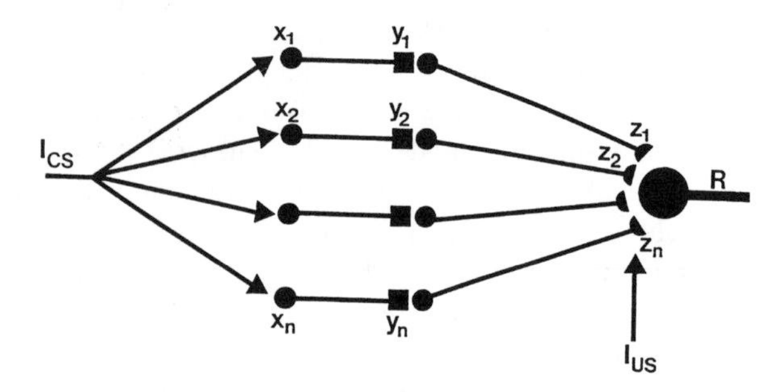

Figure 7.1. Circuit diagram of the spectral timing model. The CS activates function $I_{CS}(t)$, assumed to be a step function whose amplitude is proportional to the CS intensity, which stays on for a fixed time after CS offset. x_i denotes the activity of cells with different growth rates α_i; y_i represents the neurotransmitter gate activated by x_i; z_i denotes adaptive long-term memory traces; and $R(t)$ denotes the total circuit output.

The spectral timing model

Grossberg and Schmajuk (1989) introduced a real-time (see Box 1.4) neural network model that describes behavioral timing. The circuit diagram of the spectral timing model is schematized in Figure 7.1. Box 7.1 summarizes the most important properties of the network. Appendix 7.A offers a formal description of the model.

In the model, the CS activates function $I_{CS}(t)$, assumed to be a step function whose amplitude is proportional to the CS intensity, and which stays on for a fixed time after CS offset because it is internally stored in short-term memory (STM). Input $I_{CS}(t)$ activates all potentials x_i of the cells in its target population. The potentials x_i respond at rates proportional to α_i, $i = 1, 2, \ldots, n$. Each potential x_i generates the output signal $f(x_i)$. Figure 7.2 (bottom panel) depicts the result of a computer simulation in which $f[x_i(t)]$ is plotted at a function of time t for values of α_i ranging from 0.2 ("fast cells") to 0.0025 ("slow cells").

Each output signal $f(x_i)$ activates a neurotransmitter y_i. Neurotransmitter y_i accumulates to a constant level and is inactivated by signal $f(x_i)$. The different rates α_i at which each x_i is activated cause the corresponding y_i to become habituated at a different rate. A habituation spectrum is thereby generated. The signal functions $f[x_i(t)]$ in Figure 7.2 (bottom panel) generate the habituation spectrum of $y_i(t)$ curves shown in Figure 7.2 (middle panel).

Each signal $f(x_i)$ interacts with y_i. This process is also called the gating of $f(x_i)$ by y_i to yield a net signal g_i proportional to $f(x_i)y_i$. Each of these gated signals has a different rate of growth and decay. The set of all these curves thereby generates a gated signal spectrum, which is shown in Figure 7.2 (top panel). The curves in Figure 7.2 (top panel) exhibit the following properties: (1) Each function $g_i(t)$ is a unimodal function of time, where function $g_i(t)$ achieves its maximum value M_i at time T_i; (2) T_i is an increasing function of i; and (3) M_i is a decreasing function of i.

The curves shown in the top panel of Figure 7.2 represent neural populations whose elements are distributed along a temporal parameter and are familiar throughout the nervous system. Two examples are the size principle, which governs variable rates of responding in spinal motor centers (Henneman, 1957, 1985), and the spatial frequency-tuned cells of the visual cortex, which also react at different rates (e.g., Skrandies, 1984).

Box 7.1 Grossberg and Schmajuk's (1989) timing model

The most salient aspects of the model are:

1. A CS activates function $I_{CS}(t)$, assumed to be a step function whose amplitude is proportional to the CS intensity, which stays on for a fixed time after CS offset because it is stored in STM.
2. Function $I_{CS}(t)$ activates all potentials x_i of the cells in its target population. The potentials x_i respond at rates proportional to α_i, $i = 1, 2, \ldots, n$, ranging from fast to slow rates.
3. Each signal $f(x_i)$ activates a neurotransmitter y_i.
4. Neurotransmitter y_i accumulates to a constant level and is inactivated by signal $f(x_i)$. The different rates α_i at which each x_i is activated cause the corresponding y_i to become habituated at a different rate.
5. Each signal $f(x_i)$ interacts with y_i to yield a gated signal g_i proportional to $f(x_i)y_i$. Each of these gated signals has a different rate of growth and decay constituting a spectrum.
6. LTM associations between the spectrum of different g_i's and the US, z_i, reflect the differential ability of g_i to learn at different values of the ISI. This is the basis of the network's timing properties.
7. The output signal $R(t)$ is the sum of all $g_i z_i$.

Each long-term memory (LTM) z_i in Figure 7.1 is activated by its own temporally selective sampling signal g_i. The sampling signal g_i turns on the learning process and causes z_i to approach US during the sampling signal at a rate proportional to g_i. Each z_i thus grows by an amount that reflects the degree to which the curves $g_i(t)$ and $US(t)$ have simultaneously large values through time. The difference in the individual ability of $g_i(t)$ signals to learn at different values of the ISI is the basis of the network's timing properties.

The output signal $R(t)$ is the sum of all the doubly gated signal functions $h_i(t)$ minus the threshold F. The output signal computes the cumulative learned reaction of all the cells to the input pattern.

Computer simulations

This section presents computer simulations obtained with the spectral timing network for different temporal learning paradigms. Parameter values used in the simulations are shown in Appendix 7.B.

CR topography in classical conditioning

Figure 7.3 (top panel) illustrates how the model describes timing during classical conditioning. The CS and US are paired during four learning trials, after which the CS is presented alone on a test trial. The CS representation $I_{CS}(t)$ remains on and constant throughout the duration of each learning trial. The US input $US(t)$, is presented after 400 msec and then remains on for 50 msec. When a threshold L for responding is assumed, CR latency decreases and CR amplitude increases over trials in agreement with Smith's (1968) data. The CR latency

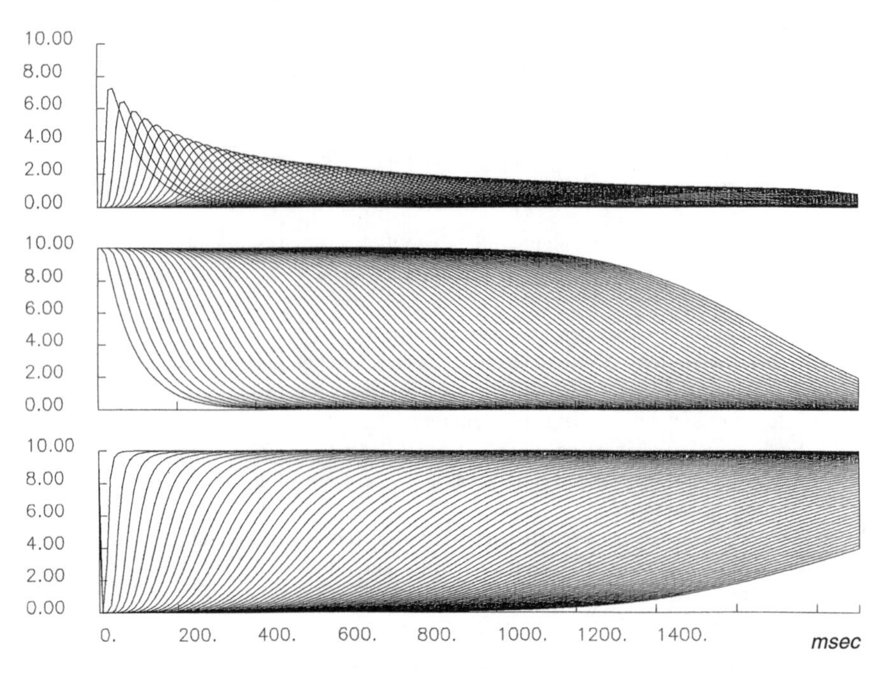

Figure 7.2. Spectrum of reactions to step input I_{CS}. *Bottom panel:* Eighty signal functions $f(x_i)$ with rates proportional to α_i, $i = 1, 2, \ldots, n$. *Middle panel:* The corresponding 80 neurotransmitter gates $y_i(t)$. *Top panel:* The corresponding gated signals $g_i = f(x_i)\, y_i$.

moves progressively forward in the CS–US interval with training. The maximal response amplitude (CR peak) is approximately located at the time of the US occurrence.

The bottom panel in Figure 7.3 displays the sampling signal functions, $g_i(t) = f[x_i(t)]y_i(t)$, generated as described in Figure 7.2. The middle panel in Figure 7.3 displays all the doubly gated signal functions, $h_i(t) = g_i(t)\ z_i(t)$ that, when summed, generate $R(t)$ on the fifth trial. Finally, the top panel in Figure 7.3 shows $R(t)$ on four succesive reinforced trials and one nonreinforced test trial. Together, these panels illustrate how function $R(t)$ generates an accurately timed response from the cumulative partial learning of all cells in the population spectrum.

ISI curves and US intensity

Figure 7.4 plots the functions R(t) that are generated by different ISIs in a series of learning experiments. These are the R(t) functions generated on the fifth trial of each experiment in response to a CS alone, after four trials of prior learning, with all time axes synchronized with CS onset. In the top panel of Figure 7.4, the US(t) was chosen twice as large as in the bottom panel of Figure 7.4. Halving US(t) amplitude reduces the R(t) amplitudes without changing their timing or overall shape. Note that the envelope of the R(t) functions increases and then decreases through time, and that the individual R(t) functions corresponding to larger ISIs are broader. The computer simulations summarized in Figure 7.4 are strikingly similar to the data of Smith (1968). The fact that conditioning occurred at ISIs much larger than CS duration implies that a trace of the CS, which we have called I_{CS}, is stored in short-term memory subsequent to CS offset.

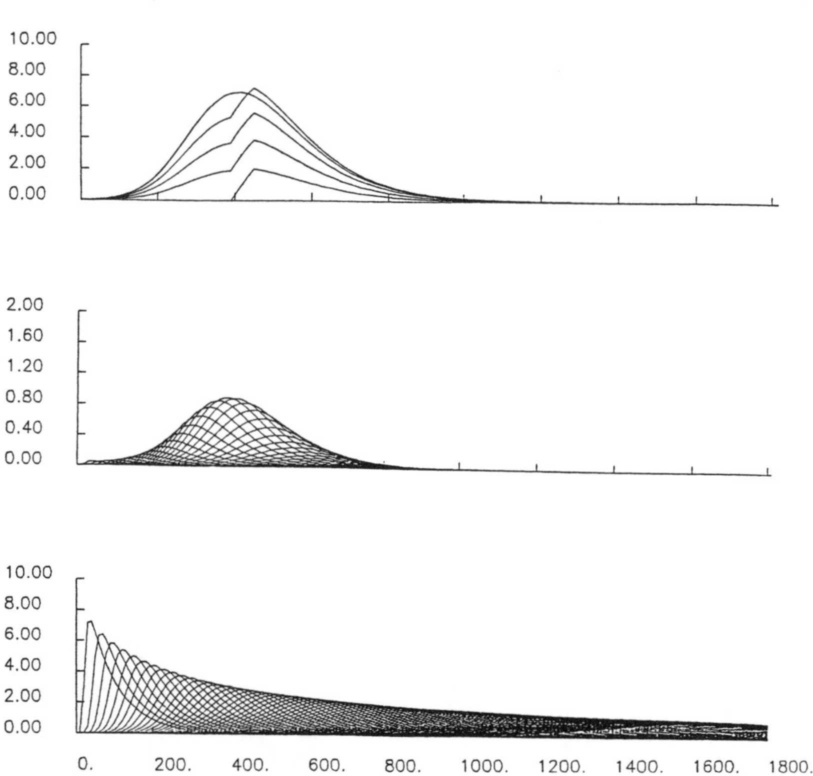

Figure 7.3. Generation of the response function R(*t*). The CS and US are paired during four conditioning trials, after which the CS is presented alone in a single trial. The CS is presented at 0 msec. The US is presented after 400 msec and then remains on for 50 msec. *Bottom panel:* Signals $g_i(t) = f[x_i(t)]y_i(t)$. *Middle panel:* All the gated signal functions $h_i(t) = g_i(t)\, z_i(t)$ that summed generate R(t) on the fifth trial. *Top panel:* R(t) on five successive trials.

A basic property of functions R(t) in Figure 7.4 is an inverted U in learning as a function of the ISI. In other words, there exists a positive ISI that is optimal for learning. In Figure 7.4, this optimal ISI is approximately 250 msec. Learning is weaker at both smaller and larger values of the ISI.

Figure 7.5 shows the similarity between Smith's (1968) experimental data and the simulated values of peak time μ, standard deviation σ, Weber fraction W, and peak amplitude A. Peak time μ was defined as the time at which the response amplitude reached its maximum value at each ISI. Standard deviation σ was estimated by approximating each response curve by a normal distribution and determining the times at which the amplitude was equal to 0.61 of the curve's peak value. This criterion was chosen because the interval between the times at which response amplitude equals 0.61 of its peak value is approximately 2σ in length. To see this, consider a normal distribution $1/\sqrt{2\pi\sigma}$ $\exp[-(t-u)^2/2\sigma_2]$. Its amplitude when $|t-\mu| = \sigma$ is $1/\sqrt{2\pi\sigma}\exp(-1/2)$. Its amplitude when $t = \mu$ is $1/\sqrt{2\pi\sigma}$. The ratio of these amplitudes is $\exp(-1/2) \approx 0.61$. The Weber fraction W was defined as $W = \sigma/\mu$. Of particular interest is the approximately constant value of the Weber fraction W as a function of ISI, in particular its tendency to reach a positive asymptote with increasing values of the ISI (Killeen and Weiss, 1987).

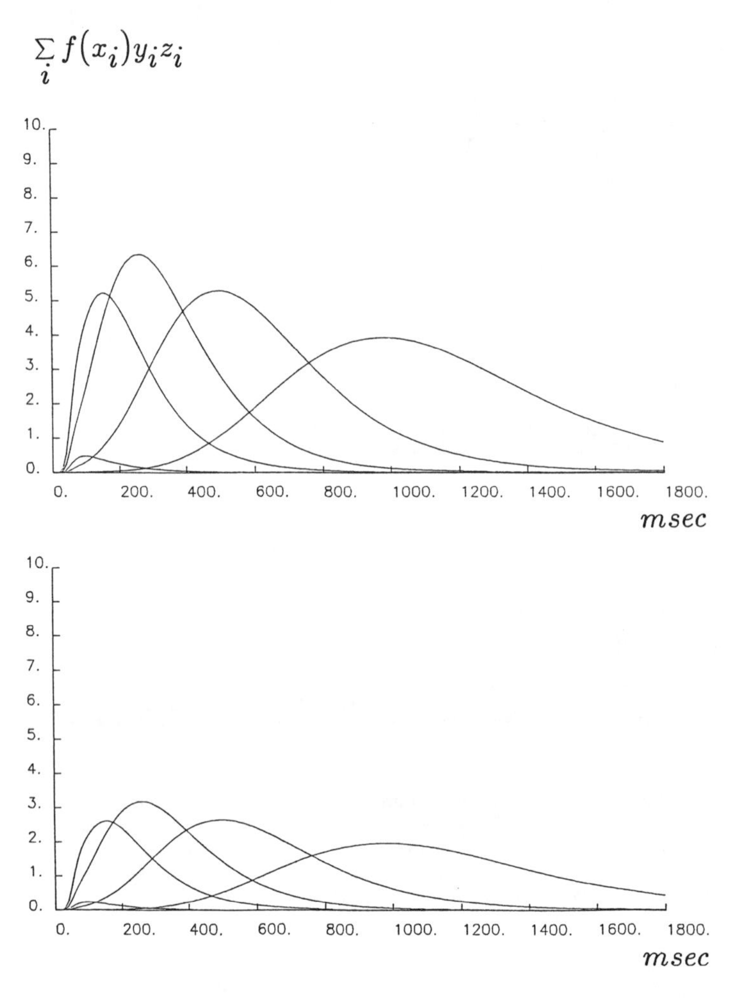

Figure 7.4. ISI curves. R(*t*) functions generated on the fifth trial in response to a CS alone, after 10 trials of prior learning with different ISIs (0, 125, 250, 500, and 1,000 msec). *Bottom panel:* US = 5. *Top panel:* US = 10.

Mixed CS–US intervals

Figure 7.6 depicts computer simulations of a CS that is paired with a US under two different ISIs in alternate trials. In agreement with Millenson, Kehoe, and Gormezano's (1977) results, when the CS is subsequently presented, the response function R(*t*) generates two peaks, each one centered at one of the ISIs.

Discussion

The model can describe different aspects of temporal discrimination in the rabbit's classically conditioned nictitating membrane response. First, in agreement with Gormezano, Kehoe, and Marshall (1983), the model indicates that CR latency decreases and CR amplitude increases over trials. Second, in agreement with Smith (1968), simulations show that in both classical conditioning the CR

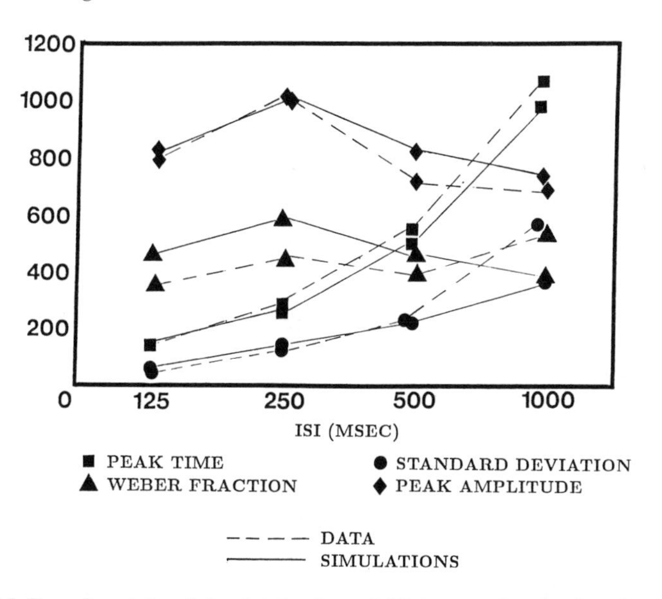

Figure 7.5. Experimental and simulated values of CR topography as a function of the interstimulus interval. Peak amplitude *A*, peak time μ, standard deviation σ, and Weber fraction *W* for simulated CRs generated with 125-, 250-, 500-, and 1,000-msec ISIs, and a 50-msec US. Peak amplitude is measured in fractions of the peak amplitude of the optimal ISI, peak time, and standard deviation in msec. Data from Smith (1968).

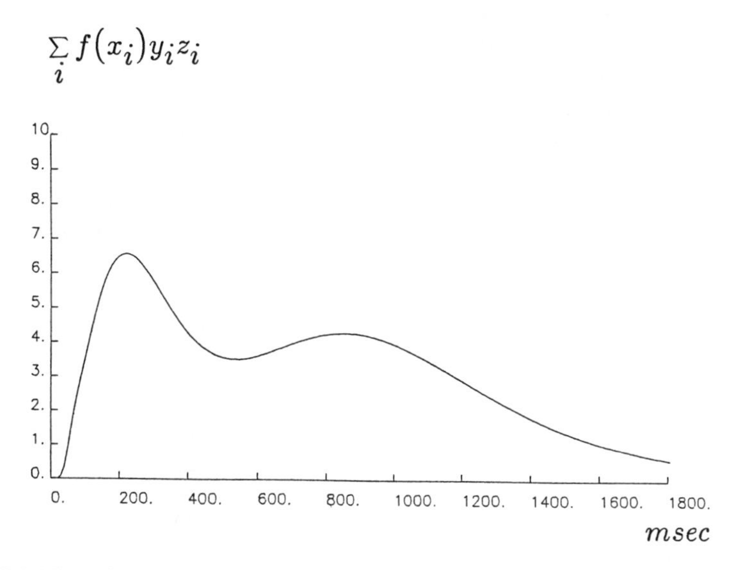

Figure 7.6. Mixed CS–US intervals. Multiple time peaks following conditioning with two ISIs. R(*t*) is plotted on a test trial after 20 conditioning trials during which a US of intensity 10 was presented alternately at an ISI of 200 and 800 msec.

peak is approximately located at the time of the US occurrence, independently of the US intensity. Third, in agreement with Smith (1968), simulations also show that the ISI curve has an inverted-U shape with an optimal ISI interval at around 250 msec. Fourth, the model describes a Weber law for temporal generalization, in agreement with Smith's (1968) data showing that the mean tempo-

Table 7.1. *Simulations obtained with the Grossberg and Schmajuk (1989) timing model.*

Paradigm	Model
Delay conditioning	Y
Trace conditioning	Y
Partial reinforcement	Y
Extinction	Y
ISI curves	Y
Mixed CS–US intervals	Y

Note. Y: The model can simulate the experimental data.

ral estimate and its standard deviation are proportional. Fifth, in agreement with Millenson et al.'s (1977) data, the model generates a two-peak CR when the CS is alternatively presented with one US at 200 msec and a second US at 700 msec. In addition to these qualitative similarities, a quantitative evaluation of the model shows a highly significant correlation between simulated ISI curves and Smith's (1968) experimental data. Table 7.1 summarizes the paradigms described by the model. The results can be extended to the description of timing in a peak procedure in appetitive operant conditioning.

The model, however, has several drawbacks. For instance, although it generates a two-peak CR when the CS is alternatively presented with one US at 200 msec and a second US at 700 msec, the model is unable to account for experimental results showing a one-peak CR when the 200-msec CS is presented and a two-peak CR when the 700-msec CS is presented (Millenson et al., 1977). In addition, since the model does not introduce the duration of the CS in its computations, it does not account for the different CR intensities obtained with different CS durations when the CS and the US do not completely overlap. For instance, Ayres, Albert, and Bombace (1987) found that extending the CS after the US weakened conditioning. For the same reasons, the model predicts the same amount of CS–US association for a given ISI with either delay or trace conditioning procedures, a result at odds with experimental data (Schneiderman, 1966).

Timing and operant conditioning

In addition to describing timing in classical conditioning, the model can be applied to the characterization of some aspects of timing in operant conditioning. For instance, Figure 7.5 also describes results obtained in a peak procedure in appetitive operant conditioning. In this procedure, rats are trained on a CS discriminated fixed-interval schedule of food reinforcement. Testing consists of a peak procedure in which the CS is turned on but the US is not delivered. The rate of responding steadily increases during the period before food is expected and then it decreases to baseline. The time at which the maximum rate occurs is the peak time (Platt, Kuch, and Bitgood, 1973). In addition, with increasing reward intervals both peak time and temporal generalization increase according to the Weber law (Roberts, 1981). In agreement with Platt et al. (1973), the rate of responding (assumed to be proportional to R) increases during the period before the US is expected and then decreases to baseline. The time at which the maximum rate occurs is the peak time. In addition, in agreement with Roberts's (1981) results, the model describes a Weber law for different intervals of reinforcement.

The model is also able to describe Wilkie's (1987) data showing that an increased CS intensity "speeds up the clock" that calibrates the response reaction time. Such a speedup is a straightforward consequence of Equation 7.1.

Appendix 7.A

Spectral timing theory

The activation spectrum

Input $I_{CS}(t)$ activates all potentials x_i of the cells in its target population by

$$d(x_i)/dt = \alpha_i\,[\,-Ax_i + (1 - Bx_i)\,I_{CS}(t)] \tag{7.1}$$

Potentials x_i respond at rates proportional to α_i, $i = 1, 2, \ldots, n$.

The habituation spectrum

Each output signal $f(x_i)$ activates a neurotransmitter y_i, according to the following equation:

$$d(y_i)/dt = C(1 - y_i) - Df(x_i)\,y_i \tag{7.2}$$

where $f(x_i)$ is a sigmoid signal function of the form:

$$f(x_i) = x_i^n/\beta^n + x_i^n \tag{7.3}$$

Neurotransmitter y_i accumulates to a constant target level 1, via term $C(1 - y_i)$, and is inactivated, or habituates, by signal $f(x_i)$, via term $-Df(x_i)y_i$. The different rates α_i at which each x_i is activated cause the corresponding y_i to become habituated at a different rate.

The gated signal spectrum

Each signal $f(x_i)$ interacts with y_i to yield sampling signals $g_i(t)$, given by

$$g_i(t) = f[x_i(t)]\,y_i(t) \tag{7.4}$$

Temporally selective associative learning

The sampling signal g_i turns on the learning process and causes z_i to approach US at a rate proportional to g_i. Each z_i thus grows by

$$d(z_i)/dt = Eg_i(t)[-z_i + \mathrm{US}(t)] \tag{7.5}$$

The doubly gated signal spectrum

The output of each spectral element is a double-gated signal given by

$$h_i(t) = g_i(t)\,z_i(t) \tag{7.6}$$

The output signal

The output signal computes the cumulative learned reaction of all the cells to the input pattern. The output signal R(t) is given by

$$R = [\Sigma_i h_i(t) - F]^+ \tag{7.7}$$

where F is the threshold and $[w]^+ = w$ if $w > 0$ and $[w]^+ = 0$ if $w \leq 0$.

Appendix 7.B

Parameter values

Eighty signal functions $f[x_i(t)]$, $i = 1, 2, \ldots, 80$ were used. Parameters are $\alpha_i = 0.2\, i^{-1}$ for $i = 1, 2, \ldots, 80$, $A = 0$, $B = 1$, $C = 0.0001$, $D = 0.125$, $\beta = 0.8$, $n = 8$, $I_{CS} = 1$ for $t > 0$. In all simulations, one time step represents 1 msec. All $f[x_i(0)] = 0$, and $y_i(0) = 1$.

II Operant conditioning

8 Operant conditioning and animal communication: data, theories, and networks

Part II describes neural network theories applied to operant conditioning. Whereas in classical conditioning, animals change their behavior as a result of the double contingency between CSs and the US, during operant (or instrumental) conditioning, animals change their behavior as a result of a triple contingency between their responses R, environmental stimuli S's, and the US (Skinner, 1938). Animals are exposed to the US in a relatively close temporal relationship with a combination of S and R_i. Although at the beginning of training, animals generate few (many) R_i when S is presented, with an increasing number of experiences with the S– R_i–US contingency, they start (stop) emitting R_i when S is presented. During operant conditioning, animals learn by trial and error from feedback that evaluates their behavior but does not indicate the correct response. Part II also suggests how neural network models of operant conditioning can be applied to the description of animal communication.

Four classes of S–R_i–S contingencies are possible: (1) positive reinforcement, in which R_i is followed by the presence of an appetitive stimulus; (2) punishment, in which R_i is followed by the presence of an aversive stimulus; (3) omission, in which R_i is followed by the absence of an appetitive stimulus; and (4) negative reinforcement (escape and avoidance), in which R_i is followed by the absence of an aversive stimulus. Although the present chapter concentrates only on escape and avoidance, the results can be generalized to positive reinforcement and omission.

Avoidance

Like other operant conditioning procedures, avoidance procedures can be divided into discrete trial procedures, in which the operant response occurs only once in a given trial (shuttle-box, running wheel), and free operant procedures, in which the operant response can occur repeatedly (bar pressing, key pecking).

Figure 8.1 illustrates the sequence of events during escape or avoidance in a shuttle-box. The experiment starts with both compartments being dark (Figure 8.1a). At time zero, the light above the compartment where the animal is located turns on a warning stimulus (WS) and the door separating both compartments opens. If the animal has not crossed to the opposite side after a given time, it escapes the US (Figure 8.1b). If the animal has crossed to the opposite side before a given time, it avoids the US (Figure 8.1b'). After a constant or an average intertrial interval, the whole sequence restarts (Figure 8.1c).

Acquisition of delay avoidance

During acquisition of avoidance, fear responses (e.g., whining, urination, shaking) decrease as the animal masters avoidance. Although initially, subjects show

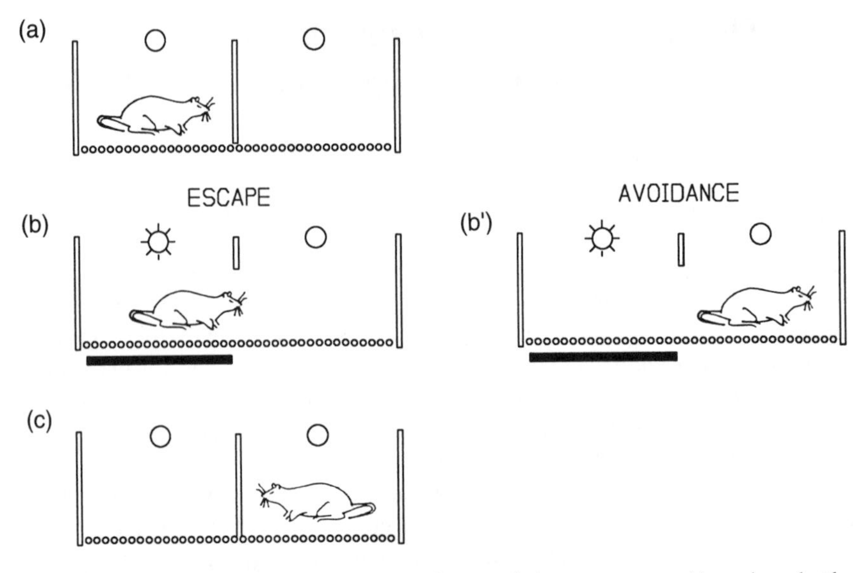

Figure 8.1. Escape and avoidance. Sequence of events during escape or avoidance in a shuttle-box. (a) The experiment starts with both compartments being dark. (b) At time zero, the light above the compartment where the animal is located turns on (WS) and the door separating both compartments opens. If the animal has not crossed to the opposite side after a given time, it escapes the US. (b') If the animal has crossed to the opposite side before a given time, it avoids the US. (c) After a constant or an average intertrial interval, the whole sequence restarts. The black solid line underneath the grid represents the US.

intense fear to the US, this fear decreases as the animal learns the avoidance response (Kamin, Brimer, and Black, 1963; Solomon and Wynne, 1953; Starr and Mineka, 1977). Subjects have been shown to substitute fear responses with stereotypical behavior: Immediately after jumping to the opposite side of the shuttle-box, dogs position themselves in a specific part of the apparatus, body and head facing a fixed direction, and maintain this position until the next WS presentation (Solomon and Wynne, 1953).

Interestingly, even after the avoidance response is established, the response latency still decreases with the increasing number of trials (Mowrer, 1947; Schoenfeld, 1950; Solomon and Wynne, 1953). Therefore, although the acquisition of avoidance is a relatively abrupt process, the gradual decrease in response latency suggests that learning continues after the initial avoidance behavior has been acquired.

Spontaneous extinction

An enigmatic feature of avoidance is that in some cases it shows very slow extinction. For example, Solomon, Kamin, and Wynne (1953) reported that latencies kept decreasing even after 200 trials without shock, and Solomon and Wynne (1954) reported no sign of extinction in experiments with up to 650 trials.

Several factors might influence the resistance to extinction of avoidance. Whereas Solomon et al. (1953) and Solomon and Wynne (1954) reported insignificant extinction using a strong (10.0–12.5 mA) US, Bolles, Moot, and Grossen (1971) found extinction of avoidance using a weaker (1.2-mA) US. In the same vein, Brush (1957) reported that resistance to extinction tends to increase with increasing shock intensities ranging from 0.7–3.1 mA. In addition to US intensity, decreasing WS–US intervals also result in an increased resistance to extinction (Kamin, 1954). Although avoidance in the shuttle-box may show negligible extinction, with certain classes of responses (like turning in a running wheel), avoidance may degenerate into escape after relatively few trials (Anderson and Nakamura, 1964; Coons, Anderson, and Myers, 1960).

Fear modulates avoidance behavior

Whereas techniques that reduce fear decrease avoidance responding, techniques that augment fear increase avoidance responding (Rescorla, 1967; Rescorla and LoLordo, 1965; Weisman and Litner, 1969).

Transfer of control

Overmier and Bull (1969) trained an animal to generate an avoidance response at the presentation of WS1. After the animal was restricted so that avoidance responses could not occur, it was presented with unreinforced WS1 presentations and WS2 reinforced presentations. Later, whereas the WS1 did not elicit avoidance responses, WS2 elicited the avoidance response. The tranfer of control to WS2 was independent of the continued responding to WS1. According to Overmier and Bull (1969), the results support the views that (1) WS1 controls fear and fear controls the avoidance response, and (2) classical fear conditioning and operant conditioning are independent sequential links.

Extinction of avoidance

Although under certain conditions avoidance might show little ordinary extinction, it can be extinguished through various alternative procedures. One procedure consists of preventing the capability of an animal to generate the avoidance response without presenting the US (Baum, 1966, 1976; Page and Hall, 1953; Solomon, Kamin, and Wynne, 1953). A second procedure consists of making the avoidance response ineffective by shocking both sides of the shuttle-box (Davenport and Olson, 1968). Finally, a third procedure consists of shocking the animal when it emits the avoidance response (Seligman and Campbell, 1965; Solomon et al., 1952).

Effects of extending the WS, terminating the WS, and classical conditioning

Kamin (1957) studied the effect that different procedures have on acquisition of avoidance. In normal avoidance, the avoidance response was followed by the immediate termination of the WS and no US was delivered in that trial. In a "terminate WS" situation, the avoidance response was followed by the immediate termination of the WS, but a US was delivered to either side of the box. In an "extended WS" situation, the avoidance response prevented the US presentation, but did not terminate the WS. Finally, in a classical conditioning situation, the avoidance response had no effect on either the WS duration or the delivery of the US. Kamin's (1957) data show that, whereas in the classical conditioning and terminate WS cases, animals did not acquire consistent avoidance, the extended WS case resulted in slower acquisition and lower asymptotic levels of avoidance.

Trace avoidance

Whereas the WS and the US temporally overlap in delay avoidance, WS offset precedes US onset in trace avoidance. Trace avoidance seems to be less resistant to extinction than delay avoidance (Brush, 1957; Church, Brush, and Solomon, 1956; Kamin, 1954). In contrast to delay avoidance, trace avoidance deteriorates with increasing interstimulus interval (ISI). Church et al. (1956) studied the effect of varying the ISI on the acquisition of avoidance under trace and delay conditions. They reported that, with the trace procedure, acquisition was faster and extinction slower with decreasing WS–US intervals. Interestingly, Brush (1957) reported that avoidance learning is not characterized by learning of the timing of the US, as is the case in classical conditioning (Smith, 1968). Recently, however, Davis, Schlesinger, and Sorenson (1989) reported that fear conditioning is also time-specific, that is, it is maximal at the time that matches the CS–US interval used in training.

Different escape and avoidance responses

Mowrer and Lamoreaux (1946) and Bolles (1969) reported that, although animals can learn an avoidance response different from that required to escape the US, in most cases avoidance performance is better if the responses are identical.

Discriminative avoidance

In a discriminative avoidance procedure, animals are taught to avoid the shock by performing one response in the presence of a warning stimulus and a different response in the presence of another warning stimulus. For instance, Overmier, Bull, and Trapold (1971) trained dogs to avoid shock by pressing a panel on the left in the presence of one stimulus and pressing a panel on the right in the presence of another stimulus. Similarly, Young (1976) trained rats to avoid shock by pressing a lever in the presence of one stimulus and licking in the presence of another stimulus.

Sidman avoidance

In a Sidman (1953) avoidance task, subjects are taught to avoid shock by pressing a lever that gives them a shock-free period. In Herrnstein and Hineline's (1966) experiment, subjects learn to choose between high and low random chances of being shocked in the absence of external WSs.

Two-process theories of avoidance

Mowrer (1947) and Miller (1948) contributed a two-process theory of avoidance that appeals to classical and operant conditioning in order to describe avoidance behavior. The classical conditioning process consists of the association between the WS and the shock US, and the consequent generation of a fear-conditioned response (CR) when WS is presented. According to drive-reduction theories popular at the time (Hull, 1951), the operant conditioning process consists of the reinforcement of the avoidance response Ra by the reduction of fear as the animal's response terminates the fear-eliciting WS.

Several experimental results support Mowrer–Miller's theory (see Levis, 1989, for a recent review). For example, ongoing avoidance responses increase in the presence of a conditioned stimulus (CS) that increases fear (Rescorla, 1967; Rescorla and LoLordo, 1965; Weisman and Litner, 1969), whereas ongoing avoidance responses decrease in the presence of a CS that alleviates fear (Mineka, Cook, and Miller, 1984; Morris, 1974; Rescorla and LoLordo, 1965; Rosellini, DeCola, and Warren, 1986; Weisman and Litner, 1972). Furthermore, Mowrer–Miller's theory is supported by data showing that avoidance extinguishes at a fast rate when animals are exposed to the WS but the US is not presented (response prevention), because the WS–US association is extinguished.

Other empirical evidence, however, presents difficulties for Mowrer–Miller's two-factor theory. Two-factor theory has problems explaining trace avoidance, in which a temporal delay is introduced between the WS offset and the presentation of the US. Although in trace avoidance the association between a WS temporal trace and the US can be explained in the same terms as trace conditioning in classical conditioning (see Chapter 3), WS termination reduces fear only slightly. To dispense with this drawback, Anger (1963), Dinsmoor (1954), and Sidman (1953) suggested that internal stimuli (e.g., proprioceptive, visual, kinesthetic feedback) that follow the WS and precede Ra become aversive and internal stimuli that follow Ra become reinforcing. Alternatively, Seligman and Johnston (1973) suggested that in trace avoidance the WS may be the passage of time.

Another major difficulty for Mowrer–Miller's theory is the sometimes almost negligible extinction of avoidance. According to classical conditioning principles, an association is formed between the WS and the US. As the animal learns to avoid the US, the WS is presented alone, and this should cause extinction of the WS–US association. A solution to the problem of extinction of the WS–US association that afflicts the two-factor theory was proposed by Soltysik (1964). According to Soltysik (see also Gray, 1975, p. 315), proprioceptive and kinesthetic feedback stimuli or environmental stimuli originated by the avoidance response become conditioned inhibitors. Inhibitory conditioning results from the presentation of the WS together with the US, alternated or followed by the presentation of the WS and Ra in the absence of the US. According to Soltysik, early in avoidance the WS becomes associated with the US and elicits fear. As the animal starts avoiding the US, Ra becomes a conditioned inhibitor that reduces fear, thereby reinforcing the instrumental response. Simultaneously, the association of the WS with the US is preserved by the presentation of Ra (Chorazyna, 1962; see also Rescorla and Wagner, 1972). Supporting his view, Soltysik (1960) showed that, in classical conditioning, nonreinforced trials might result in relatively little extinction if the excitatory CS is presented in conjunction with a conditioned inhibitor in extinction trials. Unfortunately, a direct test of the Soltysik theory was carried out in a Sidman avoidance task by LoLordo and Rescorla (1966; see also Morris, 1974) with negative results.

In line with Soltysik's view, Weisman and Litner (1969, 1972) proposed that the critical sequence in the acquisition of avoidance includes the following steps. First, some stimuli are associated with the US and become WSs and conditioned excitors of fear. Second, other stimuli are associated with the nonoccurrence of the US and become conditioned inhibitors of fear. Third, avoidance responses are scheduled to produce the conditioned inhibitor. Fourth, the conditioned inhibitors act as positive reinforcers of the avoidance responses. In contrast to other two-process theories that assume a negative reinforcement of the avoidance response by the reduction of fear as the animal's response terminates the fear-eliciting WS, Weisman and Litner suggested that responses are positively reinforced by the onset of the conditioned inhibitors predicting the nonoccurrence of the US. Similar to Weisman and Litner's (1969, 1972) approach, Denny (1971) suggested that the avoidance response is positively reinforced by the shock-free period that follows it.

Gray (1975, p. 325) suggested that avoidance responses become conditioned inhibitors of fear and, therefore, a "safety signal" that animals try to approach. According to Gray's view, the CS does become a warning signal that gradually extinguishes its association with the avoided US, but the avoidance response is maintained by the positive reinforcement of the safety signal. Several experimental results support Gray's (1975) view. For instance, Lawler (1965) found that once safety signals are acquired, their effectiveness is independent of the degree of fear demonstrated by the animal. Similarly, Mineka and Gino (1980) showed that a reduction in suppression to the WS as avoidance progresses is not accompanied by a decline in the avoidance response. According to Gray (1975, p. 327), his safety signal modification of Mowrer–Miller's two-process theory includes the two original classical and instrumental factors, the fear-inducing properties of the CS–US associations, the rewarding and fear-decreasing properties of the response, the independence of the safety signals from the level of fear, and the reduction of fear as avoidance becomes established. Gray and Smith (1969) offered a mathematical model limited to the instrumental component of a general two-process learning system.

A neural network model of avoidance

This section introduces a neural network that learns different responses by including a "response-selection" mechanism. In Dickinson's (1980) terms, the model introduced in this chapter combines declarative and procedural knowledge in order to solve environmental problems (see Chapter 2).

The diagram depicted in Figure 8.2 represents a two-process neural network of avoidance. The model assumes that through classical conditioning animals build an internal model of their environment (Sokolov, 1960), and that through operant conditioning animals select from alternative behavioral strategies (Hull, 1951). The internal model provides *predictions* (declarative knowledge) of what environmental events precede other environmental events, such as the US. Behavioral strategies refer to the *prescription* (procedural knowledge) of the responses to be generated in different circumstances. Whenever there is a mismatch between predicted and actual environmental events, (1) the internal model is modified (as in previous models), and (2) the behavioral strategies are adjusted.

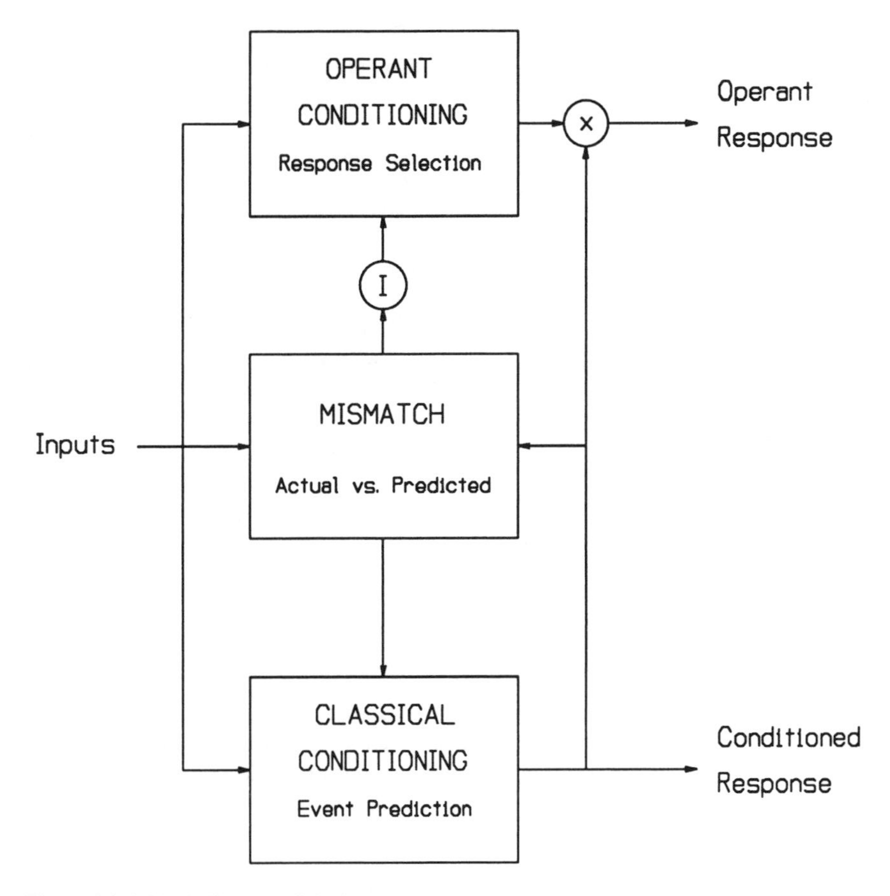

Figure 8.2. A block diagram of the interactions between the environment and the animal during operant conditioning. The internal model provides predictions of what environmental events precede other environmental events, such as the US. Behavioral strategies refer to the prescription of the responses to be generated in different circumstances. The novelty system compares observed and predicted events to compute novelty. Novelty is then used to (1) modify the internal model, and (2) to adjust the behavioral strategies.

The network

Figure 8.3 shows a real-time neural network (see Box 1.4), introduced by Schmajuk (1994; Schmajuk, Urry, and Zanutto, in press), that describes escape and avoidance. Box 8.1 summarizes the main aspects of the model. Appendix 8.A presents a formal description of the network as a set of differential equations that depict changes in the values of neural activities and connectivities as a function of time.

In simple terms, the network intimately combines classical and operant conditioning principles. The classical conditioning process involves the formation of associations between different environmental stimuli and the animal's responses with the US. These associations are used to predict the presence or absence of the US. Classical conditioning is regulated by the mismatch between the actual and the predicted intensity of the US: When the US is underpredicted, classical associations increase and decrease otherwise. The operant conditioning process entails the formation of associations between environmental stimuli with the escape or the avoidance response. These associations are

Box 8.1 Schmajuk's (1994) escape-avoidance model

The most important aspects of the model are:

1. The network includes two-processes, classical and operant conditioning.
2. Classical associations, used to generate predictions of the US by the US, WS, or R, are changed according to a real-time version of the Rescorla–Wagner rule (see Box 3.1).
3. Node M computes the mismatch (US – B) between the actual US intensity and its aggregate prediction B.
4. Classical associations V_i increase when (US – B) is greater than zero.
5. Node I computes the difference (B – US) used by the operant conditioning block.
6. Operant associations Z_{ik} increase when (B – US) is greater than zero.
7. The response most strongly activated becomes selected and is executed by the system.
8. CR amplitude, interpreted as *fear* of the aversive US, is proportional to the aggregate prediction of the US, B.
9. The strength of the output response is modulated by the amplitude of the classical CR. Therefore, similar output response strength can be obtained with multiple combinations of classical and operant associations.

used to select the adequate response in each case. Operant conditioning is controlled by a novel algorithm that mirrors the classical conditioning algorithm: when the US is underpredicted, operant associations decrease and increase otherwise.

Real-time internal representations

Environmental events are internally represented by real-time variables in the model. We assume that both the WSs and the animal's responses R can be regarded as biologically neutral CSs and might become associated with the US. In addition, we assume that the US itself is represented as a CS and as a biologically meaningful US. As mentioned in Chapter 3, Hull (1943) proposed that CSs give rise to short-term memory (STM) traces in the central nervous system that increase over time to a maximum and then gradually decay back to zero. In the same vein, we assume that WS, US, and R's activate different neural populations whose activity constitutes STM traces $\tau_{WS}(t)$, $\tau_{US}(t)$, and $\tau_{R}(t)$. The use of STM traces allows the model to describe paradigms, such as trace conditioning, in which the STM traces (but not the physical stimuli themselves) temporally overlap with the US.

Gormezano, Kehoe, and Marshall (1983) suggested that the curve representing the strength of the CS–US association as a function of the CS–US interstimulus interval (ISI) reflects the variation in the intensity of trace $\tau_{CS}(t)$ over time. In the case of the rabbit's nictitating membrane conditioned response, it has been consistently reported (Schneiderman, 1966; Smith, 1968) that CS–US associations are negligible at zero ISI, present a peak at ISIs of 100 msec, and gradually decrease for longer ISIs (see Chapter 2). In the case of fear conditioning in rats, Davis et al. (1989) recently reported that CS–US associations are negligible at zero ISI and monotonically increase from 50 msec–50 sec ISIs.

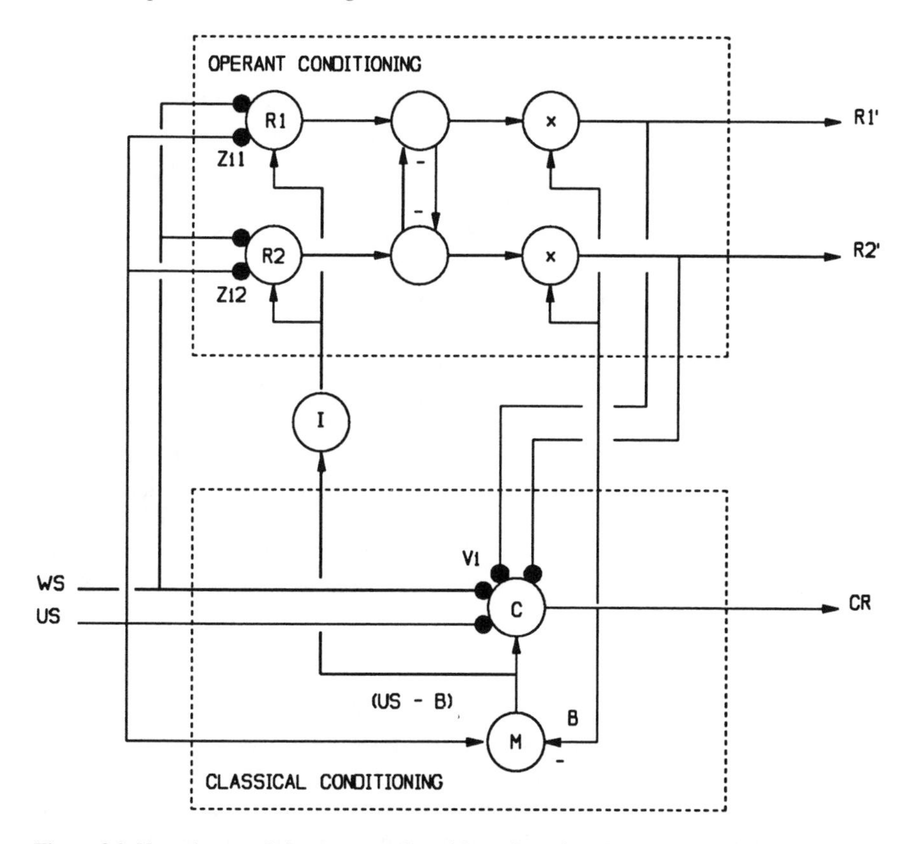

Figure 8.3. Neural network for escape and avoidance learning. Classical conditioning processes build an internal model of the environment used to generate predictions of the US. Operant conditioning processes establish the animal's behavioral strategy by selecting the adequate avoidance or escape response. Mismatch between the actual and predicted intensity of the US modulates changes in both classical and operant associations. The prediction of the US also serves to regulate the strength of the operant response. WS: warning stimulus; US: unconditioned stimulus. *Classical conditioning.* CR: conditioned response. Solid circles represent *V,* the associations of the US, WS, and R's with the US [*V*(US, US), *V*(WS, US), *V*(R1, US), and *V*(R2, US)]. $B = \Sigma_j V_j \tau_j$: aggregate prediction of the US. *Operant conditioning.* Solid circles represent *Z,* the associations of WS with R [*Z*(WS, R1) and *Z*(WS, R2)] and of the US with R [*Z*(US, R1) and *Z*(US, R2)]. Open circles with an *X* inside represent neural populations that compute the product of the inputs. Inhibitory connections indicate competition between alternative responses R1 and R2. *I:* neural population converting (US – *B*) into (*B* – US); R'1 and R'2: strength of the operant responses.

The shape of the ISI curve in fear conditioning suggests that $\tau_{WS}(t)$ grows fast after a 50-msec delay following the CS presentation, reaches its maximum value at around 50 sec, and then decays to zero.

Real-time internal representations participate in two processes, classical and operant conditioning. The classical conditioning process describes the acquisition of WS–US, R–US, and US–US associations. The operant conditioning process describes US–Re and WS–Ra associations.

Classical conditioning

The classical conditioning block of Figure 8.3 shows that node *C* receives inputs from the US (as a stimulus), WS, R1, and R2. Solid circles V_i represent the associations of these inputs with the US. Classical associations, used to generate predictions of the US by the US, WS, or R, are changed according to a real-time version of the Rescorla–Wagner rule described in Chapter 3.

Node M computes the mismatch (US – B) between the actual US intensity and its "aggregate prediction," $B = \Sigma_i V_i \tau_i(t)$ [where $\tau_i(t)$ represents the trace of either a WS, the US, or a R]. Classical associations V_i increase when (1) $\tau_{WS}(t)$, $\tau_{US}(t)$, or $\tau_R(t)$ is active, and (2) (US – B) is greater than zero. Classical associations V_i decrease when (1) $\tau_{WS}(t)$, $\tau_{US}(t)$, or $\tau_R(t)$ is active, and (2) (US – B) is less than zero. By the algorithm, USs, WSs, and R's compete with each other in order to gain association with the US. When the US is predicted but not presented, other WSs or R's present at that time might acquire inhibitory associations with the US, that is, predict the absence of the US. The algorithm provides real-time descriptions of classical conditioning paradigms that include acquisition and extinction of delay and trace conditioning, blocking, overshadowing, conditioned inhibition, and discrimination acquisition and reversal (see Chapter 3).

The aggregate prediction of the US, B, is interpreted as *fear* of the aversive US in the context of the model. Figure 8.3 shows that B is used (1) to define the strength of the conditioned response, CR = $f[\Sigma_i \tau_i(t)\ V_i]$, (2) to compute (US – B) at node M, and (3) to regulate the strength of the operant response (R1' or R2').

Operant conditioning

The operant conditioning block in Figure 8.3 shows that nodes R1 and R2 receive inputs from WS and the US (as a stimulus). Solid circles Z_{i1} and Z_{i2} represent the associations of these inputs with alternative responses R1 and R2. In Figure 8.3, node I inverts (US – B), computed by node M and used by the classical conditioning block, in order to compute (B – US), used by the operant conditioning block. Operant associations Z_{ik} increase when (1) $\tau_{WS}(t)$ or $\tau_{US}(t)$ are active together with $\tau_{Rk}(t)$, and (2) (B – US) is greater than zero. Operant associations Z_{ik} decrease when (1) $\tau_{WS}(t)$ or $\tau_{US}(t)$ are active together with $\tau_{Rk}(t)$, and (2) (B – US) is less than zero. Therefore, the model selects the adequate response by first increasing Z_{ik} of all responses active in the presence of fear alone (i.e., increasing variability) and then by decreasing the Z_{ik} of those responses active when the US is present (i.e., discarding the wrong responses). Notice that the US and WS accrue excitatory associations with responses that gain inhibitory associations with the US, that is, with responses, such as Ra or Re that predict the absence of the US.

Outputs of nodes R1 and R2, proportional to $\Sigma_i Z_{ik} \tau_i$, compete to decide which alternative response will be generated. Noise is added to the intensity of each alternative response in order to provide initial random responses at the beginning of training. The response R1 or R2 most strongly activated becomes selected and is executed by the system. This response is the one that, at a given time, predicts the minimal amount of US (either generates the strongest prediction of the absence of the US or generates the weakest prediction of its presence). Figure 8.3 shows that the selected response is combined with the error signal arriving from node I to modify its association with the active $\tau_{WS}(t)$ or $\tau_{US}(t)$.

Figure 8.3 shows that the intensity of the selected operant response R1' or R2' is proportional to the product of (1) the aggregate prediction of the US (B) and (2) the STM trace of the maximal response (R1 or R2). Because the strength of the output response is proportional to the product of WS–US and WS–R associations, so that decreases in one can be compensated by increases in the other,

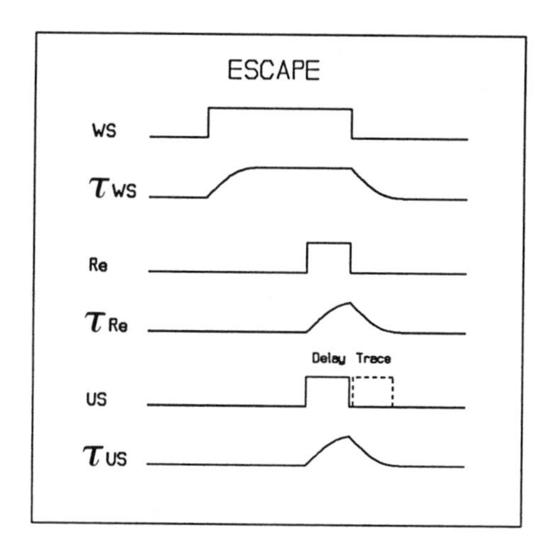

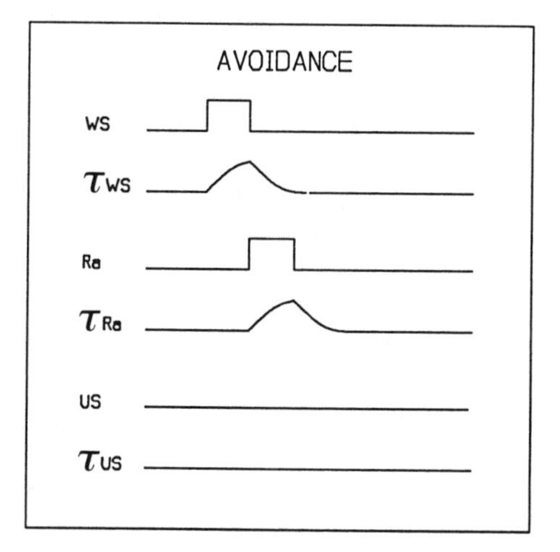

Figure 8.4. Real-time variables used in the model during escape and avoidance. WS: warning signal; τ_{WS}: trace of the WS; Re: escape response; τ_{Re}: trace of the escape response; Ra: avoidance response; τ_{Ra}: trace of the avoidance response; US: shock unconditioned stimulus; τ_{US}: trace of the US. The US represented by a solid line corresponds to delay avoidance, the US represented by a dashed line corresponds to trace avoidance.

similar output response strength can be obtained with multiple combinations of classical and operant associations.

Escape and avoidance

This section illustrates how the model processes information during acquisition of avoidance in a shuttle-box. Under similar assumptions, the model can be applied to running-wheel, leg-flexion, or lever-pressing tasks.

Figure 8.4 shows the temporal arrangement of WS, Re, Ra, US, and their respective STM traces $\tau(t)$ during escape and avoidance. Figure 8.4 shows that when WS is on, $\tau_{WS}(t)$ increases toward 1 and decreases toward 0 when WS turns off. Normally, when the animal produces the avoidance response Ra, WS turns off. Notice that the more rapidly the animal generates the Ra, the shorter the $\tau_{WS}(t)$. Also notice that $\tau_{WS}(t)$ is greater than 0 even when the animal has generated the Ra. These temporal properties of $\tau_{WS}(t)$ are essential to the functioning of the model. When the animal generates the avoidance response Ra, its trace $\tau_{Ra}(t)$ increases toward 1. As the response is terminated, $\tau_{Ra}(t)$ decreases back to 0.

At the beginning of training, when the WS is on, $\tau_{WS}(t)$ is active in the presence of the US. Therefore, $\tau_{WS}(t)$ accrues an excitatory association with the US. When the animal emits the correct avoidance response, Ra, crossing to the opposite side and terminating the WS, $\tau_{WS}(t)$ and $\tau_{Ra}(t)$ are active in the absence of the predicted US. At this time, the prediction of the US exceeds its actual value (US $< B$). Consequently, in the classical conditioning block, $\tau_{WS}(t)$ partially decreases its excitatory association and $\tau_{Ra}(t)$ acquires an inhibitory association with the US (see Equation 8.2 in Appendix 8.A). Simultaneously, in the operant conditioning block, $\tau_{WS}(t)$ accrues an excitatory association with Ra (see Equation 8.3 in Appendix 8.A).

We assume that the animal is able to avoid the US when the strength of the operant response, $Ra' = \tau_{Ra}(t)\ B$, exceeds an arbitrary threshold value. R' is a real-time function whose temporal topography is basically defined by the shape of $\tau_{max}(t)$, which exponentially grows and decays. The latency of the avoidance response is the time when R' reaches the threshold. Therefore, latency decreases for a fast-growing $\tau_{max}(t)$ and large B's. Because $\tau_{max}(t)$ grows faster with increasing values of $R_{max} = K_6\, f[\Sigma_i Z_{ik} \tau_i]$, latency decreases with increasing values of Z_{max} (WS–R_{max} and US–R_{max} operant associations). Also, because $B = \Sigma_j V_j \tau_j$, latency decreases with increasing values of V_j (WS–US, US–US, and R–US classical associations). Notice that the same amplitude and latency of the output response R' can be achieved by various combinations of operant and classical associations.

As response latency decreases, $\tau_{WS}(t)$ becomes increasingly shorter, and its shorter duration partially prevents the extinction of its excitatory association with the US. In addition, because inhibitory associations do not extinguish with WS presentations (Zimmer-Hart and Rescorla, 1974), $\tau_{Ra}(t)$ does not extinguish its inhibitory association with the US (see Equation 8.2 in Appendix 8.A). Furthermore, since $\tau_{WS}(t)$ and $\tau_{Ra}(t)$ are simultaneously active, the inhibitory association of $\tau_{Ra}(t)$ retards the extinction of the excitatory association of $\tau_{WS}(t)$ (see Equation 8.2 in Appendix 8.A). In Soltysik's (1985) terms, $\tau_{Ra}(t)$ partially protects $\tau_{WS}(t)$ from extinguishing its excitatory association with the US.

Computer simulations

The present section contrasts experimental data with computer simulations carried out with the model presented in Figure 8.3. All simulations were conducted with identical parameter values. Parameter values used in the simulations are presented in Appendix 8.B.

In Figures 8.5–8.17 (except for Figures 8.6 and 8.15), (A) displays the percentage of avoidance responses, (B) displays response latency (as the number of simulated time units), (C) displays R_i–US, US–US, and WS–US associations, and (D) represents WS–R_i and US–R_i associations. In Figures 8.18 and 8.19, (A) displays the average number of responses over a 100 time-unit period, whereas (B) displays the number of USs received during a 100 time-unit period. R_i represents responses such as running, turning, rearing, pressing, and so on. Any of these responses can be chosen to be the escape or avoidance response.

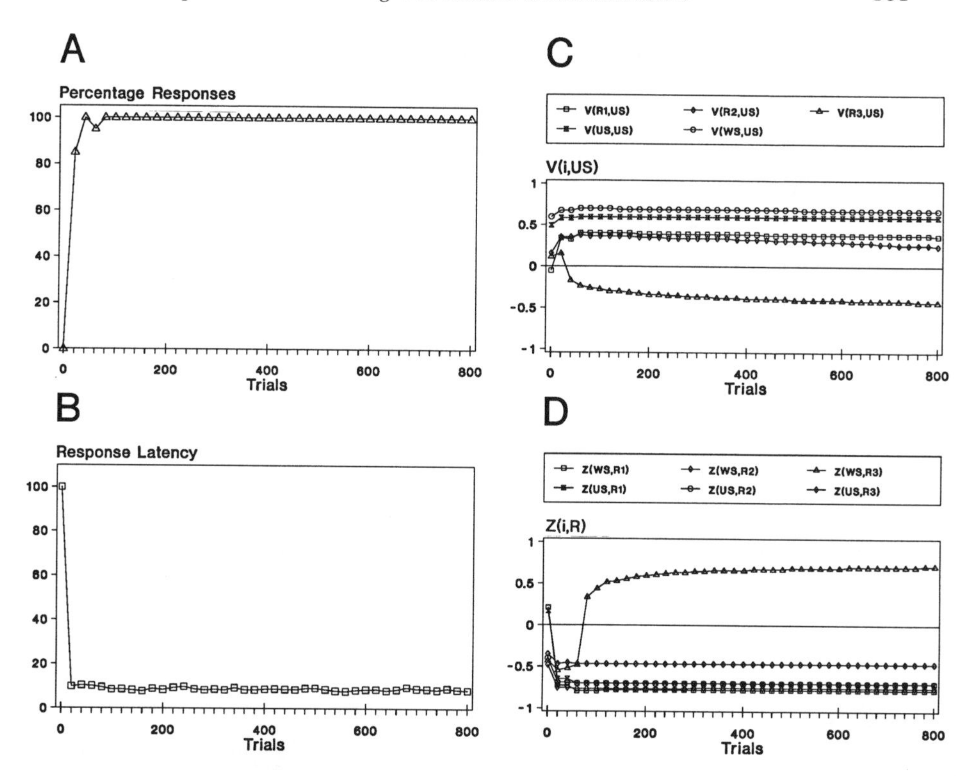

Figure 8.5. Delay avoidance. (A) Percentage of avoidance responses. (B) Response latency. (C) Associations of the US, WS, and R's with the US [*V*(US, US), *V*(WS, US), *V*(R1, US), *V*(R2, US), and *V*(R3, US)]. (D) Associations of WS and US with the R's [*Z*(WS, R1), *Z*(WS, R2), *Z*(WS, R3), *Z*(US, R1), *Z*(US, R2), and *Z*(US, R3)]. R3 is the avoidance response.

Acquisition of delay avoidance

Figure 8.5 shows computer simulations of delay avoidance. During the first trials, simulations show alternated escape and avoidance responses, followed by a long period of uninterrupted avoidance behavior (over 800 trials). Latency decreases over the first 200 trials and the subject shows no sign of extinction, even after 650 trials. These simulated results are in agreement with Solomon and Wynne's (1953, 1954) data.

Figure 8.5A shows that, in agreement with empirical results (Solomon et al., 1953; Solomon and Wynne, 1954), the percentage of avoidance responses does not decrease over trials, that is, avoidance has almost negligible extinction. This negligible extinction is explained as follows. Although protected by shorter-response latencies and the inhibitory association of Ra with the US, the WS–US association and, therefore, fear decrease over trials. As WS–US decreases, WS–Ra increases. Because the intensity of the output response is proportional to the product of the WS–US and WS–Ra associations, decreases in WS–US are compensated by increases in WS–Ra and, therefore, the percentage of avoidance responses remains constant for a long number of trials. The model's description of resistance to extinction is compatible with Baum's (1970) data, showing that response extinction is independent of fear extinction.

Figure 8.5B shows that in agreement with experimental data (Mowrer, 1947; Schoenfeld, 1950; Solomon and Wynne, 1953), response latency decreases after the avoidance response is acquired. According to the model, response latency

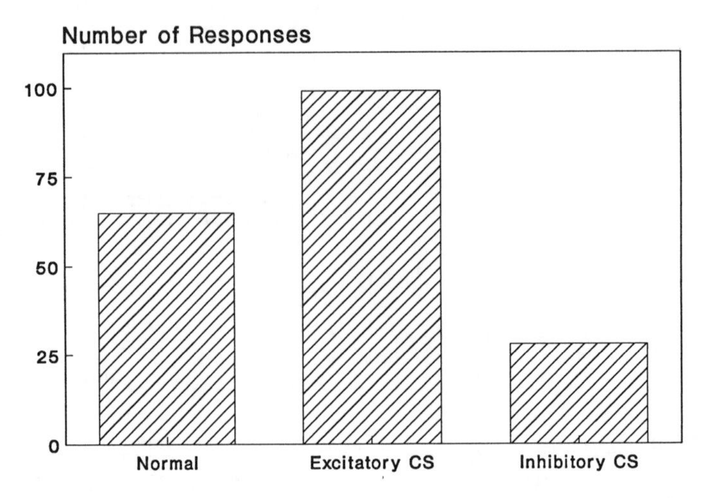

Figure 8.6. Effect of introducing an excitatory and an inhibitory CS on the rate of responding during Sidman avoidance. Average number of avoidance responses per 100 time units during trials 100–200.

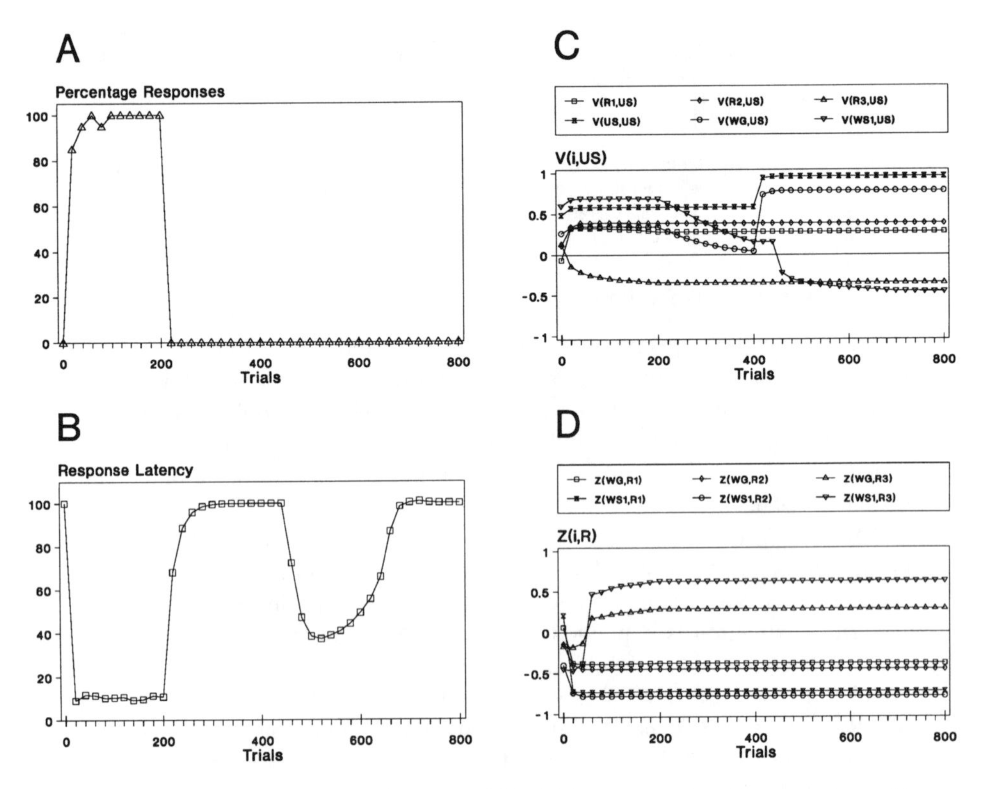

Figure 8.7. Transfer of control: responses to WS1. (A) Percentage of avoidance responses. (B) Response latency. (C) Associations of the US, WS, and R's with the US [*V*(US, US), *V*(WG, US), *V*(WS1, US), *V*(R1, US), *V*(R2, US), and *V*(R3, US)]. (D) Associations of WG and WS1 with the R's [Z(WG, R1), *Z*(WG, R2), Z(WG, R3), *Z*(WS1, R1), *Z*(WS1, R2), and *Z*(WS1, R3)]. R3 is the avoidance response.

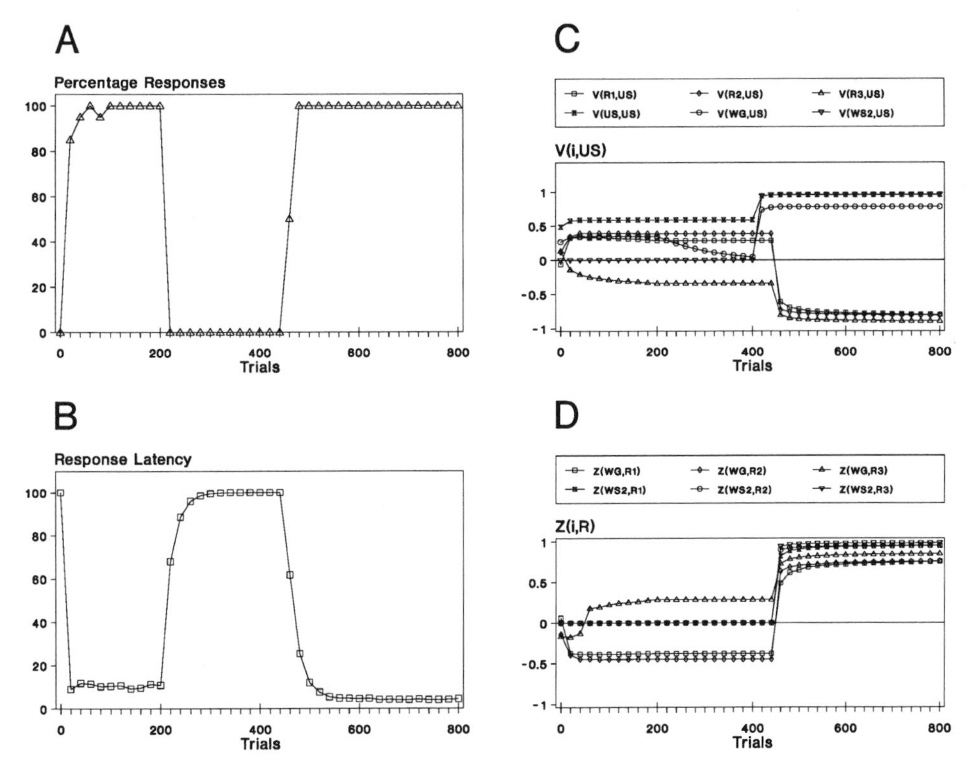

Figure 8.8. Transfer of control: responses to WS2. (A) Percentage of avoidance responses. (B) Response latency. (C) Associations of the US, WS, and R's with the US [*V*(US, US), *V*(WG, US), *V*(WS2, US), *V*(R1, US), *V*(R2, US), and *V*(R3, US)]. (D) Associations of WG and WS2 with the R's [*Z*(WG, R1), *Z*(WG, R2), *Z*(WG, R3), *Z*(WS2, R1), *Z*(WS2, R2), and *Z*(WS2, R3)]. R3 is the avoidance response.

decreases because the WS–Ra association, driven by the absence of the predicted US, keeps increasing over trials. Figure 8.5C shows that, in agreement with experimental results (Kamin et al., 1963; Solomon and Wynne, 1953; Starr and Mineka, 1977), the WS–US association and, therefore, fear continue to gradually decrease after the animal masters avoidance. Figure 8.5C also shows that, whereas WS, R1, and R2 accrue an excitatory association with the US, R3 (the avoidance response Ra) acquires an inhibitory association. Figure 8.5D indicates that the WS–R3 association increases over trials.

Fear modulates avoidance behavior

Figure 8.6 shows that, when a CS previously associated with the US (a conditioned excitor) is presented together with the WS in a Sidman avoidance paradigm, the latency of avoidance responses decreases and the percentage of avoidance responses increases. Conversely, when a CS previously associated with the absence of the US (a conditioned inhibitor) is presented together with the WS, the latency of avoidance responses increases and the percentage of avoidance responses decreases. These results are explained in terms of the modulation of the avoidance response by the intensity of the prediction of the US, CR $= f[B] = f[\Sigma_i V_i X_i]$, by all CSs present at a given time (Equation 8.4 in Appendix 8.A).

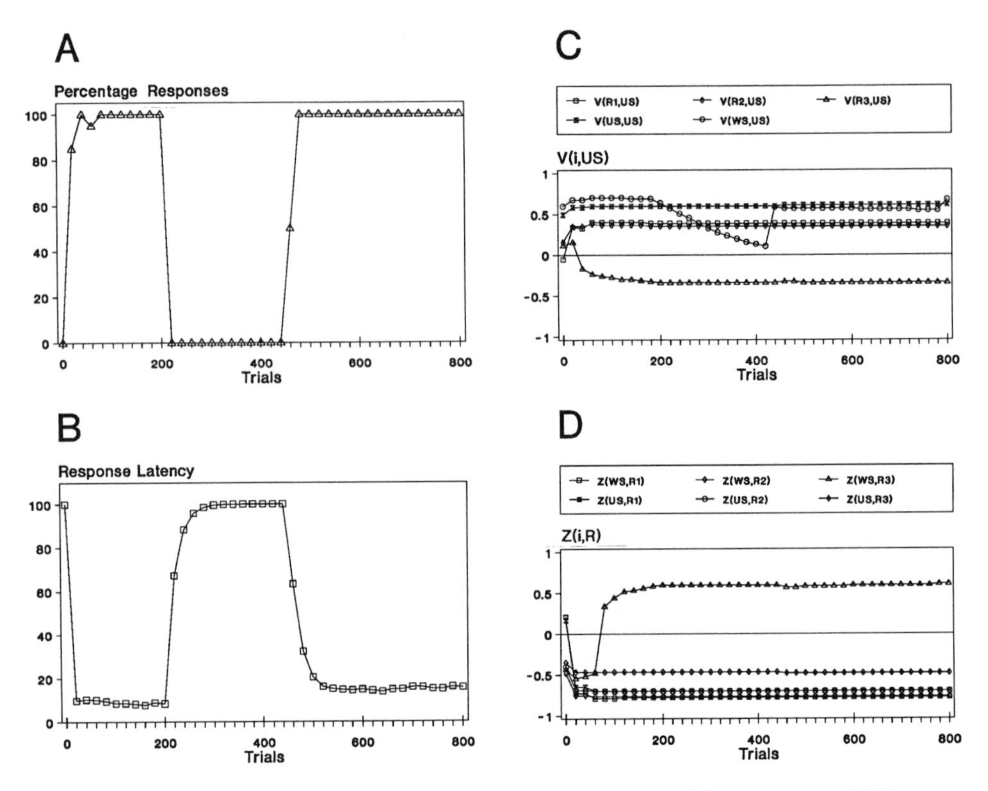

Figure 8.9. Preventing the avoidance response. (A) Percentage of avoidance responses. (B) Response latency. (C) Associations of the US, WS, and R's with the US [*V*(US, US), *V*(WS, US), *V*(R1, US), *V*(R2, US), and *V*(R3, US)]. (D) Associations of WS and US with the R's [*Z*(WS, R1), *Z*(WS, R2), *Z*(WS, R3), *Z*(US, R1), *Z*(US, R2), and *Z*(US, R3)]. R3 is the avoidance response.

Transfer of control

If we assume that WS1 and WS2 share a common element WG that represents 1 percent of the total value of both WS1 and WS2 traces, the model is able to reproduce Overmier and Bull's (1969) experimental results. Figures 8.7 and 8.8 show computer simulations of a transfer of control experiment. During the first 200 trials, simulations show avoidance acquisition to WS1. In trial 201, the simulated animal is presented with the WS1, the avoidance response is prevented, and no US is delivered. From trials 400–450, the simulated animal is presented with WS2, the avoidance response is prevented, and the US is delivered. Figure 8.7A shows that, when the barrier is removed in trial 451, the simulated animal does not generate avoidance responses to WS1. However, Figure 8.8A illustrates that, when the barrier is removed in trial 451, the simulated animal does generate avoidance responses to WS2. These results are explained in terms of the WG–Ra associations that remain unchanged during the extinction of the WS1–US association and the acquisition of the WS2–US association. During extinction of the WS1–US associations, WG–US associations are also extinguished, and animals do not avoid WS1 (Figure 8.7A). When WS2 is presented together with the US, both WS2-US and WG–US associations increase. When animals are given the opportunity to avoid the US, WS2–US and WG–US associations modulate the WG–Ra associations (see Equation 8.6), and the simulated animal avoids WS2 (Figure 8.8). In contrast to this explanation, Overmier

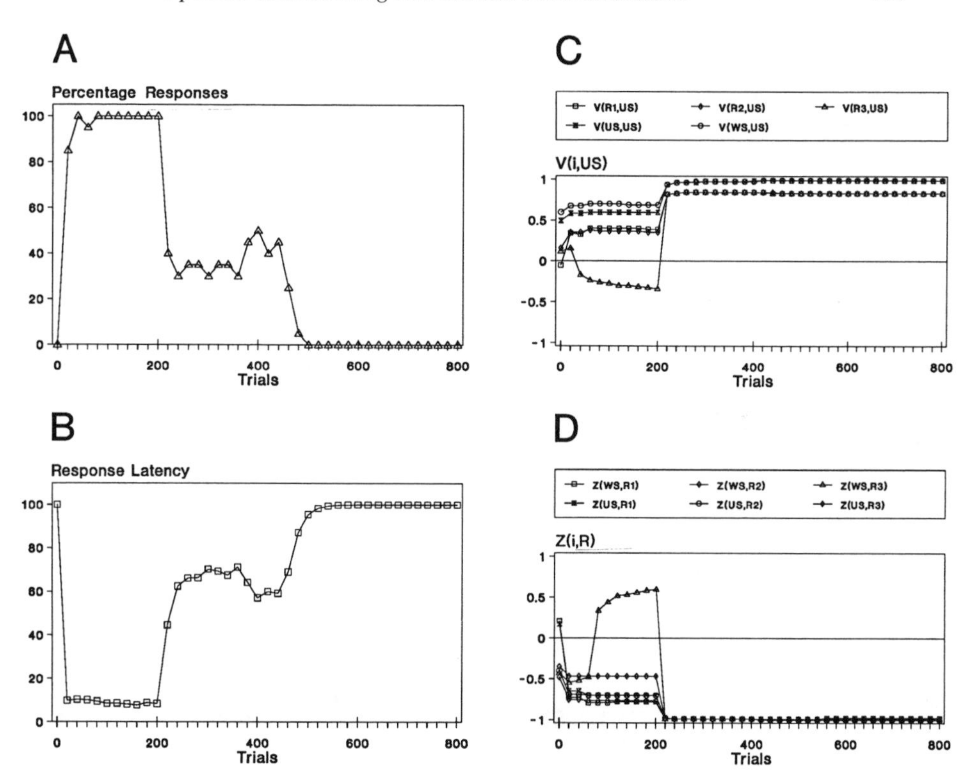

Figure 8.10. Shocking all responses. (A) Percentage of avoidance responses. (B) Response latency. (C) Associations of the US, WS, and R's with the US [*V*(US, US), *V*(WS, US), *V*(R1, US), *V*(R2, US), and *V*(R3, US)]. (D) Associations of WS and US with the R's [*Z*(WS, R1), *Z*(WS, R2), *Z*(WS, R3), *Z*(US, R1), *Z*(US, R2), and *Z*(US, R3)]. R3 is the avoidance response.

and Bull (1969) applied Mowrer–Miller's view and interpreted the results in terms of both WS1 and WS2 activating fear and fear controlling the avoidance response.

Extinction of avoidance

In agreement with experimental data (see the section on spontaneous extinction in this chapter), computer simulations show that resistance to spontaneous extinction increases with increasing US intensity and decreasing WS–US interval. In terms of the model, both procedures result in increased WS–US associations.

Preventing the avoidance response

Figure 8.9 shows computer simulations of preventing the avoidance response. During the first 200 trials, simulations show avoidance acquisition. In trial 201, the simulated animal is presented with the WS, the avoidance response is prevented, and no US is delivered. Figure 8.9A indicates that, when the barrier is removed in trial 401, the simulated animal does not generate avoidance responses. These simulated results are in agreement with Baum's (1966, 1969, 1976) and Page and Hall's (1953) data. Figure 8.9A also shows simulations of the effect of reinstating an avoidance protocol in trial 450. It illustrates that animals reacquire avoidance in few (two) shocked trials. This savings effect is

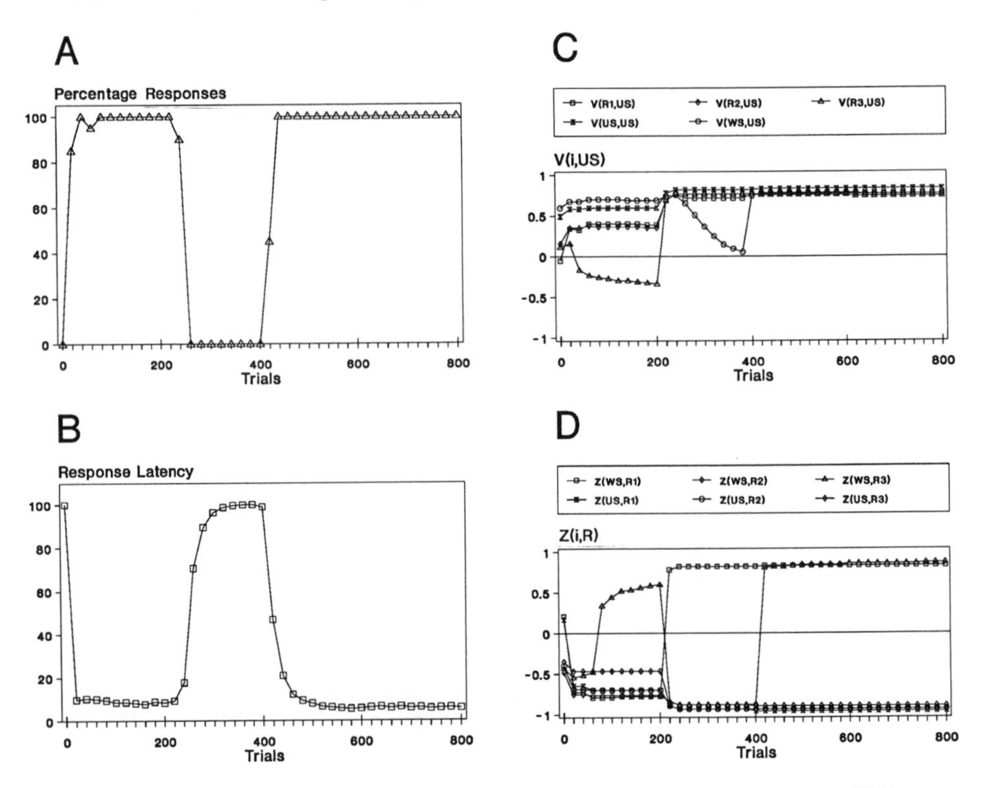

Figure 8.11. Shocking the avoidance response. (A) Percentage of avoidance responses. (B) Response latency. (C) Associations of the US, WS, and R's with the US [*V*(US, US), *V*(WS, US), *V*(R1, US), *V*(R2, US), and *V*(R3, US)]. (D) Associations of WS and US with the R's [*Z*(WS, R1), *Z*(WS, R2), *Z*(WS, R3), *Z*(US, R1), *Z*(US, R2), and *Z*(US, R3)]. R3 is the avoidance response.

based on the preserved WS–Ra association (see Figure 8.9d). This prediction of the model has not been experimentally tested.

Figure 8.9C shows that in trial 300 the WS–US association (aversion or fear to the WS) is almost entirely extinguished. According to the model, extinction of the WS–US association occurs at a faster rate than during avoidance because (1) $\tau_{WS}(t)$ is long enough to decrease its excitatory association with the US, and (2) $\tau_{Ra}(t)$ is absent and therefore does not provide protection from extinction. Based on these principles, the model is able to explain experimental data showing that weak responding [which in the model causes $\tau_{WS}(t)$ to become longer and, therefore, to decrease the WS–US association] leads to even weaker responding (Beecroft, 1967).

Shocking all responses

Figure 8.10 shows computer simulations of the effect of presenting the US 10 sec after the beginning of the trial, regardless of the animal's response. During the first 200 trials, simulations show acquisition of delay avoidance. From trials 201–600, the animal always receives the shock US independently of the response. Figure 8.10C shows that from trial 201–500, as all responses are shocked, WS–US and R3–US associations become excitatory. Therefore, the model explains extinction of avoidance because all responses become associated with the US and all WS–R associations become too small to activate a re-

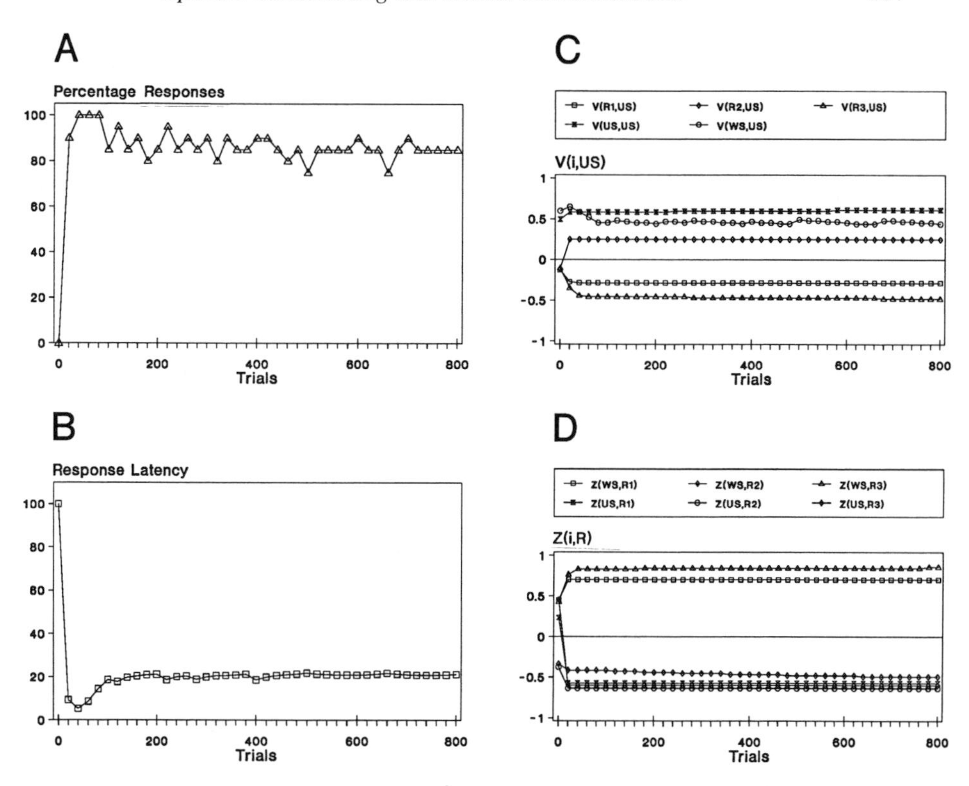

Figure 8.12. Extending the WS. (A) Percentage of avoidance responses. (B) Response latency. (C) Associations of the US, WS, and R's with the US [*V*(US, US), *V*(WS, US), *V*(R1, US), *V*(R2, US), and *V*(R3, US)]. (D) Associations of WS and US with the R's [*Z*(WS, R1), *Z*(WS, R2), *Z*(WS, R3), *Z*(US, R1), *Z*(US, R2), and *Z*(US, R3)]. R3 is the avoidance response.

sponse (Figure 8.10D). These simulated results are in agreement with Davenport and Olson's (1968) and Kamin's (1957) data.

Figure 8.10A also shows simulations of the effect of reinstating an avoidance protocol in trial 401. It indicates that, when a normal avoidance protocol is reinstated, simulated animals do not generate any response and, therefore, avoidance is never reinitiated. This predicted behavior might be related to learned helplessness (Maier and Seligman, 1976; Overmier and Seligman, 1967; Seligman, 1975), a phenomenon by which animals exposed to an inescapable shock do not attempt to escape the shock even when later the shock becomes escapable. Figure 8.10D suggests that learned helplessness is the consequence of all *Z*(US, Re) being very low for the system to try any response. Furthermore, Seligman and Maier (1967) reported that learned helplessness can be prevented by training the animal to escape-avoid in advance of exposing it to an inescapable shock. Figure 8.10D also suggests that prior training of the animal in avoidance could immunize against inescapable shocks by increasing the value of *Z*(US, Re), and thereby increasing the number of inescapable trials needed to decrease Z(US, Re) to a "helpless" value.

Shocking the avoidance response

Figure 8.11 shows computer simulations of the effect of presenting a US only when the animal generates Ra. During the first 200 trials, simulations show ac-

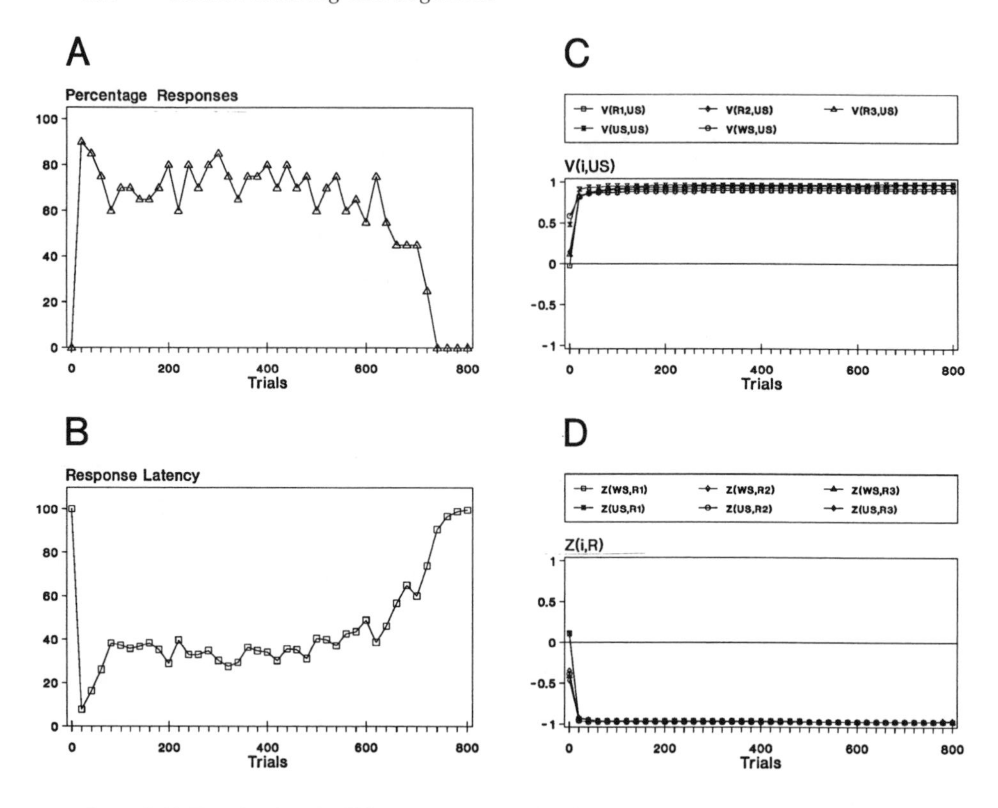

Figure 8.13. Terminating the WS. (A) Percentage of avoidance responses. (B) Response latency. (C) Associations of the US, WS, and R's with the US [*V*(US, US), *V*(WS, US), *V*(R1, US), *V*(R2, US), and *V*(R3, US)]. (D) Associations of WS and US with the R's [*Z*(WS, R1), *Z*(WS, R2), *Z*(WS, R3), *Z*(US, R1), *Z*(US, R2), and *Z*(US, R3)]. R3 is the avoidance response.

quisition of delay avoidance. From trials 201–400, the simulated animal receives a US when generating Ra. From trials 201 to approximately 320, the animal decreases its average response latency, and after that it rapidly extinguishes Ra. These simulated results are in agreement with those of Solomon et al. (1953). Figure 8.11A also shows simulations of the effect of reinstating an avoidance protocol in trial 401. Figure 8.11A illustrates that animals reacquire avoidance in eight shocked trials. As in the case of preventing the avoidance response, this savings effect is based on the preserved WS–Ra association. This prediction of the model has not been experimentally tested.

Figures 8.11C and 8.11D show that from trial 201–240, WS–US increases, WS–R3 (Ra) decreases, and WS–R1 and WS–R2 increase. The model explains extinction of avoidance in terms of the decrease in WS–R3 and increase in WS–R1 and WS–R2 associations. In agreement with Bolles et al. (1971), the simulated results show that shocking the avoidance response produces faster extinction than shocking all responses.

Effects of extending the WS, terminating the WS, and classical conditioning

Extending the WS

Figure 8.12 shows computer simulations of delay avoidance when the WS is not terminated even after the animal has generated the Ra. In agreement with

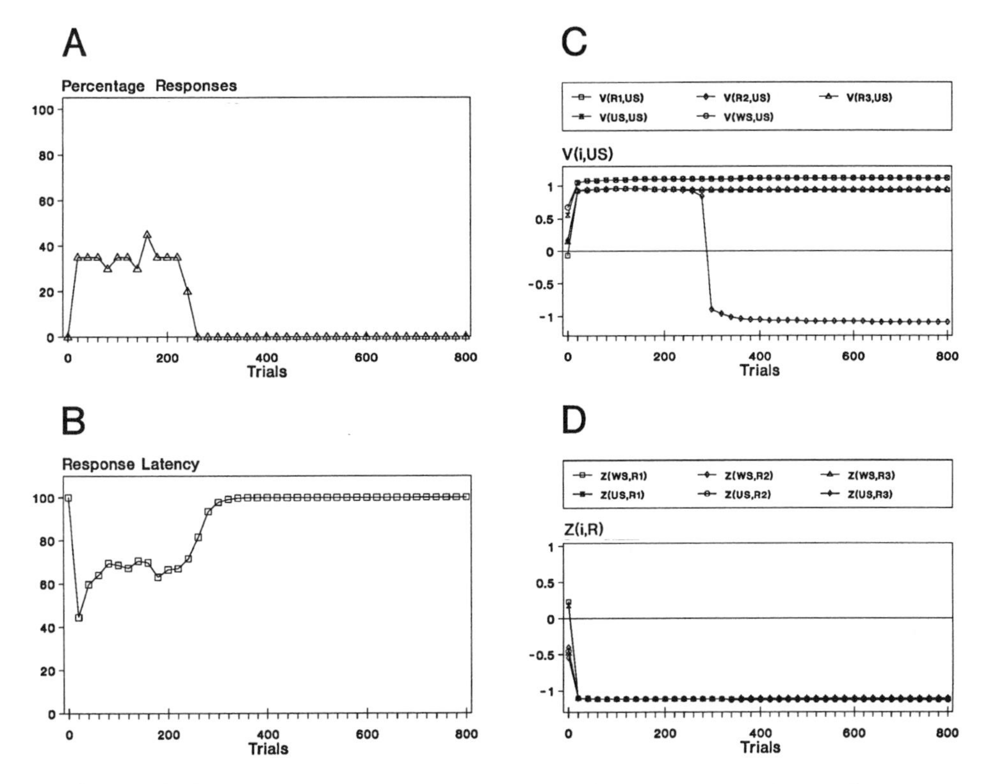

Figure 8.14. Classical conditioning. (A) Percentage of avoidance responses. (B) Response latency. (C) Associations of the US, WS, and R's with the US [*V*(US, US), *V*(WS, US), *V*(R1, US), *V*(R2, US), and *V*(R3, US)]. (D) Associations of WS and US with the R's [*Z*(WS, R1), *Z*(WS, R2), *Z*(WS, R3), *Z*(US, R1), *Z*(US, R2), and *Z*(US, R3)].

Kamin (1957), when the WS is extended after the Ra, consistent avoidance is never attained and reinforcement is intermittently needed. This alternating behavior is explained in terms of the extinction of the aversive association between the WS and the US. Figure 8.12C shows that the WS–US association extinguishes when the WS is extended. In contrast, WS–Ra is largely increased as Ra accrues an inhibitory association with the US. As mentioned, extinction of the WS–US association occurs during (1) the period when the WS is on and the US is absent, and (2) the period when WS is off but $\tau_{WS}(t)$ is still active. When the WS terminates as the animal generates Ra, $\tau_{WS}(t)$ is protected from extinction by $\tau_{Ra}(t)$ (see Equation 8.2 in Appendix 8.A). By extending the WS, the value of $\tau_{WS}(t)$ is large even when the animal has avoided the US and, therefore, extinction of WS–US increases, and periodical reinforcement is needed in order to maintain the generation of avoidance responses.

Terminating the WS

Figure 8.13 shows computer simulations of delay avoidance when the WS is terminated after Ra but the US is still delivered. Under these conditions, consistent avoidance is never attained. In agreement with Kamin (1957), simulations show alternating avoidance and escape behavior. This behavior is explained in terms of the extinction of the WS–Ra association. As mentioned, acquisition of the WS–Ra association occurs during the period when the WS and Ra are active

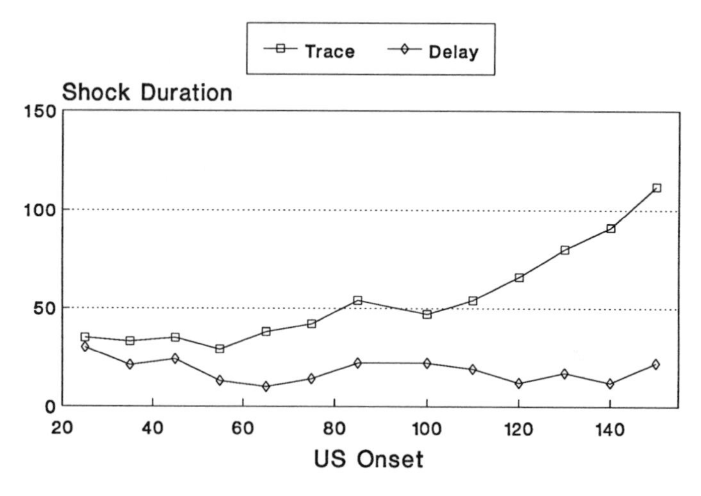

Figure 8.15. Total shock duration during delay and trace avoidance for different WS–US interstimulus intervals (ISIs).

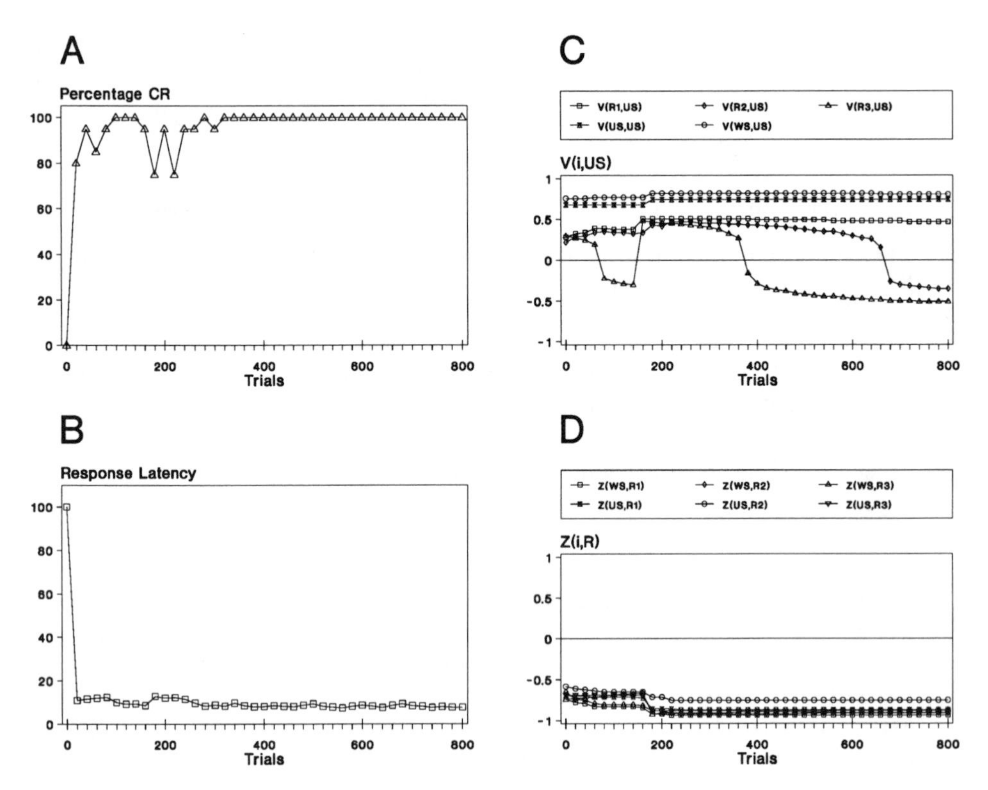

Figure 8.16. Different escape and avoidance responses. (A) Percentage of avoidance responses. (B) Response latency. (C) Associations of the US, WS, and R's with the US [*V*(US, US), *V*(WS, US), *V*(R1, US), *V*(R2, US), and *V*(R3, US)]. (D) Associations of WS and US with the R's [*Z*(WS, R1), *Z*(WS, R2), *Z*(WS, R3), *Z*(US, R1), *Z*(US, R2), and *Z*(US, R3)]. R3 is the avoidance response, R2 the escape response.

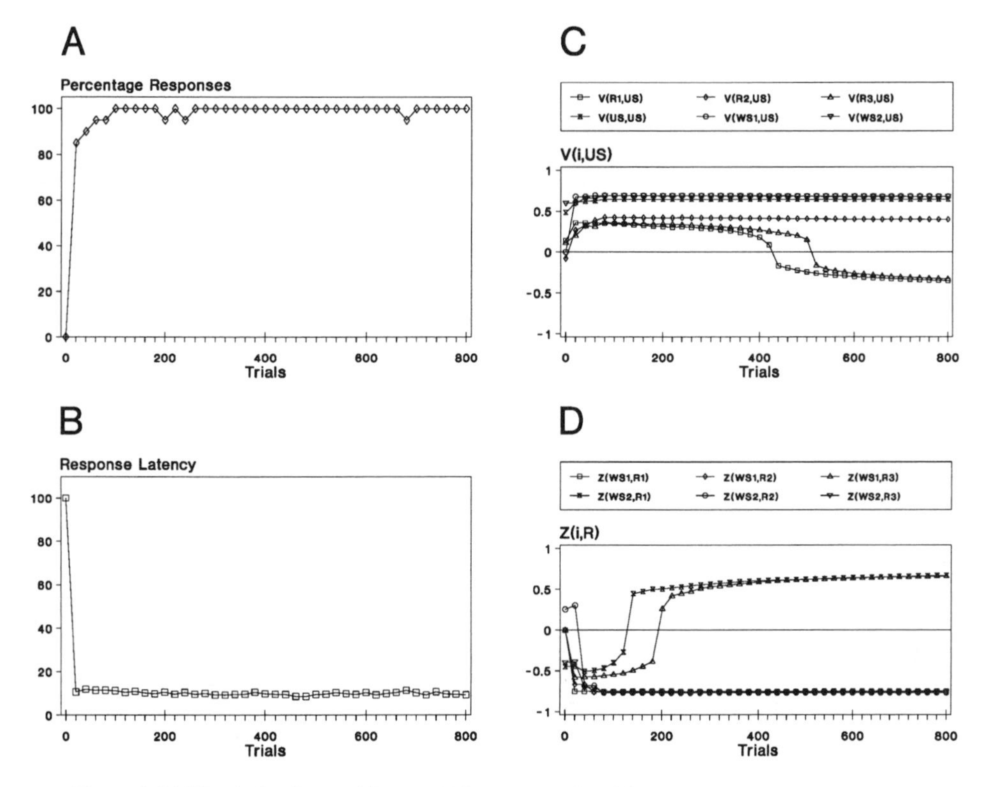

Figure 8.17. Discriminative avoidance. (A) Percentage of avoidance responses. (B) Response latency. (C) Associations of the US, WS, and R's with the US [*V*(US, US), *V*(WS1, US), *V*(WS2, US), *V*(R1, US), *V*(R2, US), and *V*(R3, US)]. (D) Associations of WS and US with the R's [*Z*(WS1, R1), *Z*(WS1, R2), *Z*(WS1, R3), *Z*(WS2, R1), *Z*(WS2, R2), and *Z*(WS2, R3)]. R1 is the avoidance response when WS2 is present, R3 the avoidance response when WS1 is present.

and the predicted US exceeds the value of the actual US. When the US is delivered as the animal generates Ra, the actual US exceeds the predicted US and, therefore, the WS–Ra association decreases. Figure 8.13C shows that the WS–US association increases as the US is consistently presented. Figure 8.13D indicates that all WS–R associations are greatly decreased.

Classical conditioning

Figure 8.14 shows computer simulations of delay avoidance when the US follows the WS at a fixed time interval, independently of the animal's response. When the WS is always followed by the US, consistent avoidance is never attained. Figure 8.14A illustrates that, in agreement with Kamin (1957), Ra increases at the beginning of training and then gradually decreases to zero. At the beginning of training, animals generate avoidance responses on a random basis, and the intensity of the responses is magnified by an increased WS–US association. As all R's, including Ra, become increasingly aversive, the animal stops responding. This result is similar to the paradigm in which all R's are punished and, therefore, avoidance is extinguished. Figure 8.14C shows that the WS–US association increases as the US is consistently presented, as in classical condi-

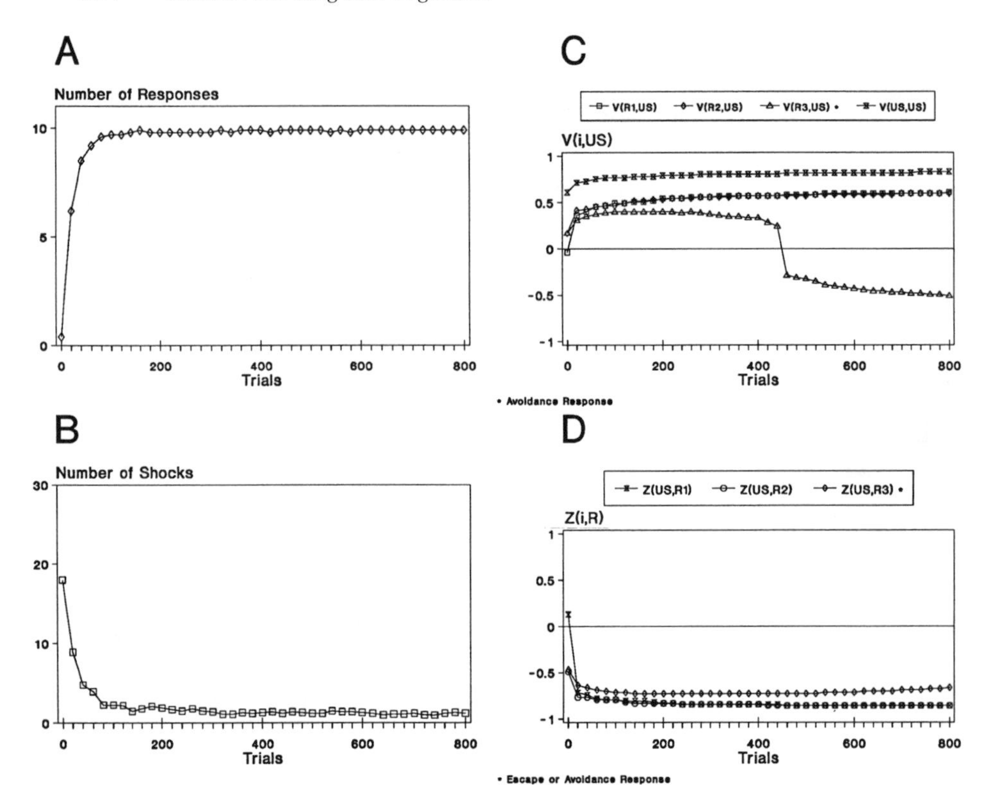

Figure 8.18. Sidman avoidance. (A) Average number of responses over a 100 time-unit period. (B) Number of US received during a 100 time-unit period. (C) Associations of the US, WS, and R's with the US [*V*(US, US), *V*(WS, US), *V*(R1, US), *V*(R2, US), and *V*(R3, US)]. (D) Associations of the US with the R's [*Z*(US, R1), *Z*(US, R2), and *Z*(US, R3)]. R3 is the escape or avoidance response.

tioning. In contrast, Figure 8.14D shows that all WS–R associations are greatly decreased.

Trace avoidance

Figure 8.15 shows the total exposure to the US in delay and trace avoidance with different ISIs. It indicates that the total exposure to the US in trace avoidance rapidly grows with increasing ISIs, whereas the total exposure to the US in delay avoidance decreases toward an asymptotic value with increasing ISIs. In trace avoidance, the intensity of Ra is determined by the classical WS–US association, which decreases with increasing ISIs. In delay avoidance, the intensity of Ra is also determined by the classical WS–US association, which also decreases with increasing ISIs. However, in delay avoidance, the trace of WS reaches a higher asymptotic value than in the trace avoidance case, thereby strongly activating the WS–Ra association, and this increased Ra compensates for decreases in the WS–US associations. This is in agreement with experimental data (Church et al., 1956). In addition, Figure 8.15 shows that, also in agreement with experimental data (Brush, 1957; Church et al., 1956; Kamin, 1954), trace avoidance is less resistant to extinction than delay avoidance and, therefore, US reinforcement is intermittently needed.

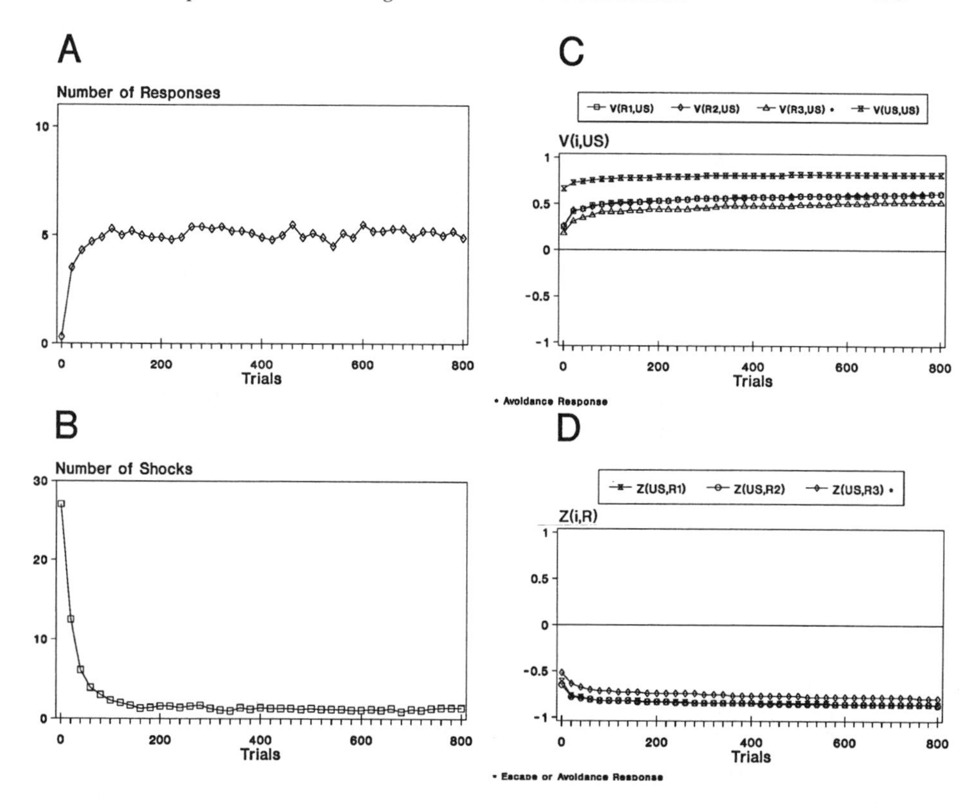

Figure 8.19. Reduction of the rate of aversive stimulation. (A) Average number of responses over a 100 time-unit period. (B) Number of US received during a 100 time-unit period. (C) Associations of the US, WS, and R's with the US [*V*(US, US), *V*(WS, US), *V*(R1, US), *V*(R2, US), and *V*(R3, US)]. (D) Associations of the US with the R's [*Z*(US, R1), *Z*(US, R2), and *Z*(US, R3)]. R3 is the escape or avoidance response.

Different escape and avoidance responses

Figure 8.16 shows that, when Ra (R3) is different from Re (R2), acquisition of delay avoidance proceeds at a slower pace than when Ra is identical to Re (see Figure 8.5). At the beginning of training, the animal generates Re either in the presence of WS or the US because both WS–Re and US–Re are the dominant responses. After 25 trials, WS–Ra becomes stronger than WS–Re, and the animal correctly discriminates avoidance from escape, that is, in the presence of WS the animal avoids, and if WS is omitted, the animal escapes when the US is presented. Because it takes more trials for the animal to learn the correct discrimination, the model predicts that WS–US and US–US associations are stronger with identical avoidance and escape responses than with different ones.

Discriminative avoidance

Figure 8.17 shows that, when two different avoidance responses R1 and R3, respectively, are required in the presence of two different warning stimuli WS2 and WS1, acquisition proceeds at a slower pace than when only one response is needed (see Figure 8.5). After 150 and 200 trials, respectively, WS2–R1 and WS1–R3 become strong and the animal reliably generates the correct response in the presence of the corresponding WS. After 400 trials, both R1 and R3 become predictors of the absence of the US.

Sidman avoidance

Figures 8.18A and 8.19A display the number of responses as a percentage over 100 time units. Figures 8.18B and 8.19B display the number of USs as a percentage over 100 time units. Figure 8.18 shows that, in agreement with Sidman (1953), the percentage of responses increases and the percentage of USs decreases with increasing number of trials. Over trials, US–US associations increase (the US predicts itself) as well as US–Ra associations (the US becomes associated with the Ra response). According to the model, animals learn the Sidman avoidance task because Ra becomes associated with the US trace. In addition, Figure 8.19 shows that in agreement with Herrnstein and Hineline's (1966) experimental data, Ra increases after the subject receives the US and gradually decreases until the next US is presented.

Discussion

We presented a novel, real-time two-process neural network model of avoidance that combines elements of classical and operant conditioning. The classical conditioning process generates associations between the WS, the US, and R's generated by the animal with the US. Whereas WS and the US become predictors of the US, the escape and the avoidance response, Re and Ra, become predictors of the absence of the US. The operant conditioning process associates WS and US with those R's that predict the absence of the US (avoidance and escape responses). The classical conditioning process (1) provides predictions of the presence or absence of the US used by the operant conditioning process to generate WS–R associations, and (2) controls the strength of the operant behavior. Because the warning WS and the US can become associated with different available responses, animals can learn different escape and avoidance responses. Since the model describes behavior in real time, it is able to capture the effects of using different temporal WS and US arrangements and to describe the latency of avoidance and escape responses.

The network comprises two processes, classical and operant conditioning. The classical conditioning process assumes that US–US, WS–US, and R–US associations increase whenever the traces of US, WS, or R are active and the *actual intensity of the US exceeds its predicted intensity (fear).* The operant conditioning process assumes that US–Re and WS–Ra associations increase whenever the traces of US and Re or WS and Ra are active together and the *predicted intensity of the US (fear) exceeds its actual intensity.* Whereas classical conditioning is regulated by a delta rule, operant conditioning is regulated by a novel algorithm that mirrors the classical conditioning algorithm. In other words, by computing the mismatch between the actual and predicted intensity of the US, and using this mismatch in opposite ways to regulate classical and operant conditioning, the network is able to describe escape and avoidance in a unified manner. Importantly, responses that acquire an excitatory association with a stimulus gain an inhibitory association with the US, that is, responses activated by a stimulus in order to escape or avoid the US come to predict the absence of the US. Furthermore, Gabriel and Schmajuk (1990) suggested that the mis-

Table 8.1. *Paradigms described by the Schmajuk (1994) escape-avoidance model.*

Paradigm	
Delay avoidance	Y
Trace avoidance	Y
Fear modulates avoidance	Y
Transfer of control	Y
Extinction of avoidance	Y
Preventing the avoidance response	Y
Shocking all responses	Y
Shocking the avoidance response	Y
Extending the WS	Y
Terminating the WS	Y
Classical conditioning	Y
Discriminative avoidance	Y
Different escape and avoidance responses	Y
Sidman avoidance	Y
Discriminative avoidance and communication	Y

Note. Y: The model can simulate the experimental data.

match between the actual and predicted intensity of the US might also control behavioral inhibition during avoidance: When a mismatch is detected, the avoidance response is inhibited. Gabriel and Schmajuk (1990) showed through computer simulations that behavioral inhibition retards acquisition and accelerates extinction of avoidance in a running wheel paradigm.

The approach introduced in this chapter for aversive operant conditioning can be extended easily to appetitive operant conditioning. As in the aversive case, in the appetitive condition classical associations V_i increase when (US – B) is greater than zero and decrease when (US – B) is less than zero. In contrast to the aversive case, in the appetitive condition operant associations Z_{ik} increase when (US – B) is greater than zero and decrease when (US – B) is less than zero.

The network was applied to the description of escape and avoidance behavior in a shuttle-box and Sidman avoidance (see Table 8.1). Computer simulations demonstrate that the model describes many of the features that characterize avoidance behavior: (1) Fear of the US decreases as the animal masters the response; (2) techniques that decrease fear decrease, and techniques that increase fear increase, ongoing avoidance responses; (3) when WS–US classical associations are extinguished, the avoidance response can be elicited by another WS classically conditioned to the US; (4) the amount of time needed to generate the avoidance response decreases with increasing number of trials; (5) in some situations avoidance undergoes negligible extinction; (6) trace avoidance is less resistant to extinction than delay avoidance; (7) acquisition of trace avoidance is slower and extinction faster with increasing WS–US intervals; (8) extinction of avoidance is obtainable by preventing the animal's capability to elicit the avoidance response without delivering the US, by shocking the animal when it emits the avoidance response, or by shocking the animal whether or not it emits the avoidance response; (9) when the avoidance response terminates the WS but does not prevent US presentation, animals show slower acquisition and lower asymptotic levels of avoidance; (10) when the avoidance response prevents US presentation but does not terminate the WS, animals learn less than in the normal avoidance situation; (11) when the avoidance response has no effect either on WS duration or the delivery of the US (classical conditioning), animals do not acquire consistent avoidance; (12) the avoidance response may be different from that required to escape the US; and (13) different avoidance responses may be associated with different discriminative stimuli.

Comparison with other models of avoidance

Two-factor theories

Like previous two-factor theories, our model incorporates a classical conditioning process by which the WS becomes associated with the US and interprets this association as fear of the WS. Therefore, as does Mowrer–Miller's two-factor theory, the model correctly predicts that presenting an aversive CS increases the rate of ongoing avoidance responses, whereas presenting a "safe" CS decreases the rate of ongoing avoidance responses. Also in agreement with Mowrer–Miller's theory, our model correctly predicts rapid extinction of avoidance when animals are exposed to the WS but the US is not presented (preventing the avoidance response), because the WS–US association is extinguished.

The operant algorithm offered in this chapter differs from Mowrer–Miller's two-factor theory in several important ways. First, Mowrer–Miller's theory suggests that fear–Ra associations are reinforced by the *reduction of fear* as Ra terminates the fear-eliciting WS. In contrast, our model assumes that WS–R associations for *all responses are reinforced by fear,* and WS–R associations for *responses other than the avoidance response are weakened by the presentation of the US.* In both models, however, when fear decreases [(US – B) is less than zero], operant associations increase [(B – US) is greater than zero]. Second, Mowrer–Miller's theory assumes that classical and operant processes are organized in a serial fashion, that is, a WS is associated with the US and activates fear, and fear is associated with the escape response and activates the (now) avoidance response. The strength of the operant response is therefore proportional to the product of the strength of classical and operant associations. In contrast, our model assumes that classical associations modulate the operant responses in a parallel fashion, that is, a WS is associated with both the US and the avoidance response, and WS–US associations multiplicatively modulate the strength of the operant response. No associations between fear and the operant responses are assumed. Third, Mowrer–Miller' theory assumes that the US and fear are associated with the same response and, therefore, escape and avoidance responses are identical. In contrast, our model establishes independent US–Re and WS–Ra associations. Fourth, whereas Mowrer–Miller's two-factor theory explains transfer of control experiments in terms of WS1 and WS2 activating fear and fear activating the avoidance response, our model describes the results in terms of a common element in WS1 and WS2.

Some empirical evidence that presents difficulties for Mowrer–Miller's two-factor theory is correctly addressed by our model. For instance, two-factor theory has problems explaining trace avoidance, in which a temporal delay is introduced between the WS offset and the presentation of the US. Although in trace avoidance the association between a WS temporal trace and the US can be explained in the same terms as classical trace conditioning, WS termination reduces fear only slightly. Our model does not need WS termination to create the reduction of fear that reinforces the avoidance response. Instead, the model explains trace avoidance in terms of the association between the WS and the avoidance response in the presence of the fear generated by the temporal trace of the WS. Another major difficulty for Mowrer–Miller's theory is the sometimes reported slow extinction of avoidance. According to the theory, WS–US associations extinguish as the animal avoids the US and, therefore, animals should cease avoiding the WS. In contrast to the two-factor theory, our model adequately addresses extinction of avoidance: Extinction of the WS–US association (supported by experiments showing a decrease in fear during learning) is perfectly compensated by the increase in the WS–R association (corroborated by data showing that response latency decreases as the animal increasingly masters the avoidance response).

Also in contrast to Mowrer–Miller's theory, our neural network contributes to the understanding of learned helplessness (Maier and Seligman, 1976; Overmier and Seligman, 1967; Seligman, 1975), a phenomenon by which animals exposed to an inescapable shock do not attempt to escape the shock even when later the shock becomes escapable. In terms of the model, learned helplessness is the consequence of all Z(US, Re) being too low for the system to try any response. Furthermore, Seligman and Maier (1967) reported that learned helplessness can be prevented by prior training of the animal to escape-avoid. The model suggests that training the animal in avoidance could immunize against inescapable shocks by increasing the value of Z(US, Re), thereby increasing the number of inescapable trials needed to decrease Z(US, Re) to a "helpless" value.

Although most two-process theories share Mowrer–Miller's account of classical conditioning, they do not all assume that the avoidance response is reinforced by the reduction of fear as the animal's response terminates the fear-eliciting WS. For example, Soltysik (1985) proposed that Ra becomes a conditioned inhibitor that reduces fear, thereby reinforcing the instrumental response. Similarly, Weisman and Litner (1969, 1972) suggested that Ra is reinforced by the onset of the conditioned inhibitor predicting the nonoccurrence of the US. In contrast to Soltysik's (1985) and Weisman and Litner's (1969, 1972) theories, in which classical conditioning processes precede operant conditioning, our model suggests that classical and operant conditioning processes occur simultaneously in a tightly integrated manner. Both classical conditioning (generating US–US, WS–US, and R–US associations) and instrumental reinforcement (generating US–R and WS–R associations) depend on the mismatch between the predicted and the actual value of the US. As Re or Ra become conditioned inhibitors of fear through a classical conditioning mechanism, they also become associated with the US or the WS in the operant conditioning process.

One-factor theory

One-factor theory (Herrnstein, 1969) assumes that only operant conditioning principles are at work during avoidance. Although one-factor theory argues that, because there is no WS to signal the US, behavior is controlled entirely by operant conditioning principles, Sidman himself recognized that the passage of time could be regarded as a classical conditioned WS. According to Mazur (1990), in Sidman's (1953) experiment, this WS becomes trace-conditioned to the US (there is a temporal gap between the WS offset and the US presentation), and in Herrnstein and Hineline's (1966) experiment, the WS becomes backward-conditioned to the US (the US presentation precedes the WS onset). In the Herrnstein and Hineline experiment, the subject's response is followed by a greater average time without a response, which might be considered a WS in a backward conditioning paradigm. Therefore, consistent with the classical conditioning process of the two-factor theory, Sidman's subjects acquired avoidance more effectively than Herrnstein and Hineline's rats because backward conditioning is weaker than trace conditioning (Mackintosh, 1974).

In our model, the Sidman avoidance task is learned because Re becomes associated with the US trace. Classical conditioning generates US–US associations, that is, presentation of the US predicts its continuation. Because Re causes the termination of the US, it predicts its absence and becomes associated with the US. Further US presentations will activate the generation of Re, until the US trace becomes too weak to sustain the response. At that point, a new US presentation is needed and a new cycle starts.

Cognitive theory

Seligman and Johnston (1973) proposed a theory of avoidance in which (1) learning occurs only in the case of a mismatch between actual and expected results of a response, and (2) a response leading to no shock is preferred to a response leading to shock. In the course of avoidance, animals learn that no US occurs if a given response is produced. Avoidance exhibits slow extinction until this expectation is disconfirmed.

Seligman and Johnston's (1973) cognitive theory is supported by data showing that avoidance extinguishes at a fast rate in experiments that prevent the generation of the avoidance response (in which animals are retained in the darkened compartment but the US is not presented), because the subjects learn that no response leads to no US. Cognitive theory is also supported by data showing that avoidance can be extinguished by presenting the US following the response (Davenport and Olson, 1968; Seligman and Campbell, 1965), because the expectation that no US follows the avoidance response is disconfirmed.

As a corollary of the cognitive theory, Seligman (Maier and Seligman, 1976; Overmier and Seligman, 1967; Seligman, 1975) suggested that animals can learn the expectation that their behavior has little effect on their environment, a phenomenon that they called *learned helplessness*. When animals are exposed to an inescapable shock US and placed later in a shuttle-box, they do not learn to escape or avoid the shock US.

Our model shares some features with Seligman and Johnston's (1973) cognitive theory of avoidance. As the cognitive theory proposes, our model assumes that learning occurs only in the case of a mismatch between actual and expected results of a response or a WS. Cognitive theory proposes that, when the animal acquires avoidance behavior, since there is no change in expectancy, there is no change in behavior and, therefore, avoidance does not extinguish. In contrast, our model suggests that avoidance does not extinguish because (1) the aversive WS becomes increasingly shorter as avoidance progresses, (2) the inhibitory Ra–US association somewhat protects the extinction of the aversive WS–US association, and (3) the strength of the WS–Ra association increases as avoidance progresses. Our model formalizes cognitive theory assumptions of a cognitive process by which animals prefer a response leading to no shock to a response leading to shock, and expresses it in the form that animals prefer a response that minimizes the shock US.

Although not incorporated in the present model, our neural network model can describe, as Seligman and Johnston (1973) proposed and Davis et al. (1989) confirmed, the presence of fear at different times in the avoidance paradigm.

Fear is considered to be timed relative to the presentation of the warning stimulus. As described in Chapter 7, Desmond and Moore (1988), Grossberg and Schmajuk (1989), and Schmajuk (1990) presented neural network models for classical conditioning that generate CRs at the time of the US presentation.

Other neural network models of operant conditioning

Two different views have been used to explain why animals approach or avoid different stimuli: a "stimulus-approach" and a "response-selection" view. According to the stimulus-approach view, subjects learn that some particular stimulus situation is appetitive and therefore it is approached. Mackintosh (1974, p. 554) proposed that in the presence of numerous intra-and extramaze cues, animals typically learn to approach a set of stimuli associated with reward and to avoid a set of stimuli associated with punishment. This approach was studied by Schmajuk (1990) (see Chapter 10), who presented a model that describes how animals navigate in the environment by approaching appetitive, and avoiding aversive, places. Only in a totally homogeneous environment (one that lacks discernible cues) will animals learn to make the correct responses that lead to the goal. According to the response-selection view, subjects learn because a particular response is reinforced in the presence of a particular stimulus situation. Different responses are learned at different places in the environment, and animals can navigate toward a goal by compounding different responses associated with different landmarks. More specifically, Mackintosh (1983) suggested that in some situations (e.g., shuttle-box), classical conditioning principles might suffice to describe avoidance in terms of stimulus-approach principles: Animals withdraw from aversive places and approach safe places. Schmajuk and Urry (unpublished results) confirmed through computer simulations the capability of stimulus-approach principles to describe avoidance in a shuttle-box. However, in situations where there is no safe place to reach (e.g., running-wheel, leg-flexion, lever-pressing), operant conditioning principles (such as those described in the present chapter) implementing response-selection are needed to provide a complete description of avoidance.

Like the model presented in this chapter, Grossberg (1972a, b), Barto and Sutton (1981), and Staddon and Zhang (1991) proposed models that incorporate "response-selection" principles. Grossberg (1972a, b) proposed a neural theory of punishment and avoidance that combines classical and operant conditioning (response-selection) mechanisms and describes avoidance in terms of an architecture that provides a rebound mechanism from fear to relief. According to Grossberg, if a WS is paired with the US, it becomes associated with fear and its termination produces a rebound that provides a relief signal that, in turn, controls the association of the WS with the avoidance response. The model describes phenomena such as the lesser effect of reducing J units of shock to $J/2$ units than reducing $J/2$ to 0 units, persistent nonspecific fear that biases the interpretation of specific cues, different effects of gradual and abrupt shock on response suppression, reduction of pain in the presence of a loud noise, influence of drugs on conditional emotional and avoidance responses, among others.

Barto and Sutton (1981) characterized a model that describes spatial learning. They suggested that the association between a stimulus (one of four spatial landmarks) and a response (a movement in one of four spatial directions) increases whenever the response moves the animal toward a more appetitive location.

Staddon and Zhang (1991) proposed a nonassociative (no S–R associations are formed) model that describes the context-free properties of operant conditioning, selection, delay-of-reward, contingency, and unsignaled avoidance. The model assumes that different responses are represented by a filtered noise V that is augmented or decremented by pleasant or unpleasant events in an amount proportional to its value. The response of the highest V value is the one that actually occurs. Although Staddon and Zhang's (1991) model cannot describe discriminated escape and avoidance, the model is able to explain some aspects of Sidman avoidance.

Maki and Abunawas (1991) presented a neural network that, employing a generalized delta rule (see Box 2.1), describes matching-to-sample tasks (MTS). In a MTS paradigm, animals are first presented with the sample stimulus and later presented with the comparison stimulus. Animals are rewarded for choosing the comparison stimulus that matches the preceding sample. The network correctly describes coding of sample stimuli in terms of anticipated events (prospective coding), improvement of delayed MTS with delay training (rehearsal), and decrements in matching accuracy when compound samples are presented (shared attention).

Donahoe, Burgos, and Palmer (1993) proposed a three-layer "selection network" that implements a single positive reinforcement principle for both classical and operant conditioning. The network includes (1) a response-selection component that mediates environment–behavior relation, and (2) a stimulus-selection component that mediates environment–environment relations. The model describes acquisition, extinction, reacquisition, conditioned reinforcement, blocking, and stimulus discrimination.

Ethological validity of the theory

Hull (1929) suggested that principles similar to those found in the laboratory may be applied to the description of how animals in the wild learn to avoid predators. According to Hull, when vulnerable animals are attacked by a predator but luckily manage to escape and survive, they learn that stimuli present at the time of the assault predict an impending assault. Therefore, when potential prey perceive similar stimuli on future occasions, they flee to avoid the predator. According to Bolles (1970), this account is wrong, because predators do not announce their presence and do not give their potential prey enough trials for learning to occur. However, supporting Hull's view, Curio (1976) reported that predators may be effective in less than 43 percent of their strikes, thereby revealing their presence and providing their prey with plenty of opportunities to learn about their enemies and their circumstances. In sum, it is possible that animals in the wild learn to avoid predators utilizing mechanisms similar to those described in the present chapter.

Applications to animal communication

Recently, Schmajuk and Axelrad (1995) applied the model presented in this chapter to the description of some theoretical aspects of animal communication. For instance, an avoidance task, in which two animals have to communicate in order to avoid shock presentations. In this task, one of the animals is presented with the warning stimuli, but cannot generate responses to avoid the

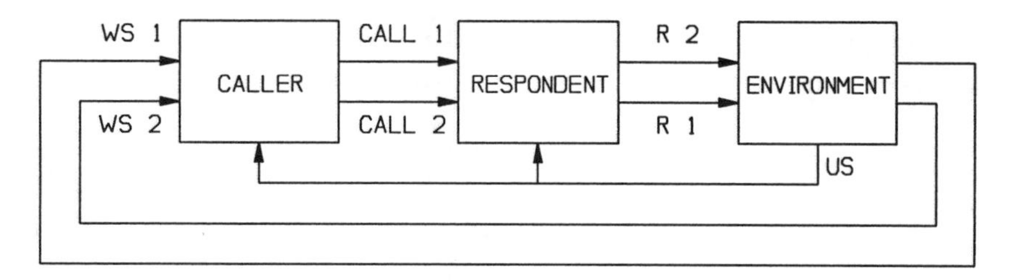

Figure 8.20. A block diagram of the interactions between the environment, a caller, and a respondent during discriminative avoidance. The environment establishes that for both animals, R2 is the correct response when WS1 is present, and R1 the correct response when WS2 is present. The simulated animals can solve the problem with different combinations of CALLs and R's. In this case, the caller emits CALL1 when WS1 is presented and CALL2 when WS2 is presented. In turn, the respondent emits R2 when CALL1 is emitted by the caller and R1 when CALL2 is emitted by the caller.

shock. A second animal is presented with the first animal's calls but not the warning stimuli, and is able to generate responses to avoid the shock for both animals.

Figure 8.20 shows two identical neural network models of avoidance that were allowed to interact with each other (through some of their responses assumed to be calls) and with the environment (through potential avoidance responses). In the simulations, a simulated caller was assumed to perceive the warning stimuli WS1 and WS2, and a simulated respondent was assumed to generate the avoidance responses R1 and R2. As in the discriminative avoidance paradigm, two different avoidance responses, R1 and R2, respectively, were required from the respondent when one of two different warning stimuli, WS2 and WS1, were presented to the caller. When the adequate response was emitted by the respondent, both the caller and the respondent avoided the US.

Figure 8.21 illustrates computer simulations of the communication avoidance paradigm. Figure 8.21A shows that acquisition proceeds at a slower pace for two animals than for one animal trained in a discriminative avoidance paradigm (see Figure 8.17). Figure 8.21B shows that as the percentage of correct responses increases, response latency decreases over trials.

Figure 8.21C displays R1–US, R2–US, R3–US, US–US, CALL1–US, and CALL2–US associations for the respondent. Whereas R3, US, CALL1, and CALL2 strongly predict the US, R2 predicts the absence of the US (R2–US is inhibitory). After 300 trials, R1 still predicts the US, but with more trials will predict the absence of the US as R2 does. Figure 8.21D displays CALL1–R1, CALL1–R2, CALL1–R3, CALL2–R1, CALL2–R2, CALL2–R3 associations for the respondent. In the presence of CALL1, the respondent will choose R2; in the presence of CALL2, the respondent will choose R1.

Figure 8.21E displays CS1–US, CS2–US, US–US, CALL1–US, and CALL2–US associations for the caller. All associations are excitatory. Figure 8.21f displays WS–CALL1, WS1–CALL2, WS2–CALL1, WS2–CALL2 associations for the caller. Although it is clearly represented in Figure 8.21C–F, in the presence of WS1, the caller will choose CALL1, and in the presence of WS2, the caller will choose CALL2.

In sum, because the caller emits CALL1 for WS1 and CALL2 for WS2, and the respondent emits R2 for CALL1 and R1 for CALL2, the simulated animals show adequate discrimination, responding with R2 to WS1 and R1 to WS2.

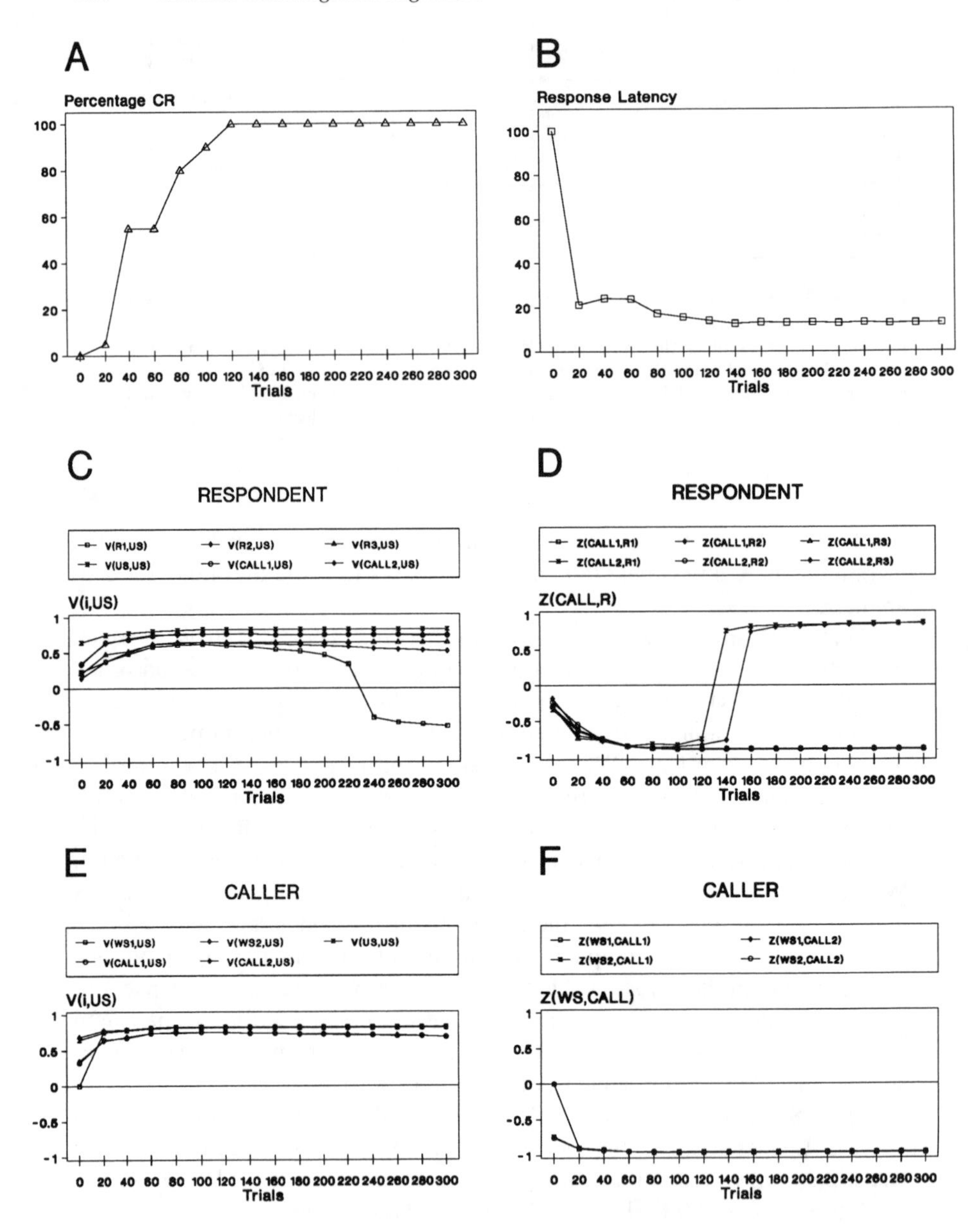

Figure 8.21. Animal communication during discriminative avoidance. (A) Percentage of avoidance responses for the both the caller and the respondent. (B) Response latency for both the caller and the respondent. (C) Respondent's classical associations of the US, CALLs, and R's with the US [*V*(US, US), *V*(CALL1, US), *V*(CALL2, US), *V*(R1, US), *V*(R2, US), and *V*(R3, US)]. (D) Respondent's operant associations of CALLs with the R's [*Z*(CALL1, R1), *Z*(CALL1, R2), *Z*(CALL1, R3), *Z*(CALL2, R1), *Z*(CALL2, R2), and *Z*(CALL2, R3)]. R2 is the selected avoidance response when CALL1 is present, and R1 the selected avoidance response when CALL2 is present. (E) Caller's classical associations of the US, WS, and CALLs with the US [*V*(US, US), *V*(WS1, US), *V*(CALL1, US), and *V*(CALL2, US)]. (F) Caller's operant associations of WSs with the CALLs [*Z*(WS1, CALL1), *Z*(WS1, CALL2), *Z*(WS2, CALL1), and *Z*(WS2, CALL2)]. CALL2 is the selected call when WS1 is present, and CALL1 the selected call when WS2 is present.

Appendix 8.A

A formal description of the model.

This section formally describes the model depicted in Figure 8.3. The network presented is able to describe in real time many classical conditioning paradigms, such as acquisition, extinction, blocking, overshadowing, and conditioned inhibition (see Chapter 3). As shown by Schmajuk and DiCarlo (1992) (see Chapter 6), with the addition of a configural system, it also describes positive and negative patterning. Importantly, in agreement with experimental data (Zimmer-Hart and Rescorla, 1974), the network characterizes conditioned inhibition as not extinguishable by nonreinforced presentations of a conditioned inhibitor.

Classical conditioning process

Classical conditioning is described with a real-time delta rule model similar to that presented in Chapter 3. We assume that $WS_i(t)$, $R_i(t)$, and the US(t) [generically represented by $X_i(t)$] generate a STM trace $\tau_i(t)$ according to

$$d[\tau_i(t)]/dt = -K_1\, \tau_i + K_2\, [K_3 - \tau_i(t)]\, X_i(t) \tag{8.1}$$

where $-K_1\, \tau_i(t)$ represents the passive decay of the STM of $WS_i(t)$, $R_j(t)$, or the US(t), K_2 represents the rate of increase of $\tau_i(t)$, and constant K_3 is the maximum possible value of $\tau_i(t)$.

Changes in the association of $\tau_i(t)$ with the US, are given by

$$d[V_i(t)]/dt = K_4 \tau_i\, [\mathrm{US}(t) - \mathrm{B}(t)]\, (1 - |V_i(t)|) \tag{8.2}$$

where $\tau_i(t)$ represents the trace of $WS_i(t)$, $R_i(t)$, or the US(t). $B(t) = \Sigma_j\, V_j(t)\, \tau_j(t)$ represents the aggregate prediction of the US by all τ's active at a given time. $K_4 = K_4'$ if $\mathrm{US}(t) > \Sigma_j\, V_j(t)\, \tau_j(t)$, $K_4 = K_4''$ if $\mathrm{US} < \Sigma_j\, V_j(t)\, \tau_j(t)$. By Equation 8.2, $V_i(t)$ increases whenever $\tau_i(t)$ is active and $\mathrm{US}(t) > \Sigma_j\, V_j(t)\, \tau_j(t)$ and decreases when $\mathrm{US}(t) < \Sigma_j\, V_j(t)\, \tau_j(t)$. In order to prevent the extinction of conditioned inhibition or the generation of an excitatory CS by presenting a neutral CS with an inhibitory CS, we assume that, when $\Sigma_j\, V_j(t)\, \tau_j(t) < 0$, then $\Sigma_j\, V_j(t)\, \tau_j(t) = 0$. The term $[1 - |V_i(t)|]$ bounds $V_i(t)$ $[-1 \leq V_i(t) \leq 1]$.

Operant conditioning process

We adopt a response-selection view of operant conditioning by which WS or the US become associated with different alternative responses. Changes in the association of τ_i of WS_i with the R_j are given by

$$d\,[Z_{ik}(t)]/dt = K_5 \tau_i(t)\, \tau_k(t) [B(t) - \mathrm{US}(t)]\, [1 - |Z_{ik}(t)|] \tag{8.3}$$

where $\tau_i(t)$ is the STM trace of WS(t) or the US(t), and $\tau_k(t)$ is the STM trace of the response that has the maximum output at a given time. $K_5 = K_5'$ if $\mathrm{US}(t) < \Sigma_j\, V_j(t)\, \tau_j(t)$, $K_5 = K_5''$ if $\mathrm{US} > \Sigma_j\, V_j(t)\, \tau_j(t)$. According to Equation 8.3, WS(t) or the US(t) become associated with $R_k(t)$ when they are active simultaneously and $\Sigma_j\, V_j(t)\, \tau_j(t) > \mathrm{US}(t)$, and decrease if $\Sigma_j V_j(t)\, \tau_j(t) < \mathrm{US}(t)$. Because WS and the US may become associated with different responses, Ra and Re might be different. More generally, the system can learn operant discriminations by generating

R_1 in the presence of WS_1 and R_2 in the presence of WS_2. The term $[1 - |Z_{ik}(t)|]$ bounds $Z_{ik}(t)$ $[-1 \leq Z_{ik}(t) \leq 1]$.

Performance rules

The strength of the conditioned response is given by

$$CR(t) = f[B(t)] = f[\Sigma_j V_j(t)\ \tau_j(t)] \tag{8.4}$$

The strength of $R_k(t)$ is given by

$$R_k(t) = K_6 f[\Sigma_i Z_{ik}(t)\ \tau_i(t) + r_i(t)] \tag{8.5}$$

where $f[x] = 1/1 + e^{-(x)}$, *and* $r_i(t)$ is a random number. $R_k(t)$ is always positive, even when x may vary between –1 and 1. We assume that the animal selects and tries the response with the maximum $R_k(t)$ every five time units. When $R_{max}(t)$ is selected, it activates $\tau_{max}(t)$ according to Equation 8.1.

The amplitude of the operant response $R'(t)$ is given by

$$R'(t) = \tau_{max}(t)\ B(t) \tag{8.6}$$

where $B(t) = \Sigma_j V_j(t)\ \tau_j(t)$. Equation 8.6 implies that the same operant response $R'(t)$ can be achieved by various combinations of $\tau_{max}(t)$ and $B(t)$.

Let $t_{latency}$ denote the time at which the animal crosses to the opposite side in a shuttle-box and thereby avoids or escapes the shock. Then $t_{latency}$ is the earliest time such that

$$R'(t_{latency}) = \tau_{max}(t_{latency})\ B(t_{latency}) \geq K_7 \tag{8.7}$$

where K_7 is the threshold to escape or avoid. Equation 8.7 implies that latency decreases for a fast-growing $\tau_{max}(t)$ and large B's. Because $\tau_{max}(t)$ grows faster with increasing values of $R_{max}(t) = K_6 f[\Sigma_i Z_{ik}(t)\ \tau_i(t)\ ]$, latency decreases with increasing values of $Z_{imax}(t)$ (WS–R_{max} and US–R_{max} operant associations). Also, because $B(t) = \Sigma_j V_j(t)\ \tau_j(t)$, latency decreases with increasing values of $V_j(t)$ (WS–US, US–US, and R–US classical associations).

Appendix 8.B

Simulation procedures and parameters

In our computer simulations, each trial is divided into 250 time units for shuttle-box avoidance and 100 time units for Sidman avoidance. Each time unit represents approximately 500 msec. At time zero, the WS is presented. If the avoidance response has not been performed after 25 time units, a shock US is applied. Three alternative responses are considered and no initial hierarchy (see Hull, 1951) is assumed, that is, $Z_{ik} = 0$.

Parameter values are $K_1 = 2\ 10^{-2}$, $K_2 = 25\ 10^{-3}$, $K_3 = 1$, $K_4' = 1\ 10^{-5}$, $K_4'' = 6\ 10^{-4}$, $K_5' = 5\ 10^{-3}$, $K_5'' = 3\ 10^{-1}$, $K_6 = 5\ 10^{4}$, $K_7 = 5\ 10^{-2}$, $r_k = \pm\ 2\ 10^{-4}$.

III Cognition

9 Animal cognition: data and theories

Cognitive views of animal and human behavior suggest that behavior is purposive and can be described in terms of how different goals are pursued (Tolman, 1932b). These goals are determined by internal motivational states. When pursuing these goals, animals display remarkable adaptability and adopt alternative behavioral strategies that are independent of any specific set of responses. In accomplishing their motivational objectives, animals learn to predict the occurrence of future events and foretell the consequences of their own behavior. Either as passive spectators or as active participants, they anticipate and prepare for forthcoming circumstances.

Tolman (1932b) assumed that organisms learn about their environment, the location of goals, and how to move from one place to another. According to Tolman, animals learn to expect (predict) that an event is followed by another event. In contrast to reinforcement theories popular at the time, Tolman suggested that no reward is needed for animals to learn this sequence of events, that is, associations are learned by simple temporal contiguity.

Tolman (1932) proposed that multiple expectancies can be integrated into larger units through a reasoning process called inference. Tolman hypothesized that a large number of expectancies can be combined into a cognitive map (see Box 2.2). Cognitive maps allow organisms to predict when and where temporally or spatially remote environmental events might be expected.

When seeking a reward in space, organisms compare the expectancies evoked by different alternative movements. For Tolman, *vicarious trial-and-error behavior,* that is, the active scanning of alternative pathways, reflects the animal's generation and comparison of different expectancies. Animals sample different stimuli before making a decision and approach the one that best predicts the desired goal.

Part III describes neural network theories applied to cognitive behavior. Animal cognition results from the interaction between a goal-seeking mechanism and a cognitive system. The goal-seeking mechanism allows the animal to pursue different goals with great adaptability and by adopting alternative behavioral strategies that are independent of any specific set of responses. Chapter 8 presented a neural network that learns specific responses. In contrast, a cognitive system generates expectancies of the consequences of alternative behavioral strategies. Examples of cognitive systems are temporal maps (Chapters 3 and 5), spatial maps (Chapter 10), and cognitive maps (Chapter 11). A temporal map is a cognitive system that defines the proximity in time of when environmental events take place, that is, the temporal expectancy of environmental events. A spatial map is a cognitive system that defines the spatial proximity to the location where environmental events occur, that is, the spatial expectancy of environmental events. Finally, a cognitive map is a cognitive system that permits the generation of new expectancies through the combination of multiple old temporal or spatial expectancies.

Cognitive networks differ from the classical conditioning and operant conditioning networks presented in Parts I and II in several aspects. In classical conditioning networks, the animal's response is fixed and mostly determined by the US. In operant conditioning networks, the response is selected from a set of alternative responses. In contrast, in cognitive networks, the goal can be reached by multiple alternative responses. In classical conditioning networks, behavior is changed as a result of the double contingency between the CS and the US. In operant conditioning networks, behavior changes as a result of a triple contingency between responses R, environmental stimuli S's, and the US. In contrast, in a cognitive network, behavior changes as a result of changes in the internal representation of the environment. In general, behavior described by cognitive nets is more flexible than behavior described by operant nets, and behavior described by operant nets more flexible than that described by classical networks.

Goal-directed paradigms

Cue learning

In a cue-learning task, animals have to approach a visible object in order to receive reward (Jarrard, 1983; Morris et al., 1982; Okaichi, 1987). During a cue task, usually extramaze cues are minimized either by illuminating the maze but not the room (Jarrard et al., 1984), or by surrounding the maze with a wall (Jarrard, 1983).

Appetitive place learning

In place-learning experiments, a rewarded location is defined by distal landmarks and animals are found able to learn the location of reward independently of the responses used to approach the goal. During a place task, animals are trained to approach a spatial location in a well-lighted room and a number of distinctive extramaze cues are present.

Morris (1981) trained rats to escape cool water by locating a hidden platform in a tank filled with opaque water. A diagram of Morris's water tank is shown in Figure 9.1. He found that if rats were trained to navigate to the (appetitive) invisible platform from a given starting point, they could swim directly to the goal from any other starting location in the pool perimeter during testing.

Brown (1942) studied the approach gradient to an appetitive place in an alley. He trained a group of hungry rats to obtain a food reward at the end of an alley. After the rats were trained, Brown measured the magnitude of the approach tendency by the force with which rats pulled against a restraining harness at different distances from the place where they were previously rewarded. The rats pulled harder as the distance to the rewarded place decreased.

Whishaw (1991) studied the effect of different training procedures on performance in Morris's tank task. He found that one very brief exposure to a platform location (goal exposure) is effective in improving performance. Performance is further improved following a single swimming trial (goal approach), and platform placement combined with a swim resulted in the best execution of the task.

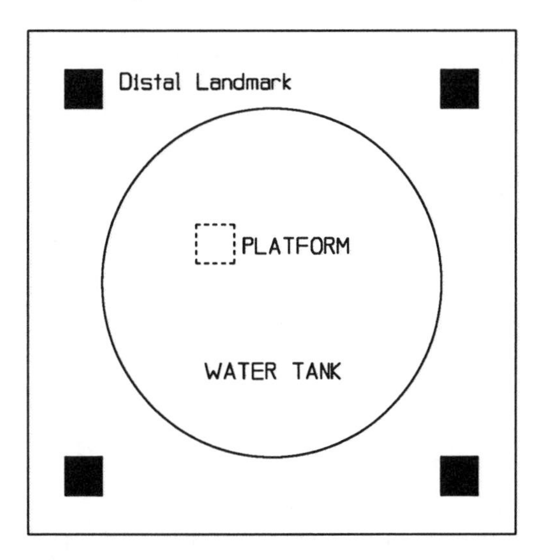

Figure 9.1. Place learning. Diagram of a Morris tank employed to study place learning.

Aversive place learning

Brown (1942) gave a group of rats two brief electric shocks at the end of an alley. After receiving the shocks, when animals were placed at that end of the alley without shock, they showed a tendency to avoid the place. He measured the magnitude of the avoidance tendency by the force with which rats pulled against a restraining harness at different distances from the place where they had received the shock. The rats pulled harder as the distance from the shocked place decreased. In addition to aversive stimuli, barriers, walls, and objects may be also regarded as mildly aversive objects, since animals avoid colliding with them.

Aversive-appetitive interactions in a runway

Miller, Brown, and Lipovsky (1943) studied the behavior of albino rats in situations in which different intensities of avoidance and approach tendencies were in conflict. Miller et al. trained hungry rats to run down an alley to obtain food, and afterwards they gave the rats a brief shock while they were eating. After training with different combinations of shock intensity and hunger level, the rats were placed at the start of the alley and their behavior was observed. Figure 9.2 shows that, under both strongly appetitive and aversive drives, the animals stop close to the goal (A). With both weakly appetitive and aversive drives, animals stop close to the goal (B). With a weakly appetitive and a strongly aversive drive, animals stop far from the goal (C).

Navigational dynamics in a runway

Wagner (1961) and Amsel, Rashotte, and MacKinnon (1966) recorded the running speed at different points in a runway. Both groups found that, under continuous reinforcement, animals accelerate in the start box, reach peak velocity in

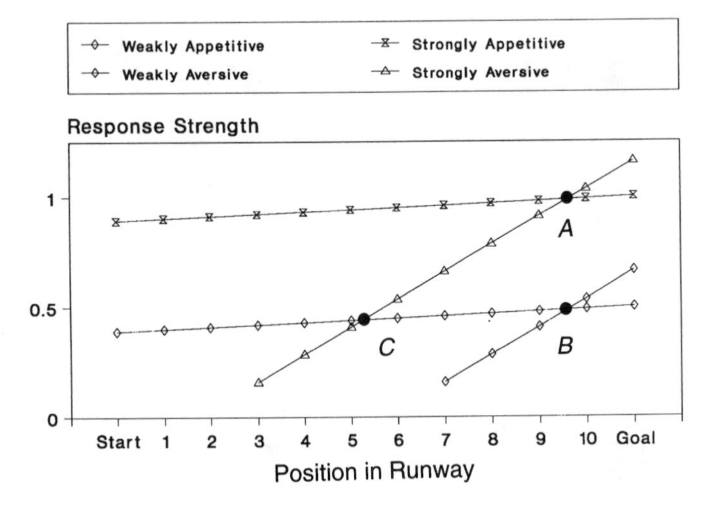

Figure 9.2. Appetitive-aversive interactions in a runway (after Miller, 1944). Tendencies to approach and avoid the goal box as a function of the distance to the goal box, when the goal box is made strongly or weakly appetitive and strongly or weakly aversive.

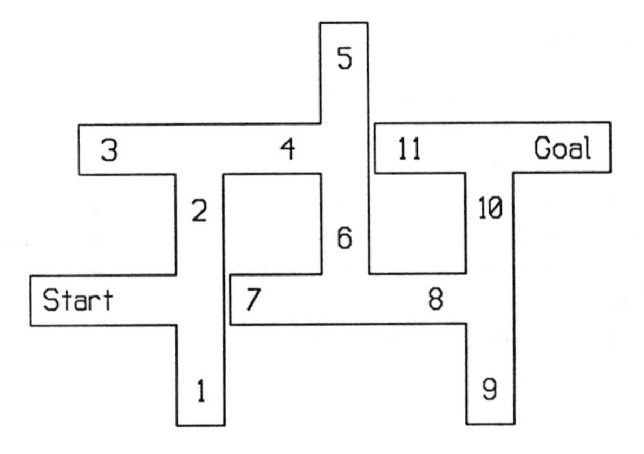

Figure 9.3. Latent learning. Diagram of a multiple-T maze employed to study latent learning.

the alley, and decelerate in the goal box. Velocities at all points in the runway increase over acquisition trials.

Latent learning and extinction

Blodgett (1929) studied how nonrewarded trials affect performance when reward is later introduced in a multiple T-maze. A diagram of Blodgett's maze is shown in Figure 9.3. Food-deprived rats received one trial a day in which they were placed in the start box and retrieved after reaching the goal box. Several groups of rats were trained. One group received food reward in the goal box starting in the first trial. Two other groups were rewarded after three or seven nonrewarded trials. Blodgett found that, after only one rewarded trial, the performance of the initially nonrewarded groups improved to nearly the level of the group rewarded in all trials. In agreement with Blodgett's (1929) results, Haney (1931) exposed rats to a maze without reward for an extended period of time (instead of trials) and observed that preexposed animals made fewer errors

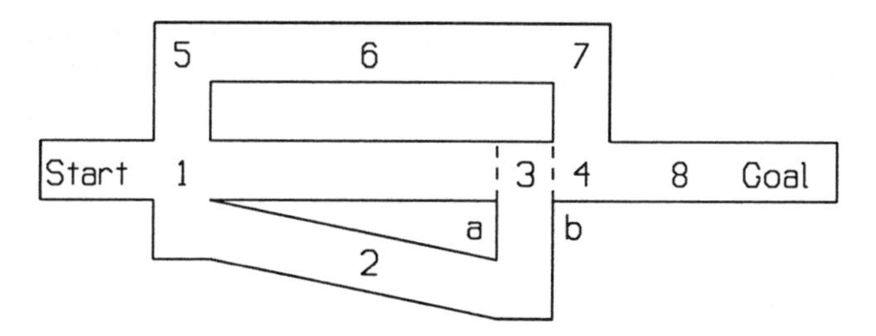

Figure 9.4. Detour problem. Diagram of the maze employed to study a detour problem. Path *A* is formed by places 1, 3, 4, 8, and goal, path *B* is formed by places 1, 2, 3, 4, 8, and goal; path *C* is formed by places 1, 5, 6, 7, 4, 8, and goal.

when reward was later introduced. Parallel to this latent learning phenomenon, Seward and Levy (1949) described latent extinction as the phenomenon observed when animals receive nonrewarded exposures to the goal after being trained in a rewarded maze.

Detour learning

In order to study detour performance, Maier (1929) used a complex experimental room, which included tables, ring stands, and elevated paths, to show that rats swiftly take an alternative known path if the usual path to the goal is blocked with a wire mesh. Tolman and Honzik (1930) used a three-path maze with three pathways (*A, B,* and *C*) to analyze how rats solve a detour problem. A diagram of the maze is shown in Figure 9.4. Path *A* is the shortest, path *B* the next shortest, and path *C* the longest. Path *A* is formed by places 1, 3, 4, 8, and 9; path *B* is formed by places 1, 2, 3, 4, 8, and 9; path *C* is formed by places 1, 5, 6, 7, 4, 8, and 9. Place 9 is the goal. Paths *A* and *B* share a common place, place 3, which path *C* does not have. During preliminary training, rats are forced to take alternatively paths *A, B,* and *C*, by introducing blocks in different segments of the maze. During testing, a block is placed in the common section of paths *A* and *B* (block *b* between places 3 and 4 in Figure 9.4). In Tolman and Honzik's experiment, most rats, after backing out of path *A*, took the longest path, path *C*, instead of the second shortest path, path *B*.

Using a different apparatus than that employed by Tolman and Honzik (1930), Thompson, Harmon, and Yu (1984) studied the effects of different lesions in detour problem-solving behavior. The apparatus consisted of a start box and a choice chamber separated from the goal box by a partition. During preliminary training, animals reach the goal through a door in the partition. After several training trials, the door in the partition is blocked and the goal can be reached only through a new pathway. Thompson et al. (1984) reported that animals showed fast detour behavior.

Mechanistic theories of goal-directed behavior

For many years, Tolman's ideas were considered "mentalistic" concepts that could not be mechanistically implemented. Gallistel (1980) proposed that Tolman's view of purposive behavior can be derived from a "biological machine" with two distinct subsystems: a cognitive system that forms representations of the environment and an action system that employs the representations to control the navigation through the external world. In the same vein, Schmajuk and Thieme (1992) showed that (1) simple feedback systems can implement Tolman's concept of goal-seeking behavior, and (2) neural networks can mechanistically embody Tolman's concept of expectancy and cognitive mapping.

Goal-directed mechanisms

According to Tolman's stimulus-approach view, subjects learn that a particular stimulus situation is appetitive and therefore it is approached. Because Tolman never specified how action was related to cognition, Guthrie (1935) criticized Tolman's approach by suggesting that Tolman's rat was left "buried in thought."

Tolman's concept of goal-seeking behavior can be accurately interpreted in cybernetical terms as that governed by the comparison between an internal representation of the goal and the present situation of the animal (Toates, 1986). Walter (1951) built a cybernetic "animal" that displayed goal-seeking behavior by combining two simple actions: exploration and approach. This robotic animal explored the environment until a photocell detected a light and approached the light when the photocell was aimed at it. Walter's cybernetic animal demonstrated that the stimulus-approach view can be easily implemented in cybernetic terms: Exploratory behavior receives a negative feedback, and approach behavior a positive feedback, from the stimulus. When the goal is reached, the animal stops. Miller, Galanter, and Pribram (1960) used a similar feedback system to describe plans. A TOTE (test-operate-test-exit) system continually monitors the difference between its current and desired states and performs operations to reduce that difference. According to Miller et al. (1960), the behavior of an organism reflect the plans implemented by an internal TOTE system. Interestingly, McFarland and Bosser (1993) proposed that tradeoff theories, which select the alternative that maximizes utility, should be preferred to stimulus-approach or goal-seeking theories.

In sum, Tolman's concept of a purposive, goal-directed behavior can be implemented by negative feedback systems. Chapters 10 and 11 introduce neural networks that incorporate goal-directed mechanisms.

Expectancies in neural networks

According to Tolman, animals become operantly conditioned by learning an expectancy that the performance of response R1 in a situation S1 will be followed by S2. Similarly, Tolman proposed that animals become classically conditioned by forming an expectancy that a CS will be followed by the US. In line with Tolman's cognitive view, Part I introduced several neural networks that use CS–US associations to generate the prediction of the US by the CS.

In many ways, Tolman's idea that animals learn expectations about the external world is similar to Sokolov's (1960) idea of a neural model of the environment presented in Parts I and II. Chapter 6 described a model of stimulus configuration in which two CSs can be combined to generate the expectation of the US. Such a model is also capable of combining a S1 and the sensory representation of a response, R1, to originate the expectation of S2 (S1–R1–S2 expectancy). Chapters 3 and 5 described a model of classical conditioning that incorporates CS–CS associations. As in the case of CS–US associations,

CS1–CS2 associations are learned by temporal contiguity and control the expectation (prediction) elicited by CS1 that CS2 is going to occur in the immediate future.

Therefore, Tolman's concepts of CS–CS, CS–US, and S–R–S expectancies can be mechanistically implemented by neural networks that describe classical conditioning. Chapter 10 introduces a neural network that incorporates a goal-directed mechanism and an internal representation of the environment that stores classical conditioning associations in order to navigate through space.

Cognitive mapping

Tolman hypothesized that a large number of expectancies can be combined into a cognitive map (see Box 2.2). Chapters 3 and 5 presented a network of classical conditioning that integrates short-term predictions, generated by reading out CS–CS associations, into a cognitive map. Chapter 3 showed how, in sensory preconditioning, CS1–CS2 associations can be combined with CS2–US associations to infer the presence of the US based on the presentation of CS1.

To the extent that the layout of a maze can be accurately portrayed in terms of the connections between its contiguous places, maze connectivity can be described in terms of the associations between the representations of adjacent places. However, because associations between spatially adjacent places are insufficient to predict remote places, navigation to remote goals requires the building and reading of a cognitive map that combines information about spatially adjacent places to infer by reasoning the connections to remote places.

Therefore, Tolman's concept of cognitive maps can be mechanistically implemented by networks able to combine different associations between temporally contiguous stimuli or spatially contiguous places. Chapter 11 presents a network that makes use of a goal-directed mechanism and a cognitive map to navigate through mazes.

10 Place learning and spatial navigation

As mentioned in the previous chapter, Tolman's (1932b) concept of purposive, goal-seeking behavior can be implemented by feedback systems, and his notion of expectation can be implemented by a model of the environment that stores classical conditioning associations.

Tolman (1932b) proposed that place learning illustrates animals' cognitive abilities to the extent that animals are able to learn the location of reward independently of the responses used to approach a spatial location. This chapter explains how a neural network is capable of guiding the animal's navigation, independently of the responses used, to approach the location of a specific goal in the environment. The network describes aversive and aversive-appetitive place learning as well as the physical dynamics (acceleration, velocity, and position as a function of time) of spatial navigation.

Place learning: response-selection versus stimulus-approach theories

Two different views have been used to explain place learning: a "response-selection" and a "stimulus-approach" view. According to the response-selection view, introduced in Chapter 9, subjects learn because a particular response is reinforced in the presence of a particular stimulus situation. Different responses are learned at different places in the environment, and animals can navigate toward a goal by compounding different responses associated with different landmarks. According to the stimulus-approach view, subjects learn that some particular stimulus situation is appetitive and therefore it is approached. Mackintosh (1974, p. 554) proposed that, in the presence of numerous intra- and extramaze cues, animals typically learn to approach a set of stimuli associated with reward and to avoid a set of stimuli associated with punishment. Only in a totally homogeneous environment, that is, one that lacks discernible cues, will animals learn to make the correct responses that lead to the goal.

Several theories have been proposed to account for spatial learning in terms of response-selection mechanisms: Tolman (1932a, b), Milner (1960), Hull (1952), and Barto and Sutton (1981). Tolman (1932a, b) advocated the idea that, in place-learning experiments, animals can learn the location of the reward independently of the responses used to approach the goal. Surprisingly however, Tolman includes the response performed by the animal when describing S_1–R_1–S_2 expectations, which represent the belief that the performance of response R_1 in a situation S_1 will be followed by a change to another situation S_2. Milner (1960) proposed a system capable of building a Tolmanian spatial map and of using it to control the animal's movements in a spatial environment. The model has elements that are active only when a particular response R_i has been made in a particular location S_j. These elements can be associated with the location S_k that results from making response R_i at location S_j. When the organism is next placed in location S_j, random responses are generated. When response R_i appears, element S_j–R_i is active and, in turn, activates S_k. If S_k is associated with an appetitive stimulus, it activates a mechanism that holds R_i in the response generator, and S_k can be reached.

Hull (1938, 1952) also described spatial behavior in animals. According to Hull, animals may withdraw from and approach different regions in space in order to reduce their drives. Hull suggested that an object whose image has been conditioned to a response at one distance will tend to evoke the response at any distance and direction from which the image may be perceived. He proposed that this response generalization curve, which he called a "reaction potential gradient," decreases as a function of distance and direction. These gradients create a field of behavior potentiality that extends over the two-dimensional space. Hull also derived equations describing how the interaction between two appetitive action potential gradients, as well as one appetitive and one aversive potential gradient, determines the spatial trajectory undertaken by an animal.

Elaborating on Hull's ideas, Miller (1944) proposed four fundamental principles to describe approach and avoidance interactions: (1) The tendency to approach a goal increases with decreasing distances to the goal (approach gradient), (2) the tendency to avoid a place increases with decreasing distances to the place (avoidance gradient), (3) the avoidance gradient is steeper than the approach gradient, and (4) the strength of approach and avoidance tendencies varies with the strength of the drive on which they are based.

Barto and Sutton (1981) described a system capable of controlling the locomotor behavior of an organism in a spatial environment. Five spatial landmarks (north, south, east, west, and a tree) are the inputs to the system, and responses in four possible directions (north, south, east, west) constitute its output. Associations between landmarks and a particular response increase when the response brings the organism closer to the appetitive goal. Barto and Sutton assumed that the organism determined its closeness to the appetitive goal using an odor gradient emitted by the reward. After learning, the organism can determine the correct combination of responses that will bring it to the goal from each point in the environment. Barto and Sutton's (1981) model is an example of the response-selection view. As mentioned before, whereas in the presence of numerous intra- and extramaze cues animals typically learn to approach a set of stimuli associated with reward and to avoid a set of stimuli associated with punishment (stimulus-approach), only in a totally homogeneous environment without intra- and extramaze cues do animals learn to make the correct responses that lead to the goal (response-selection). In contrast to response-selection theories, Daly and Daly (1982), Zipser (1986), Wilkie and Palfrey (1987), Arbib and House (1987), Schmajuk (1990), and Schmajuk and Blair (1993) have proposed theories that describe goal-seeking behavior in terms of stimulus-approach principles.

Daly and Daly (1982) incorporated the major assumptions of Amsel's (1958, 1992) frustration theory into the Rescorla and Wagner (1972) model. As in the Rescorla–Wagner model, when reward is larger than expected, the approach strength increases. Following Amsel (1958), when reward is smaller than expected, the avoidance strength increases, and rewarding goal events increase "courage" (counterconditioning.) Behavior is determined by the sum of the approach, avoidance, and counterconditioning strength values. Daly and Daly (1982) demonstrated that the resulting model yields correct descriptions of numerous appetitive instrumental learning situations.

Zipser (1985, 1986) proposed a neural model capable of recognizing a given spatial location. The model comprises two layers of neural elements. Each unit in the first layer is specific for a given spatial landmark and fires maximally when the current visual angle of the landmark equals a previously stored visual angle. The second layer in Zipser's (1985, 1986) model generates an output proportional to the sum of the outputs of all units in the first layer. Since Zipser's model is "not concerned with how information is represented or transmitted to the processors in the first layer" (1986, p. 439), Schmajuk (1990) suggested that tuned neural detectors constitute a possible mechanism for the first layer in his model. In addition, Schmajuk (1990) proposed that the second layer in Zipser's model might be regarded as equivalent to a delta rule used to describe the associations between the tuned detectors and the reinforcing events.

Wilkie and Palfrey (1987) proposed a perceptual memory-matching model that describes place learning in the Morris water maze. The model assumes that, when reaching the platform, the rat learns the distance to extramaze landmarks and stores this perceptual information in memory. When placed in the maze on subsequent trials, the simulated rat attempts to match the learned with the perceived distances to the landmarks.

Arbib and House (1987, p. 153) suggested that spatial trajectories that characterize detour behavior could be described in terms of the interaction between vector fields that represent the preferred direction of motion at different spatial locations. One appetitive field represents motion toward the location of the goal, a second aversive field represents motion away from the position of a fence, and a third field represents motion from the location of the animal. When the goal appetitive field is combined with the animal's position field, the resulting field shows various paths starting at the position of the animal and ending at the location of the goal. When the three fields are combined, the net field shows a set of paths that are diverted around the fence ends.

The neural network

Schmajuk (1990) and Schmajuk and Blair (1993) presented a real-time neural network (see Box 1.4) capable of describing place learning and the dynamics of spatial navigation. Figure 10.1 represents a block diagram of the model showing the interaction between the goal-seeking system, a cognitive system (spatial map), and the environment. The neural network is capable of guiding the animal's search for a specific goal in the environment with the assistance of the spatial map.

The goal-seeking system in the model shown in Figure 10.1 implements a stimulus-approach view of behavioral control by which animals approach appetitive stimuli and avoid aversive stimuli. Appetitive and aversive stimuli are defined by the animal's motivational state. A given motivation elicits a search behavior that receives negative feedback from the appropriate goal. When the goal is encountered, it activates approach responses and inhibits search responses.

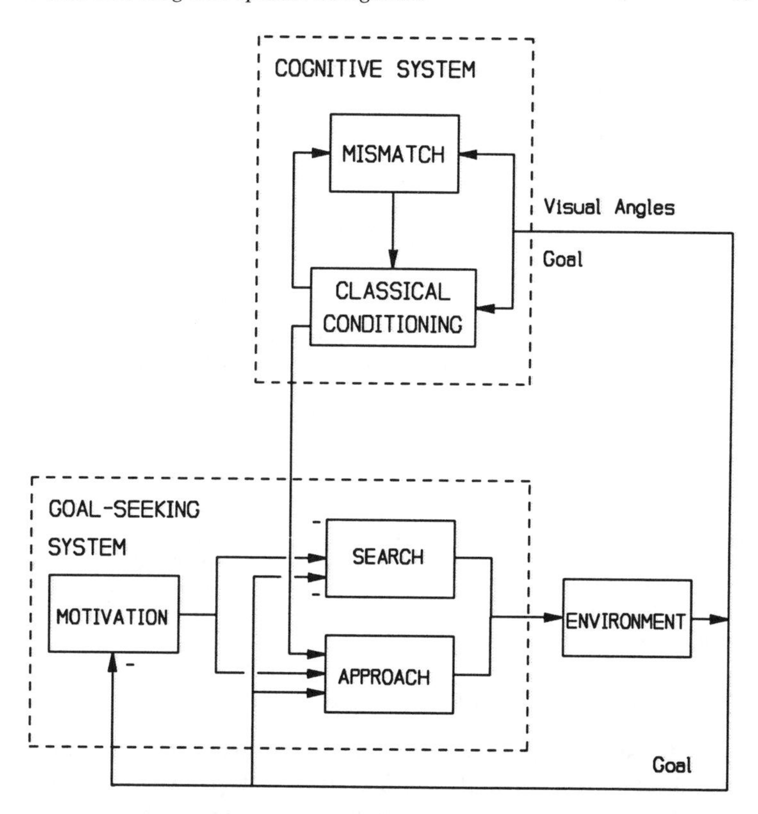

Figure 10.1. A block diagram of system capable of spatial navigation. Block diagram of the model showing the interaction between the goal-seeking system, spatial map, and the environment. The goal-seeking system approaches appetitive stimuli defined by the motivational state. A given motivation elicits a search behavior that receives negative feedback from the appropriate goal. When the goal is encountered, it activates approach responses and inhibits search responses. As the animal explores the environment under the control of the goal-seeking system, it builds a spatial map. Once the spatial map is available, the search behavior is guided by a hill-climbing technique. The spatial map is updated whenever a mismatch between predicted and actual inputs is detected.

The spatial map in the model shown in Figure 10.2 incorporates detectors that can be tuned to the values of visual angles of different landmarks as perceived from the spatial location where a reinforcing event is encountered. After a detector has been tuned, its output generates an effective stimulus that peaks at the distance from the landmark where positive or negative reinforcement was encountered before. The outputs of the tuned detectors become associated with the reinforcing event. The network generates spatial generalization surfaces that can guide navigation from any location that is within view of familiar landmark cues, even if that location has never been visited before, and independently of the responses used to approach the goal. Spatial navigation is accomplished by adopting a stimulus-approach principle, that is, by approaching appetitive places and avoiding aversive places. When generalization surfaces are assumed to represent forces exerted by the animal, the dynamics of spatial movements can be described. Schmajuk and Blair (1993) showed that the network correctly describes the navigational trajectories and dynamics of many spatial learning tasks.

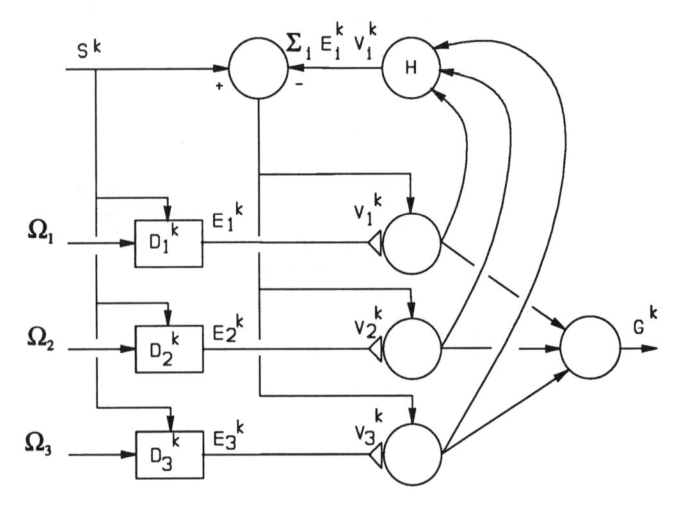

Figure 10.2. Diagram of the adaptive neural network. Effective stimuli E_i^k generated by different tuned detectors D_i^k are connected to a classical conditioning network. Visual angles Ω_i are the inputs to detectors D_i^k. US_k is the tuning signal for detectors D_i^k. V_i^k are modifiable synaptic weights. G^k is the output of the network. H computes the aggregate prediction of S^k. Arrows represent nonmodifiable connections, triangles variable synapses.

Spatial maps and place learning

Figure 10.2 shows a neural network that accomplishes spatial mapping through spatial discrimination. Briefly, the first stage of the network consists of a layer of detectors that receive a combination of visual angles, the second stage consists of a layer of neural elements that associate the output of the detectors with another event (e.g., an unconditioned stimulus or US), and the third stage adds the output of the neural elements of the second stage and generates the output of the system. An additional neural element (H) computes the aggregate prediction of S^k used to control the associations between the output of the detectors and S^k.

In the first stage of the network shown in Figure 10.2, a detector tuned by S_k to a particular value of the visual angle Ω_i is denoted by D_i^k. In the spatial domain, a detector can be tuned to the value of the visual angle of spatial landmark L_i, Ω_i, at the spatial location where S_k is encountered. When landmark L_i is encountered after the detector has been tuned, the output of the detector, effective stimulus E_i^k, peaks at the distance from L_i where S_k was encountered before.

In the second stage of the network depicted in Figure 10.2, the outputs of the tuned detectors E_i^k become associated with S_k. The association between E_i^k and S_k is designated V_i^k. According to Figure 10.2, each effective stimulus increases its association V_i^k until $\Sigma_i V_i^k(t)\, E_i^k$ equals the intensity of stimulus k, λ^k. Each E_i^k activates the final common path in proportion to V_i^k, thereby generating an output O^k also proportional to $\Sigma_i V_i^k(t)\, E_i^k$. Notice that because the effective stimuli E_i^k, but not the visual angle Ω_i, is associated with the S_k, the output of the network O^k peaks at the place where S_k was encountered before. This feature allows the network to generate appropriate spatial topographies. The second stage of the network shown in Figure 10.2 is identical to the real-time delta rule network illustrated in Figure 3.2.

Box 10.1 summarizes the main properties of the model. Appendix 10.A formally describes how detectors are tuned and generate effective stimuli, how

BOX 10.1 Schmajuk and Blair's (1993) place-learning model: goal-seeking and spatial map systems

1. The system is composed of (1) a goal-seeking system and (2) a spatial map.
2. The goal-seeking system implements a stimulus-approach view of behavioral control by which animals approach appetitive stimuli and avoid aversive stimuli.
3. The spatial map incorporates detectors D_i^k that are tuned by S_k to the particular value of the visual angle Ω_i of landmark L_i at the spatial location where S_k is encountered.
4. When landmark L_i is encountered after the detector has been tuned, the output of the detector, effective stimulus E_i^k, peaks at the distance from L_i where S_k was encountered before.
5. Outputs of the tuned detectors E_i^k become associated with S_k. The association between E_i^k and S_k is designated V_i^k. Associations V_i^k are controlled by a real-time delta rule (see Box 3.1).
6. The output of the network O^k peaks at the place where S_k was encountered before. This feature allows the network to generate appropriate spatial generalization surfaces that can guide navigation from any location that is within view of familiar landmark cues, even if that location has never been visited before, and independently of the responses used to approach the goal.

effective stimuli become associated with other stimuli and generate the output of the network, and how the efference copy is used to control associative learning.

Selection of the most accurate effective stimulus

Figure 10.3 illustrates how the network presented in Figure 10.1 learns to discriminate points in the spatial dimension. The lower panel in Figure 10.3A shows the visual angle Ω_1 generated by landmark L_1 (a tree) at different spatial locations. When the US is presented at the point indicated by the asterisk, detector $D_1{}^{US}$ stores the value of Ω_1. The middle panel in Figure 10.3A shows that L_1, when encountered in a following trial, generates an effective stimulus $E_1{}^{US}$ that peaks at the place of the US presentation and generalizes over previous and subsequent spatial locations. The upper panel in Figure 10.3A shows the output G^{US} generated by Equation 10.5 in Appendix 10.A. Activity at locations different than that of the US presentation represents spatial generalization.

Figure 10.3B illustrates how the network selects and enhances the effective stimulus of the landmark that is the most accurate predictor of the US. As in Figure 10.3A, the lower panel in Figure 10.3B shows visual angle Ω_1, generated by L_1, and visual angle Ω_2, generated by L_2 (another tree). The middle panel in Figure 10.3B shows effective stimuli E_1^{US} and E_2^{US} when a US is presented at the point indicated by the asterisk. The upper panel in Figure 10.3B depicts the individual outputs (continuous lines) and compound outputs (dotted line) G^{US}. Due to the competitive mechanism of the delta rule described by Equation 10.4 (Appendix 10.A), although the peak values of both effective stimuli E_1^{US} and E_2^{US} are identical, the peak values of the individual outputs generated by the effective stimulus with the smallest generalization (effective stimulus E_2^{US}) are increased relative to that of the stimulus with the greatest generalization (effective stimulus E_1^{US}).

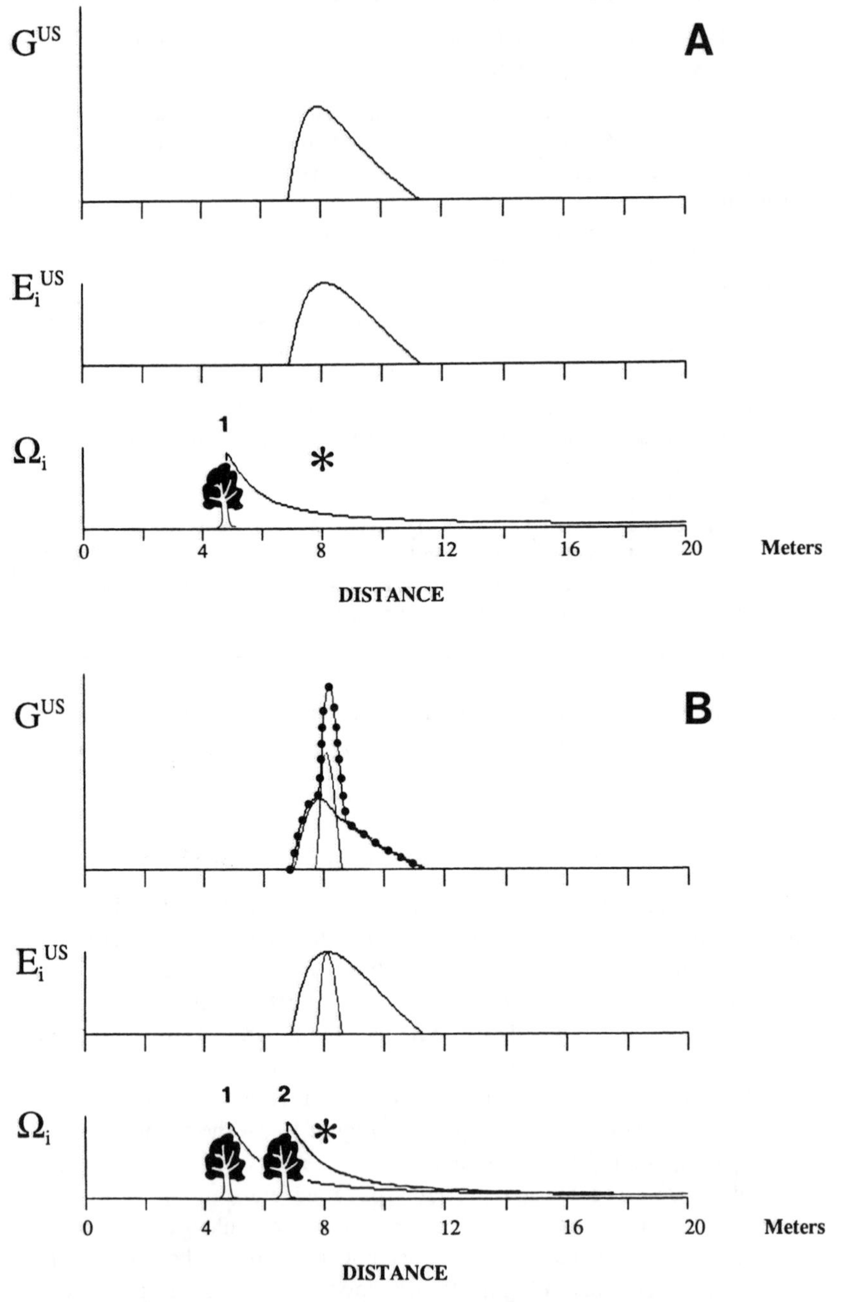

Figure 10.3. Spatial generalization curves. (A) Spatial generalization curve generated by one landmark. Lower panel: visual angle Ω_i as a function of distance to the origin. Middle panel: spatial generalization curve of the effective stimulus E_i^{US}. Upper panel: generalization curve on a test trial in which landmark L_i is presented in the absence of the US, after four training trials in which L_i is paired with the US. The asterisk indicates the place where the US is encountered during training. (B) Spatial generalization curve generated by two landmarks. Lower panel: visual angles Ω_1 and Ω_2 as a function of distance from the origin. Middle panel: spatial generalization curves of effective stimuli E_1^{US} and E_2^{US}. Upper panel: CRs on a test trial in which L_1 and L_2 are presented together in the absence of the US, after four training trials in which both L's are paired with the US. The asterisk indicates the place where the US is encountered during training.

Multiple effective stimuli

According to Equation 10.3 (Appendix 10.A), if a US is encountered at two different locations, a detector D_i^k would become tuned to the average of the values of the visual angles subtended by landmark L_i at the places where the US is encountered. Therefore, the detector would generate only one peak at a place located between the two rewarded places. At odds with this prediction, Olton (1978) found that animals can learn multiple rewarded places in space instead of an average rewarded place. Schmajuk (1990) suggested how Equation 10.3 (Appendix 10.A) can be modified in order to generate multiple effective spatial stimuli that allow the model to correctly describe multiple rewarded places. For simplicity, however, the present chapter only addresses single rewarded places.

Goal-seeking system

Navigational trajectories

We adopt a stimulus-approach view of behavioral control, by which animals approach appetitive and avoid aversive stimuli. Different appetitive stimuli will be approached according to the animal's motivational state (hunger, thirst, etc.). Therefore, at each point in space the animal should move from less to more appetitive regions of the motivationally selected generalization surface, G^{USnet}. In order to accomplish this result, we assume that the animal is able to produce small movements around a given point, staying in the next point if it is more appetitive than the previous one, and going back to the previous one in the other case. This behavior is similar to Tolman's (1932a, b) *vicarious trial-and-error behavior*. According to Tolman, at choice points in a maze, animals sample different stimuli before making a decision. This active scanning of alternative next places reflects the animal's generation and comparison of the appetitiveness of different places. Appendix 10.B formally defines how the model describes navigational trajectories.

Navigational dynamics

As explained in the previous section, animals move on either the appetitive or aversive generalization surface, depending on the value of the net surface $G^{\text{US net}}$. In addition to determining on which surface the animal moves, the net generalization surface also determines the dynamics of the movement. In order to derive the dynamics of spatial navigation, the behavioral meaning of the generalization curve $G^{\text{US}}(x, y)$ needs to be specified.

According to Brown (1942), the magnitude of the approach tendency to a goal can be measured by the force with which rats pull against a restraining harness at different distances from the rewarded place. Therefore, a possible behavioral interpretation of a generalization surface $G^{\text{US}}(x, y)$ is that it represents the force that drives the animal's movements toward or away from point (x, y). This view is coherent with the stimulus-approach position adopted in the previous section and assumes that appetitive stimuli "pull" the animal toward them, and aversive stimuli "push" the animal away from them. Appendix 10.C formally describes how the model describes navigational dynamics.

BOX 10.2 Schmajuk and Blair's (1993) place-learning model: navigational trajectories and dynamics

Navigational trajectories and dynamics are defined by the following performance rules:

1. When the net generalization curve, G^{USnet}, is appetitive, animals will move toward the most appetitive adjacent location (Equation 10.7).
2. When the net generalization curve is aversive, animals will move toward the least aversive adjacent location (Equation 10.8).
3. At locations removed from the goal, the velocity of the approach to the most appetitive adjacent place, or the least aversive adjacent place, is a function of the value of the net generalization surface at the adjacent place and a retarding factor (Equation 10.10).
4. Appetitive goal locations accrue decelerative associations that prevent collisions with the rewarding objects (Equations 10.15 and 10.16).

Box 10.2 summarizes the performance rules that define the navigational trajectories and dynamics.

Computer simulations

This section presents computer simulations obtained with the tuned network for different spatial learning paradigms: cue learning, appetitive place learning, effects of goal exposure in place learning, aversive place learning, combined appetitive-aversive place learning, and appetitive-aversive interactions.

In every case, similar results were obtained by assuming that the simulated animal (1) is free to navigate through the training environment and to find appetitive and aversive locations, (2) is forcibly exposed to every spatial position including appetitive and aversive locations, and (3) is forcibly exposed only to the appetitive and aversive locations.

Parameter values are presented in Appendix 10.D.

Cue learning

Figure 10.3A explains how the model describes cue learning. During training trials, the simulated animal perceives the tree placed inside the field (intramaze landmark or cue) but not extramaze landmarks. The simulated animal is exposed at different points and rewarded only at a point in the maze (indicated with an asterisk) close to the cue. After training, the simulated animal builds a generalization surface based on the cue and finds a maximum at the location of the reward. It should be noted that when represented in a two-dimensional space, the generalization surface shows an annular maximum around the cue rather than the peak shown in place learning (see Figure 10.4).

Appetitive place learning

Figure 10.4 shows simulated testing of place learning after 1, 3, and 10 trials. The simulated environment consisted of a rectangular matrix (20×20) representing a

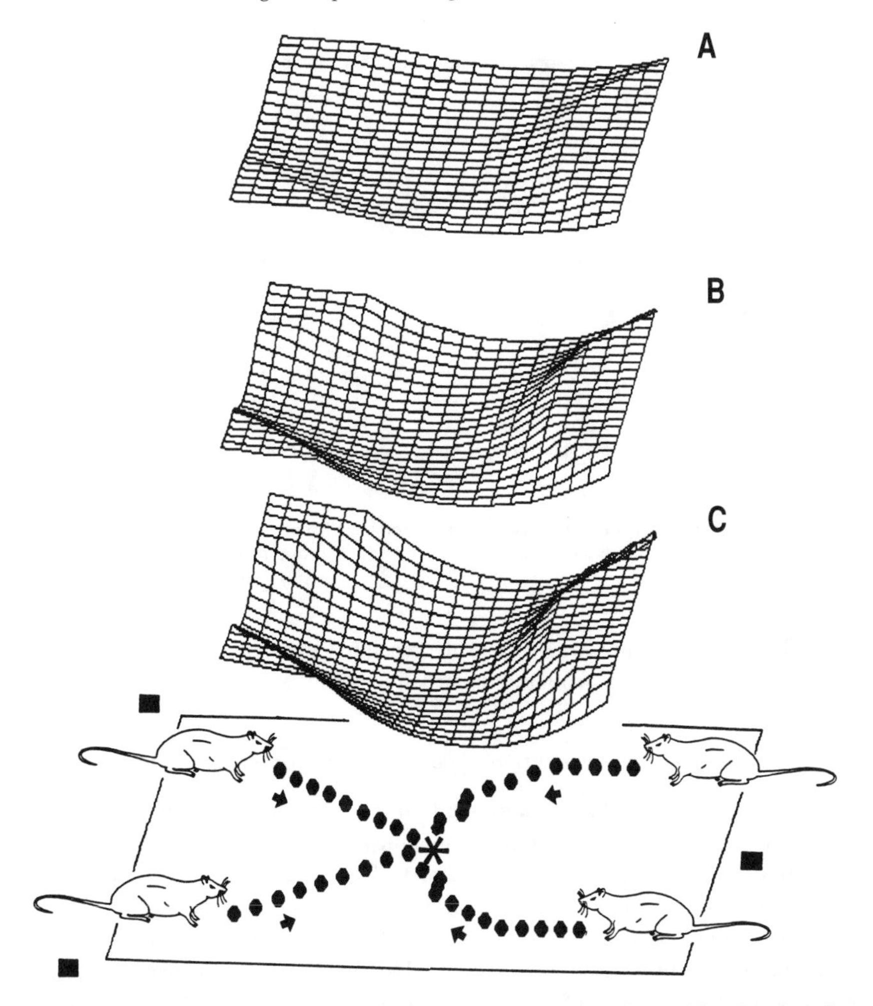

Figure 10.4. Appetitive generalization surfaces during acquisition of appetitive place learning. Appetitive generalization surfaces $G^{US}(x, y)$ for an appetitive US encountered at the point indicated with the asterisk, as a function of trials (a: trial 1, b: trial 3, c: trial 10). Squares indicate the location of the spatial landmarks, circles different routes to the goal from different starting places, arrows the direction of movement.

pool. In an environment like Morris's water tank, the animal is motivated to escape the water and, therefore, the platform becomes an appetitive US. Therefore, during simulated acquisition, the animal is exposed to 36 equally spaced points in the pool and rewarded only at the goal. It is assumed that during training the animal perceives the three landmarks (distal cues) from every point in the field. The subject builds a generalization surface by sampling the three landmarks and finds its maximum at the place of the goal. In agreement with Morris's (1981) data, the animal can find the goal from any start point.

Figure 10.4 shows how the appetitive generalization surface is shaped in successive trials. When the animal is first placed in the tank, the spatial map is flat. Figure 10.4A–C indicates that over trials, the generalization surface becomes deeper and more defined. Because velocity is proportional to the depth of the generalization surface, the animal finds the platform faster over trials. In agreement with Brown (1942), the appetitive force shows a maximum at the re-

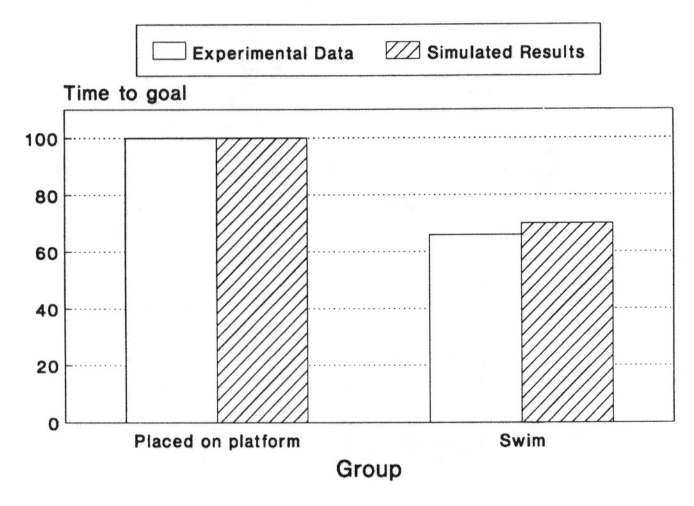

Figure 10.5. Morris water maze. Experimental and simulated latency to reach the submerged platform in a Morris water maze after (1) placing the rat on the platform and (2) a single swim to the platform.

warded place (minimum height) and decreasing values (increasing elevation) with increasing distances from the rewarded location.

Effects of goal exposure and goal approach on place learning

Figure 10.5 shows simulated latency to find the goal (hidden platform) for different training procedures. It seems reasonable to assume that the appetitiveness of the platform is related to the aversiveness of the water in the pool. Therefore, the intensity of the appetitive US when the animal swims to the platform from the pool is greater than that of the appetitive US when the animal is simply exposed to the platform. Consequently, simulations of "goal exposure" applied a weak US at the goal ($\lambda^k = 0.5$ in Equation 10.4 in Appendix 10.A), whereas simulations of "goal approach" made use of a strong US at the goal ($\lambda^k = 1$ in Equation 10.4 in Appendix 10.A). In agreement with Whishaw (1991), Figure 10.5 shows that (1) animals are able to find the goal after a single exposure to it, and (2) the latency to find the goal (hidden platform) after goal exposure is longer than after swimming to the goal.

Aversive place learning

Figure 10.6 shows a simulation of aversive place learning. The simulated environment consisted of a rectangular matrix (20 × 20), representing the open field. Simulated acquisition consisted of 10 trials, in which the animal is exposed to 36 equally spaced points on the open field and punished only at one location. During training the animal perceives three landmarks in the environment. The subject builds an aversive generalization surface. In agreement with Brown (1942), the surface representing the aversive force has its maximum at the place of the aversive location and decreasing elevation with increasing distances from the aversive location.

Figure 10.6 shows that when placed at the aversive place, the animal moves away from it. As the shape of the aversive generalization surface changes over

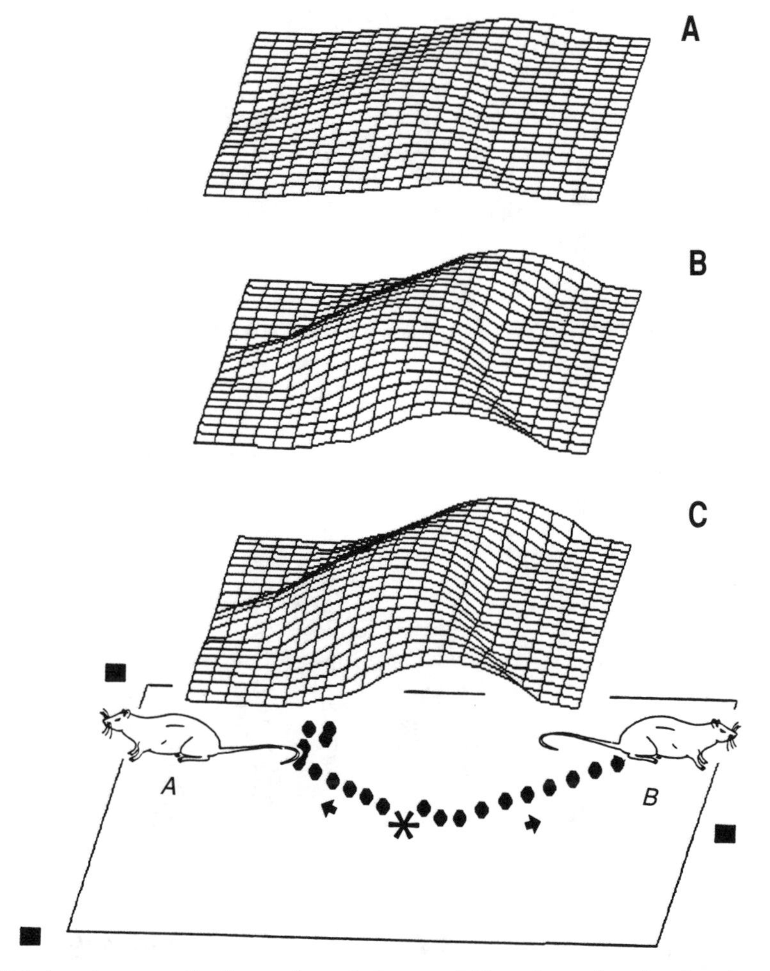

Figure 10.6. Aversive generalization surfaces during acquisition of aversive place learning. Aversive generalization surfaces $G^{aversive}(x, y)$ for an aversive US encountered at the point indicated with the asterisk, as a function of trials (a: trial 1, b: trial 3, c: trial 10). Squares indicate the location of the spatial landmarks, circles different routes to the goal from different starting places, arrows the direction of movement.

trials 1, 3, and 10, the simulated animal takes route *A* during trial 1 and route *B* afterwards. In agreement with Brown (1942), the aversive force shows a maximum at the punished location (maximum height) and decreasing values (decreasing elevation) with increasing distances from the punished location.

Combined appetitive-aversive place learning

Figure 10.7A shows an appetitive place-learning simulation. The simulated environment consisted of a rectangular matrix (20 × 20) representing a pool. Simulated acquisition consisted of 10 trials, during which the animal is rewarded at one location. During training the animal perceives three landmarks in the environment. The animal builds a generalization surface for the appetitive location. Using this appetitive generalization surface, the subject finds the goal from any start point.

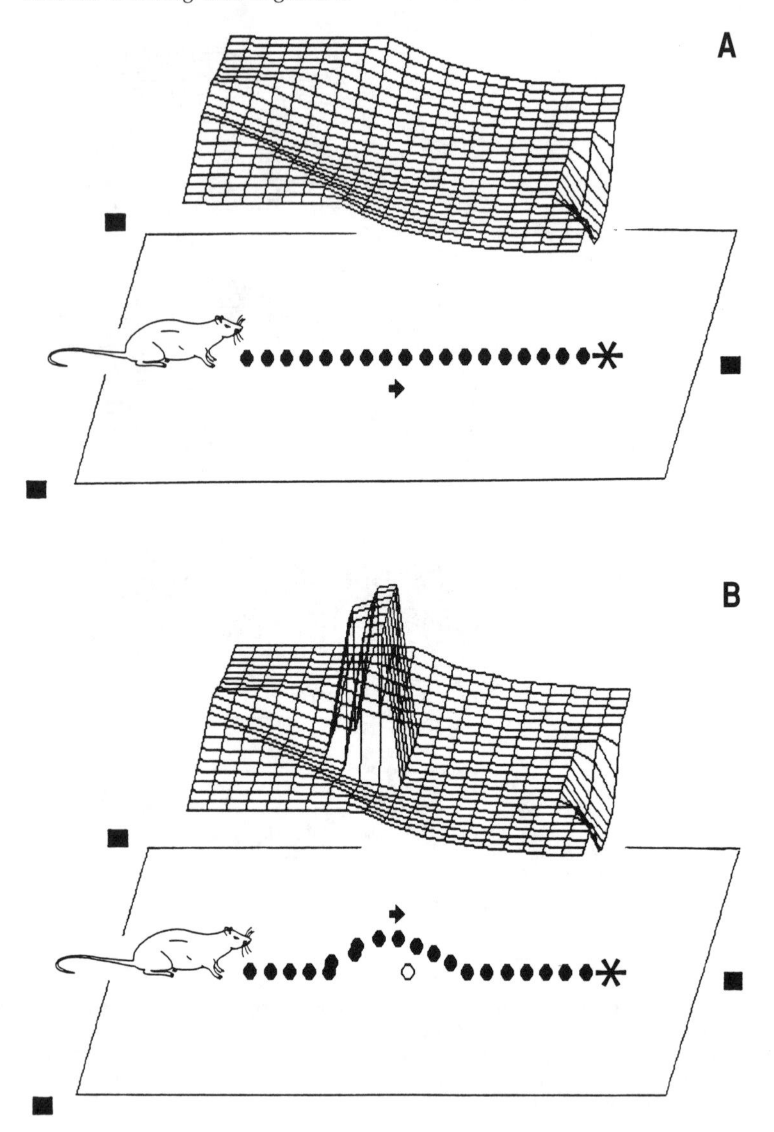

Figure 10.7. Appetitive generalization surface for combined appetitive and aversive place learning. (A) Appetitive generalization surface $G^{\text{appetitive}}(x, y)$ for an appetitive US encountered at the point indicated with the asterisk. (B) Composite generalization surface $G^{\text{aversive}}(x, y)$ for an aversive US encountered at the point indicated with the asterisk and an aversive US encountered at the point indicated with the open circle. Squares indicate the location of the spatial landmarks, solid circles different routes to the goal from different starting places, arrows the direction of movement.

Figure 10.7B shows the simulated testing of a goal-directed behavior in which an invisible aversive object has been incorporated. Simulated acquisition consisted of 10 trials, during which the animal is rewarded at the same location of Figure 10.7A and punished at another location. During training the animal perceives three landmarks in the environment. The animal builds two generalization surfaces, one for the appetitive and another for the aversive location. Figure 10.7B shows the appetitive surface where the net generalization surface is appetitive, that is, $G^{\text{US appetitive}}(x, y) > G^{\text{US aversive}}(x, y)$, and the aversive generalization surface where the net generalization surface is aversive, that is, $G^{\text{US appetitive}}(x, y) < G^{\text{US aversive}}(x, y)$.

Using these generalization surfaces, the subject finds the goal from any start point, avoiding the aversive place. It should be noted that, because velocity is defined by the net generalization surface according to Equation 10.10 in Appendix 10.A, the animal slows down as it approaches and detours around the aversive location.

Navigational dynamics in a runway

In order to evaluate the performance of the model, we computed velocity at different positions in the runway. A point at one end of the alley was chosen as the location of an appetitive US. Because a narrow alley is essentially a unidimensional environment, one landmark suffices to define a point. We assumed that only this landmark is behind the rewarded place, at the end of the runway. Navigation under the guidance of a single landmark has been called "cue" learning (Jarrard, 1983).

In a unidimensional environment, generalization surfaces become generalization curves. We assume that a rewarded place acquires both appetitive and decelerating generalization curves (see Appendix 10.C). Whereas the appetitive curve represents the accelerating force driving the animal toward the goal, the decelerating curve represents a decelerating force (which, opposing the accelerating force, ensures that the goal is reached with a small velocity, that is, without collision with the rewarding object).

Figure 10.8 shows aversive (decelerative), appetitive, and net generalization curves, together with velocity, as a function of trials and position in the runway. Positive values of the net generalization curve represent the force driving the animal in the direction of the goal. Positive values of the velocity curve represent the velocity with which the animal moves toward the goal. Both the appetitive accelerating curve and the aversive decelerating curve peak at the position of the goal and increase in amplitude over trials. Both net driving force and velocity increase at the start point, peak midway down the runway, and decrease toward the goal. Although peak net driving force and velocity increase over trials, the net driving force and final velocity remain relatively constant at the goal. Figure 10.8 shows that the appetitive curve generalizes over the whole runway and, therefore, is able to drive the animal from the start box to the goal. The aversive curve generalizes over half of the runway, thereby decelerating the animal from the middle of the runway, starting at the brake point, until it reaches the goal with a small velocity. In agreement with Wagner's (1961) results, Figure 10.8 indicates that velocities at different points in the runway increase over acquisition trials.

In agreement with Brown's (1942) results, Figure 10.8 (see Equation 10.20 in Appendix 10.A) shows that the force measured by the harness equals $G^{\mathrm{USappetitive}}(x_0 + \Delta x, y_0 + \Delta y)$, which increases as the animal is located closer to the goal.

Figure 10.9 shows the simulated velocity profile in a runway after 10 simulated trials. As illustrated in Figure 10.9, simulation results are in close agreement with Amsel, Rashotte, and MacKinnon's (1966) empirical data.

Appetitive-aversive interactions in a runway

In order to evaluate the performance of the model, we computed the final position in a runway under different combinations of appetitive and aversive reinforcements at the goal box. Simulated animals were trained to run down an

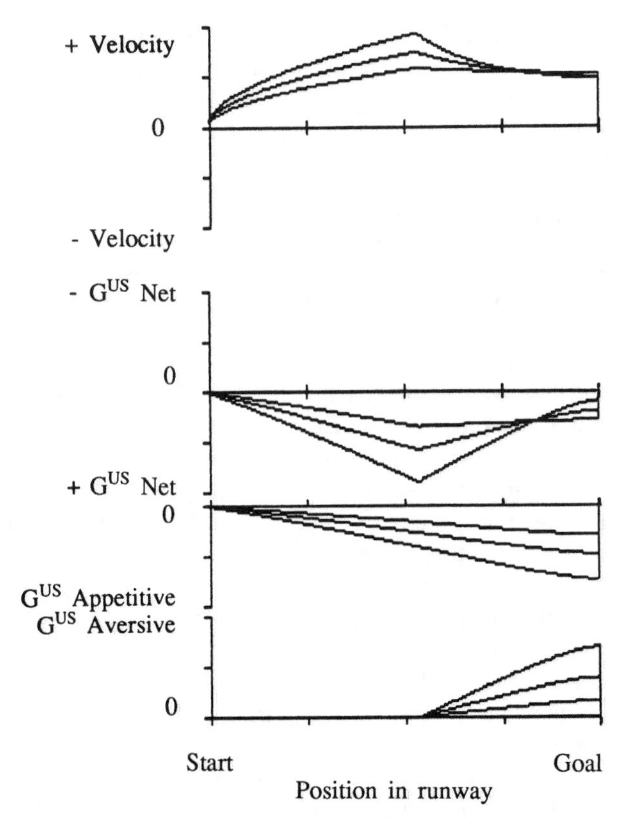

Figure 10.8. Navigational dynamics in a runway. Simulated velocity, net generalization curve, appetitive generalization curve, and aversive (decelerating) generalization curve at different places in a runway, for acquisition trials 1, 3, and 10. The goal acquires both appetitive and aversive generalization curves. Whereas the appetitive generalization curve accelerates the animal toward the goal, the aversive generalization curve decelerates the animal and ensures a smooth arrival to the goal.

alley to obtain either a strong or a weak reward for 10 trials. Subsequently, animals received several trials with either strong or weak punishment at the goal box, in addition to appetitive reinforcement. After training in different combinations of punishment and reward intensities, the simulated rats were placed at the start of the alley and their behavior was observed: Movement proceeded in the direction of the goal until a certain point was reached, where navigation stopped. From the start of the alley up to the point where movement stops, simulated results were computed in real time as the animal moves toward the goal. Because beyond the stop point velocity becomes negative, indicating that the animal moves away from the goal, simulated results were computed in the direction of the movement.

As in Miller et al.'s (1944) experiment, Figure 10.10 shows that the place at which animals stop is determined by the relative strength of the appetitive and aversive stimuli. After four trials with both strongly appetitive and aversive stimuli, the animals stop close to the goal (point A). After four trials with both weakly appetitive and aversive stimuli, animals stop close to the goal (point B). Finally, after only one trial with a strongly aversive stimulus and a weakly appetitive stimulus, animals stop far from the goal (point C). Figure 10.10 indicates that the simulations accurately reflect Miller et al.'s results.

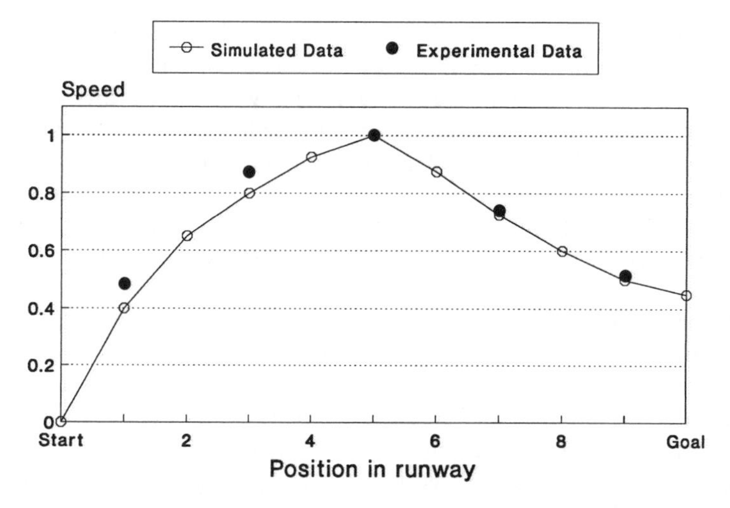

Figure 10.9. Velocity profile in a runway. Experimental and simulated velocity as a percentage of peak velocity at different places in a runway. Experimental data from Amsel, Rashotte, and MacKinnon (1966).

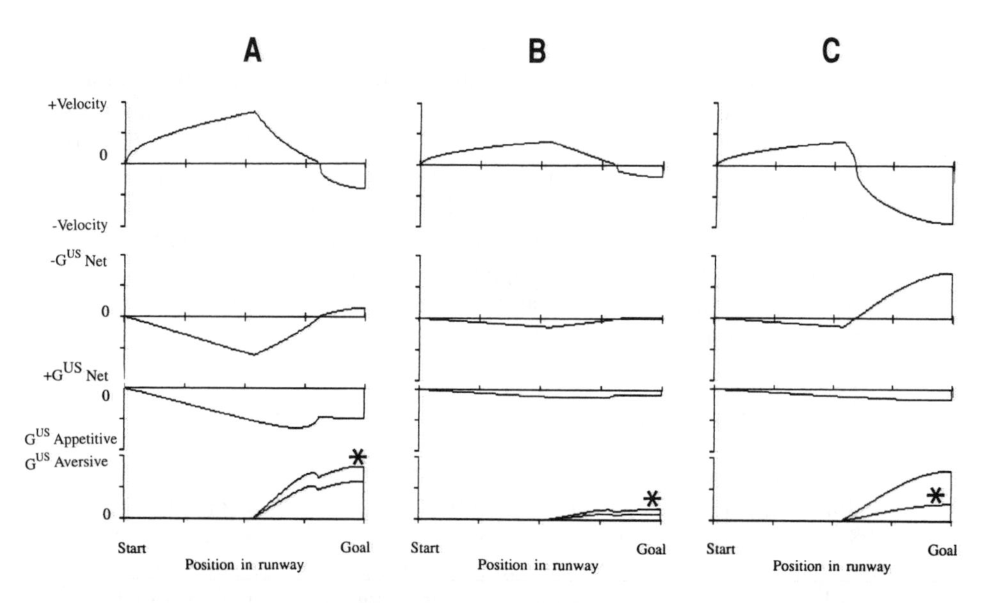

Figure 10.10. Appetitive-aversive interactions in a runway. Velocity, net, appetitive, decelerating, and aversive generalization curves for (A) a strong appetitive US and a strong aversive US, (B) a weak appetitive US and a weak aversive US, and (C) a weak appetitive US and a strong aversive US encountered at the goal. Decelerating curves are graphed as an aversive force and indicated by asterisks.

Discussion

The dynamics of spatial navigation

Schmajuk (1990) introduced a real-time neural network capable of describing both temporal discrimination and spatial learning in a unified fashion. When applied to discrimination in the temporal domain, the neural network tunes detectors that generate their maximum response when a continuous temporal variable assumes a particular value and generalizes over temporal dimensions. Computer simulations were carried out for temporal discrimination in classical conditioning and instrumental learning, classical conditioning under different interstimulus intervals, and classical conditioning with mixed ISIs. Schmajuk (1990) showed that quantitative evaluation of the model indicates a highly significant correlation between simulated and experimental data obtained in classical conditioning experiments.

When applied to place discrimination in the spatial domain, the neural network tunes detectors to respond to the values of the visual angles of different landmarks as they are perceived from the spatial location where a reinforcing event is encountered. The pattern recognition mechanism that provides each detector with information concerning a specific landmark is not described in the present rendition of the model. After a detector has been tuned, its output generates an effective stimulus that peaks at the distance from the landmark where positive or negative reinforcement was encountered before. The outputs of the tuned detectors become associated with the reinforcing event. This feature allows the network to generate appropriate spatial generalization surfaces that are used to guide spatial navigation in situations in which appetitive places are approached and aversive places avoided. Under the assumption that appetitive generalization gradients are gentler than an aversive generalization gradient (Miller, 1944), the present chapter shows that the model successfully depicts appetitive-aversive place and cue learning.

When generalization surfaces are assumed to represent forces driving animal behavior, the model allows the description of the dynamics of navigation toward the goal. We assume that animals are subject to a retarding force that opposes the movement of the animal. At locations removed from the goal, velocity is a function of the value of the net generalization surface and the retarding factor. We propose that appetitive places accrue both appetitive (accelerating) and decelerating associations. Whereas appetitive associations accelerate the animal toward the goal, the decelerating associations slow down the animal to ensure proper velocity on arrival at the goal.

Simulation results

In an appetitive place-learning task, computer simulations show that the animal can find the goal from any start point, a result that is in agreement with Morris's (1981) data. In addition, in agreement with Brown (1942), simulations demonstrate that the force driving the animal toward the goal shows a maximum at the rewarded place and decreases with increasing distances from the rewarded location. The model accounts for data showing that, when placed in a Morris tank, (1) animals are able to find the hidden platform after a single exposure to it, and (2) their latency to find the goal after goal exposure is longer than after swimming to the goal (Whishaw, 1991). Coinciding with Sutherland, Whishaw, and Kolb's (1987) results, simulations of place learning with restricted views of distal cues during training show that, in many (but not all) cases, place navigation from a novel starting location is seriously impaired. In an aversive place-learning task, simulations demonstrate that the force driving the animal away from the goal shows a maximum at the punished place and decreases with increasing distances from the punished location, a result in agreement with Brown's (1942). When combined appetitive and aversive place learning was

Table 10.1. *Simulations obtained with Schmajuk and Blair's (1993) place-learning model*

Paradigm	Model
Water maze	Y
Latent learning in the water maze	Y
Detour problems in the water maze	Y
Velocity profile in a runway	Y
Appetitive-aversive interactions in a runway	Y
Multiple-T maze	
Latent learning in the multiple-T maze	
Detour problems in the multiple-T maze	

Note. Y: The model can simulate the experimental data.

simulated, results show that animals are able to move toward the goal and detour around aversive places.

When the model was applied to the computation of the velocity at different positions in a runway, simulation results offered a proper quantitative description of Amsel et al.'s (1966) data. Finally, computer simulations of the final position in a runway under different combinations of appetitive and aversive reinforcement at the goal box are in close agreement with Miller et al.'s (1943) results.

Table 10.1 summarizes the paradigms described by the model. In brief, the model provides (1) accurate qualititative descriptions of appetitive and aversive place learning and (2) good quantitative characterizations of speed profiles in a runway.

Stimulus configuration and spatial learning

Chapter 6 describes a multilayer network that portrays stimulus configuration. The network incorporates a layer of "hidden" units positioned between input and output units, and includes inputs that are connected to the output directly as well as indirectly through the hidden-unit layer. In a multilayer extension of the network shown in Figure 10.2, the information coming from the tuned detectors would be recoded by the hidden units into internal representations that can be regarded as configural stimuli. Therefore, the outputs of the tuned detectors would not only accrue a *direct* association with the US at the output, as illustrated in Figure 10.2, but would also become *configured* with other detector outputs in hidden units that, in turn, would become associated with the US at the output.

Configural stimuli computed in the hidden-unit layer might be regarded as a representation of the spatial area where the animal is situated. Therefore, the addition of the hidden-unit layer helps to make not ambiguous cases in which the same landmark is visible from different local areas, that is, to disambiguate between different spatial maps. For instance, the visual angle of a given landmark might define the place where the US is encountered in one spatial area (specified by the configural stimulus of the visual angles of a given set of landmarks), but not in a different local area (specified by the visual angles of the configural stimulus of another set of landmarks).

Comparison with other models of goal-directed behavior

The model presented in this chapter describes the acquisition of goal-directed and goal-avoidance behaviors independently of the responses that are necessary to reach the goals. In contrast, as mentioned at the beginning of this chapter, several theories have been proposed to account for spatial learning in terms of response-selection mechanisms: Tolman (1932a, b), Milner (1960), Hull (1952), and Barto and Sutton (1981).

In addition to describing the animal's positions during goal-seeking behavior, our model also describes the dynamics (acceleration and velocity) of motion toward the goal. Recently, Killeen and Amsel (1987) presented a mathematical model that describes animal speed at different points in a runway. Killeen and Amsel assume that a decelerating force is introduced at a point in the alley (the brake point), and that this force is proportional to the accelerating force that draws the animal toward the goal. Killeen and Amsel employ two separate equations for computing velocity, one for points between the start box and the brake point, and another for points between the brake point and the goal. The equation used after the brake point includes a deceleration term. Killeen and Amsel found that this approach has difficulty describing data from long runways in which animals reach and maintain a maximum "cruising velocity." To describe such data, they must assume an additional brake point at the location where cruising velocity is reached, and thus a third equation for velocity is required. Killeen and Amsel suggested that an alternative to the separate equations approach is to describe, as our model does, deceleration as a force proportional to velocity.

Appendix 10.A

The model

The visual angle hypothesis

Hull (1952, p. 259) proposed that *optical convergence* in visual fixation yields an indication of *distance,* and that the size of the optical image (*visual angle*) in conjunction with the degree of optical convergence (coding distance) indicates the *size* of the object. According to Hull, animals may use information regarding the distance to an object in order to generalize responses that have been associated with the optical image at a given distance.

In line with Hull's approach, Zipser (1985, 1986; Wilkie and Palfrey, 1987; see also Levine and Shefner, 1981) suggested that animals learn a set of visual angles subtended by different spatial landmarks at the place where reward is encountered. Because the visual angle subtended by a given landmark at the rewarded place is a function of the distance between that landmark and the place, and the location of a point in space is uniquely determined by its distance to three different landmarks, it follows that the visual angles to three different landmarks completely specify the position of the rewarded place.[1] When the animal is placed again in the environment on a subsequent trial, the current set of

[1] It should be noted that although the distances to two landmarks define two symmetric spatial locations, two landmarks might suffice to define one spatial location if the straight line connecting both landmarks does not cross the testing area. In this case, one of the two symmetric locations will always lie outside the testing area.

visual angles is compared with the set of visual angles that characterizes the rewarded location and, therefore, the animal is capable of approaching the place where reward was encountered before.

Following Zipser (1985, 1986), Schmajuk (1990) assumed that animals use the visual angle subtended by spatial landmarks in order to define spatial locations. A landmark L_i of size O_i at a distance d subtends a unique visual angle $\Omega_i\,(d)$. Formally, the visual angle $\Omega_i\,(d)$ of spatial landmark L_i is defined by

$$\Omega_i\,(d_i) = \arctan(O_i/d_i) \tag{10.1}$$

where O_i is the size of landmark L_i and d_i the distance to L_i. The visual angle $\Omega_i(d)$ of L_i is equal to 90 deg when $d_i = 0$. $\Omega_i(d_i)$ gradually decays to zero deg as distance d_i from L_i increases. As Zipser (1985) pointed out, visual angle $\Omega_i(d_i)$ remains constant if both distance d_i and size O_i "dilate" simultaneously.

Tuned detectors

Detectors are tuned by S_k to the value of the visual angle Ω_i of landmark L_i. When L_i is encountered after the detector has been tuned, effective stimulus $E_i^k(d_i)$ peaks at the place where S_k was encountered before. The tuning of a detector is accomplished by storing in memory the value of the visual angle Ω_i at the place S_k is encountered. For example, consider the case of stimulus S_k encountered in the spatial vicinity of landmark L_i. T_i^k represents the memory of the value of the visual angle Ω_i stored at the place where S_k is encountered. Memory T_i^k is updated according to

$$\Delta T_i^k(t) = B\, X_i(d_i)\, S_k(d_i)[\Omega_i(d_i) - T_i^k(t)\, X_i(d_i)] \tag{10.2}$$

where B is the learning rate and $X_i(d_i)$ is a representation of L_i that, unlike the visual angle $\Omega_i(d_i)$, does not decrease over distance. $X_i(d_i)$ is 1 whenever $\Omega_i(d_i) > \epsilon 1$ (where $\epsilon 1$ is arbitrarily small) and 0 otherwise. S_k equals 1 when stimulus k is encountered and 0 when it is absent. By Equation 10.2, changes in $T_i^k(t)$ occur only when $S_k(d_i)$ is experienced, $X_i(d)$ is 1, and $T_i^k(t)\, X_i(d_i)$ differs from $\Omega_i(d_i^k)$. Therefore, $T_i^k(t)$ stores the value of $\Omega_i(d_i^k)$, that is, the value of the visual angle subtended by L_i at the place where S_k is encountered.

Once a spatial detector has been tuned according to Equation 10.2, when the L_i is encountered again, the value of the visual angle $\Omega_i(d_i^k)$ is recognized and the detector generates an output with a peak at a distance from the landmark L_i where S_k was encountered before. The effective spatial stimulus i for S_k based on L_i, $E_i^k(d_i)$, is

$$E_i^k(d_i) = f\{X_i(d_i) - C_k[T_i^k(t)\, X_i(d_i) - \Omega_i(d_i^k)]^2\} \tag{10.3}$$

where $f[x] = x$ if $f[x] > 0$ and $f[x] = 0$ otherwise. By Equation 10.3, at a distance d_i^k from L_i, when $T_i^k(t)\, X_i(d) = \Omega_i(d_i^k)$, $E_i^k(d)$ has value $f[X_i(d_i)] = 1$. Equation 10.3 describes spatial generalization with a peak at the distance where S_k, d_i^k, was encountered before. In agreement with Miller's (1944, 1959) suggestion that the slope of the generalization of an aversive stimulus is steeper than the generalization of an appetitive stimulus, C_{aversive} is assumed to be greater than $C_{\text{appetitive}}$. Schmajuk (1990) presented a neural architecture for detector D_i^k.

Tuned detectors can be regarded as feature detectors abstracting the amplitude of the signal to which they are tuned. Experimental data (e.g., Cynader and Chernenko, 1976; Hirsch and Spinelli, 1970; Bakin and Weinberger, 1990) support the idea that although many neurons are genetically tuned, the tuning of these detectors can be modified by experience.

Associations between effective stimuli and S_k

Once spatial effective stimuli E_i^k are defined, they can be associated with stimulus S_k. Changes in the associative value between effective stimulus $E_i^k(d)$ and S_k, V_i^k, are defined by

$$\Delta V_i^k(t) = D\, E_i^k(d_i)[\lambda^k - \Sigma_j\, V_j^k(t)\, E_j^k(d_i)] \qquad (10.4)$$

where $E_i^k(d_i)$ is the magnitude of the effective stimulus of L_i for S_k, D is the learning rate parameter ($0 < D \leq 1$), λ^k is the intensity of S_k, and $\Sigma_j V_j^k(t)\, E_j^k(d_i)$ is the combined prediction of S_k made on all spatial effective stimuli. In order to have a net positive V_i^k in a given trial, $D = D'$ whenever $\lambda^k > \Sigma_j\, V_j^k(t)\, E_j^k(d)$, and $D = D''$ otherwise, with $D' > D''$. It is important to notice that in Equation 10.4, whereas $E_i^k(d_i)$ is a function of the distance d from a given location to landmarks L_i, $V_i^k(t)$ is a function of time. That is, $E_i^k(d_i)$ changes only when the animal moves, but $V_i^k(t)$ changes continuously even in the absence of movement.

For simplicity, Equation 10.4 assumes that V_i^k exhibits both positive and negative values. However, in the network presented in Figure 10.2, V_i^k, whether excitatory or inhibitory, assumes only positive values. Therefore, in order to obtain inhibitory effects, excitatory and inhibitory activities are summed with their corresponding signs in the output unit. This unit determines the output of the system.

As in the case of the simple delta rule (see Box 3.1), Equation 10.4 implies that each E_i^k competes with all other E_i^ks' in order to gain association with S_k, and only the best predictor of S_k becomes associated with it. In associative learning the best predictor is a stimulus highly correlated with S_k, that is, a stimulus that is present every time S_k is present and absent every time S_k is absent. In the case of effective stimuli, the best predicting E_i^k is that one most precisely predicting S_k, that is, other things being equal, the E_i^k with the smallest generalization. In addition, the use of Equation 10.4 implies that when equally good predictors are used, all of them should be present in order to predict the full intensity of S_k, $\lambda^k(t)$.

Generalization surfaces

Equation 10.3 defines the value of effective stimulus $E_i^k(d_i)$ at a given spatial location as a function of the Euclidian distance d_i from that location to landmark L_i. A more general expression is obtained when distance d_i is specified in terms of Cartesian coordinates x and y. In that case, the output of the neural network at a spatial position (x, y), $G^k(x, y)$, is given by

$$G^k(x, y) = \Sigma_i V_i^k(t)\, E_i^k(x, y) \qquad (10.5)$$

$G^k(x, y)$ represents the height of a three-dimensional generalization surface at a point (x, y). In the case of combined appetitive and aversive generalization surfaces, the net generalization surface is given by

$$G^{\text{US net}}(x, y) = G^{\text{US appetitive}}(x, y) - G^{\text{US aversive}}(x, y) \qquad (10.6)$$

Different appetitive USs (food, water, etc.) are assumed to have different generalization surfaces that the animal will use to approach the alternative goals according to their motivational states. We adopt the convention that generalization surfaces of appetitive goals increase downward, whereas generalization surfaces of aversive goals increase upward.

Appendix 10.B

Navigational trajectories

When the *net generalization surface is appetitive,* that is, $G^{US\ appetitive}(x, y) > G^{US\ aversive}(x, y)$, animals will move from a point defined by coordinates x_0 and y_0 to a point defined by coordinates $x_0 + \Delta x$ and $y_0 + \Delta y$, if

$$G^{US\ appetitive}(x_0 + \Delta x, y_0 + \Delta y) > G^{US\ appetitive}(x_0, y_0) \quad (10.7)$$

where Δx and Δy are randomly generated.

When the *net generalization surface is aversive,* that is, $G^{US\ appetitive}(x, y) < G^{US\ aversive}(x, y)$, animals will move from a point defined by coordinates x_0 and y_0 to a point defined by coordinates $x_0 + \Delta x$ and $y_0 + \Delta y$, if

$$G^{US\ aversive}(x_0 + \Delta x, y_0 + \Delta y) < G^{US\ aversive}(x_0, y_0) \quad (10.8)$$

where Δx and Δy are randomly generated. Movement continues until the net generalization surface equals zero, $G^{US\ net}(x, y) = 0$.

Under the adopted convention that generalization surfaces of appetitive goals increase downward whereas generalization surfaces of aversive goals increase upward, by Equations 10.7 and 10.8 animals always move in the direction of the steepest descent on the generalization surface. As in any gradient traversal method, movement along the generalization surfaces might be halted by local minima. However, because $V_i^k(t)$ changes if the animal does not find the US (see Equation 10.4), generalization surfaces $G^{US}(x, y)$ undergo modifications over time, and these variations usually permit escape from local minima.

Appendix 10.C

Navigational dynamics

Retarding forces

Equations 10.7 and 10.8 define the trajectory an animal will follow when navigating in an environment with aversive and appetitive locations. We assume that (1) animals move from a point defined by coordinates x_0 and y_0 to a point defined by coordinates $x_0 + \Delta x$ and $y_0 + \Delta y$ driven by a motivational force defined by the value of the net generalization surface at the next point $G^{US\ net}(x_0 + \Delta x, y_0 + \Delta y)$, and (2) animals are subject to a retarding force $\mathrm{R}v(t)^2$ that tends to slow down their movement. Whereas the motivational driving force may be understood as the appetitiveness or aversiveness of different environmental locations, the retarding force may be interpreted as a resistance to movement linked to physical factors (muscle and joint friction, terrain in which the animal is moving). For instance, the retarding force will be larger when the animal is swimming in the water than when the animal is walking or running on the ground.

Applying Newton's second law, we obtain

$$a(t)M = G^{US\ net}(x_0 + \Delta x, y_0 + \Delta y) - \mathrm{R}v(t)^2 \quad (10.9)$$

where M represents the animal's mass and R a retarding coefficient.

By Equation 10.9, as velocity increases, $a(t)$ will gradually decrease and the animal will move at a constant velocity, called *terminal velocity,* v_t. The value of the animal's terminal velocity is obtained by setting $a(t)$ equal to zero in Equation 10.9:

$$v_t(x_0, y_0) = \sqrt{[G^{\mathrm{US\ net}}(x_0 + \Delta x, y_0 + \Delta y)/\mathrm{R}]} \qquad (10.10)$$

By Equation 10.10, terminal velocity increases with increasing driving forces (e.g., a more appetitive location) and decreasing retarding coefficient R (e.g., running vs. swimming). If the initial velocity exceeds the value of the terminal velocity, the velocity will decrease. When the initial velocity is much smaller or much larger than the value of the final velocity computed by Equation 10.10, the animal might not be capable of reaching terminal velocity between points (x_0, y_0) and $(x_0 + \Delta x, y_0 + \Delta y)$. In this case, Equation 10.9 must be used to describe the transition from initial to terminal velocity. However, when the difference between initial and terminal velocity is small enough, the animal is capable of reaching terminal velocity between points (x_0, y_0) and $(x_0 + \Delta x, y_0 + \Delta y)$, and Equation 10.10 may be used to approximate the actual velocity at each point. For a simple treatment of motion with a retarding velocity-dependent force, the reader is referred to Tipler (1982, p. 144).

Navigation toward a goal: decelerating forces

Hull (1931) argued that, in order to obtain reward at an appetitive location, animals should reach the place where reward is encountered at an appropriate final velocity. This final velocity should be small in order to apprehend the reward and avoid colliding with the rewarding object or the wall behind it. In order to combine a fast movement toward the goal with a slow final velocity, animals should decelerate as they near the goal. The dynamics of this combined accelerated–decelerated movement are readily captured by the assumption that *an appetitive location accrues a decelerating association that prevents collisions with the rewarding object.*

Therefore, at the location of the goal, an *appetitive and a decelerating detector is tuned by the appetitive US,* according to Equation 10.2, yielding

$$\Delta T_i^{\mathrm{appetitive}}(t) = B\, X_i(d)\, \mathrm{S}_{\mathrm{US\ appetitive}}(d)[\Omega_i(d) - T_i^{\mathrm{appetitive}}(t)\, X_i(d)] \qquad (10.11)$$

and

$$\Delta T_i^{\mathrm{decelerating}}(t) = B\, X_i(d)\, \mathrm{S}_{\mathrm{US\ appetitive}}(d)[\Omega_i(d) - T_i^{\mathrm{decelerating}}(t)\, X_i(d)] \qquad (10.12)$$

Each detector will provide an effective stimulus given by

$$E_i^{\mathrm{appetitive}}(d) = f\{X_i(d) - C_{\mathrm{appetitive}}[T_i^{\mathrm{appetitive}}(t)\, X_i(d) - \Omega_i(d)]^2\} \qquad (10.13)$$

and

$$E_i^{\mathrm{decelerating}}(d) = f\{X_i(d) - C_{\mathrm{decelerating}}[T_i^{\mathrm{decelerating}}(t)\, X_i(d) - \Omega_i(d)]^2\} \qquad (10.14)$$

Since $C_{\mathrm{decelerating}}$ is assumed to be larger than $C_{\mathrm{appetitive}}$, the decelerating generalization surface has a steeper gradient than the appetitive generalization surface, and thus falls to zero faster (closer to the goal). Therefore, the net generalization surface is only affected by the decelerating force within a certain distance from the goal. The points in space where the animal comes under the influence of the decelerating force are called *brake points,* since that is where the animal begins slowing down to avoid collision with the goal.

In order to obtain a near-zero velocity at the goal, we assume that the decelerating effective stimulus $E_i^{decelerating}$, which has been tuned by $S_{US\ appetitive}$, is also associated with $S_{US\ appetitive}$. That is, we assume that the decelerative force preventing the collision with the goal increases together with the appetitiveness of the goal. Changes in the associative value between effective stimulus $E_i^{appettive}$ (x,y) or $E_i^{decelerating}(x, y)$ and $S_{US\ appetitive,}$ are defined by

$$\Delta V_i^{appetitive}(t) = D\, E_i^{appetitive}[\lambda^{appetitive} - \Sigma_j V_j^{appetitive}(t)\, E_j^{appetitive}(x, y)] \quad (10.15)$$

and

$$\Delta V_i^{decelerating}(t) = D\, E_i^{decelerating}[\lambda^{appetitive} - V_j^{decelerating}(t)\, E_j^{decelerating}(x, y)] \quad (10.16)$$

As explained before, $G^{appetitive}(x, y)$ represents the motivational force driving the animal toward the goal, and the final force driving the animal is equal to the motivational force minus a retarding force. In addition to the retarding force introduced in Equation 10.9, a decelerative force must now be added, yielding

$$a(t)M = G^{US\ appetitive}(x_0 + \Delta x, y_0 + \Delta y) - S\, G^{decelerating}(x_0, y_0)v(t) - Rv(t)^2 \quad (10.17)$$

The term S $G^{decelerating}(x_0, y_0)\, v(t)$ represents the *decelerating* force opposing the accelerating force in proportion to the velocity of the animal, where $G^{decelerating}\,(x,y) = V_i^{decelerating}(t)\, E_i^{decelerating}$. Like the retarding force, the decelerative force is also a function of velocity. This is to be expected because the animal only needs to decelerate when moving too fast in the direction of the goal. However, unlike the frictional force, the decelerative force is sensitive to the direction of movement. Whereas the frictional force opposes movement in any direction, the decelerative force only opposes movement in the direction of the goal object.

The value of the animal's terminal velocity in the vicinity of the goal is given by

$$v_t(x_0, y_0) = 1/2R\,[-S\, G^{decelerating}(x_0, y_0) + \sqrt{[S\, G^{decelerating}(x_0,y_0)]^2 + 4R[G^{US\ appetitive}(x_0 + \Delta x, y_0 + \Delta y)]} \quad (10.18)$$

By Equation 10.18, the terminal velocity in the vicinity of the goal increases with increasing appetitive values and decreases with increasing decelerating forces and retarding coefficient R. Prior to the brake point, when $G^{decelerating}$ $(x_0,y_0) = 0$, Equation 10.18 is identical to Equation 10.10.

In summary, in order to obtain reward at an appetitive location, animals should reach the place where reward is encountered at an appropriately small final velocity. This condition can be achieved by assuming that the dynamics of movement toward the goal are controlled by a learned decelerating force that opposes the learned accelerating force driving the animal toward the goal.

Notice that, although a decelerating force prevents the animal from colliding with or overshooting the goal, this force is not measured by the harness in Brown's (1942) experiment. Applying Newton's second law when the animal is restrained by the harness with force F_h in the vicinity of the goal, we have

$$a(x_0, y_0)\, M = G^{US\ appetitive}(x_0 + \Delta x, y_0 + \Delta y) - F_h - G^{decelerating}(x_0, y_0)\, v(x_0, y_0) - Rv(x_0,y_0)^2 \quad (10.19)$$

When the animal does not move, $a(x_0, y_0) = 0$ and $v(x_0, y_0) = 0$, and therefore,

$$F_h = G^{\text{US appetitive}}(x_0 + \Delta x, y_0 + \Delta y) \tag{10.20}$$

Navigation toward a goal: aversive stimulation at the goal

As described, any appetitive goal location acquires a decelerating aversive association, which is proportionate to its accelerating appetitive association. If the goal location is negatively reinforced by an additional aversive stimulus, then an added aversive association will be developed. When the appetitive generalization curve is replaced by the net generalization curve G^{net}, the terminal velocity is given by

$$v_t(x_0, y_0) = 1/2\text{R}[-\text{S}G^{\text{decelerating}}(x_0, y_0) + \sqrt{[\text{S}\ G^{\text{decelerating}}(x_0,y_0)]^2 + 4\text{R}[G^{\text{US net}}(x_0 + \Delta x, y_0 + \Delta y)]} \tag{10.21}$$

When the initial velocity is large compared to the value of the final velocity computed by Equation 10.21, the animal might not be capable of reaching terminal velocity between points (x_0, y_0) and $(x_0 + \Delta x, y_0 + \Delta y)$. In this case, Equation 10.17 must be used to describe the transition from initial to terminal velocity. However, when the difference between initial and terminal velocity is small, and therefore the animal is capable of reaching terminal velocity between points (x_0, y_0) and $(x_0 + \Delta x, y_0 + \Delta y)$, Equation 10.21 may be used to approximate the actual velocity.

From a physical standpoint, the problem of deceleration involves determining the proper strength of the decelerative force and the proper place to apply it (the brake point) in order to stop a moving object at the goal. The strength of the decelerative force increases with increasing accelerating force $G^{\text{US net}}$ and decreases with increasing resistance coefficient R and decreasing distance from the brake point to the goal. In our model, the brake point is predetermined by the value of the parameter $C_{\text{decelerating}}$.

Notice that as $G^{\text{decelerating}}$ increases, thereby decreasing terminal velocity, the animal takes longer to reach the goal. Because the associations V_i^k are a function of time in Equations 10.4 and 10.5, longer latency to reach the goal results in greater extinction of the association V_i^k, and, thus, a decreased $G^{\text{appetitive}}$. Therefore, decelerating force $G^{\text{decelerating}}$ decreases the terminal velocity at the goal by (1) actively opposing the appetitive force $G^{\text{appetitive}}$ and (2) by contributing to the extinction of the appetitive force $G^{\text{appetitive}}$.

Appendix 10.D

Parameter values

Computer simulations generate values of the relevant variables only at discrete space distances. Simulation parameters were $B = 1$, $D_{\text{appetitive}}' = 0.1$, $D_{\text{appetitive}}'' = 0.0013$ $D_{\text{aversive}}' = .1$, $D_{\text{aversive}}'' = 0.0013$, $R = 0.5$, $\text{S} = 2$, $C_{\text{appetitive}} = 160{,}000$, $C_{\text{aversive}} = C_{\text{decelerative}} = 400{,}000$.

11 Maze learning and cognitive mapping

Chapter 10 explains how the combination of (1) a goal-directed mechanism and (2) a spatial map is capable of guiding the animal's navigation in a relatively limited territory. This chapter explains how a goal-directed system is capable of guiding the animal's navigation through complex mazes.

Tolman (1932a, b) proposed that latent learning and detour learning illustrate the animals' capacity for cognitive mapping. In latent learning, animals are exposed to a maze without being rewarded at the goal box. When a reward is later presented, animals demonstrate knowledge of the spatial arrangement of the maze, which remains "latent" until the reward is introduced. Detour problems are maze problems that can be solved by integrating separately learned pieces of information into a comprehensive depiction of the environment. The present chapter presents a real-time, biologically plausible neural network capable of latent learning and solving detour problems through cognitive mapping (Schmajuk and Thieme, 1992; Schmajuk, Thieme, and Blair, 1993). This network is an extension of the cognitive map models presented in Chapters 3 and 5.

Maze learning: response-selection versus stimulus-approach views

As in the case of place learning, two different views have been used to explain maze learning: "response-selection" and "stimulus-approach" views. Whereas some theories advocate a response-selection strategy to describe maze learning (Hampson, 1990; Klopf, Morgan, and Weaver, 1993; Tolman, 1932a, b), others adopt a stimulus-approach scheme (Hampson, 1990; Libliech and Arbib, 1982; Schmajuk and Thieme, 1992).

As mentioned, Tolman (1932a, b) proposed that multiple expectancies can be combined into a cognitive map (see Box 2.2) and tried to demonstrate the existence of cognitive maps through latent learning and detour experiments.

In contrast to Tolman's cognitive approach, Hull (1951) proposed a stimulus–response theory to explain maze learning. According to Hull's view, maze stimuli that usually precede the goal are comparable to CSs, the reward is comparable to a US, and the goal response comparable to an unconditioned response (UR), as defined in classical conditioning. Because the goal elicits an approach response, stimuli in initial sections of the maze will also be approached as they become associated with the reward. In Hull's theory, the association between a maze stimulus and the UR is a function of the intensity of a temporal trace generated by the stimulus. Because the trace decays as a function of time, the strength of the association between stimuli in a given place in a maze and the US encountered at the goal decreases as the *time* required to travel from the place to the goal increases. Because traveling time is proportional to the distance between the place and the goal, place–US associations also code the distance between places and the goal. Therefore, stimuli in portions of the maze distant from the goal form weaker associations with the US than stimuli close to the goal. Also, stimuli in sections of the maze that do not lead to the goal are weakly associated with the US. Sections of the maze that lead to the goal are selected because their association with the reward (US) is stronger than those sections that do not lead to the goal (Hull, 1951, p. 281). Hull (1951) explained latent learning in the following terms. When preexposed to the maze without being rewarded at the goal box, animals still receive some minimal reinforcement when they are removed from the goal box. When a reward is later presented, there is an abrupt change in "incentive motivation" and performance is suddenly improved.

Deutsch (1960) presented a formal description of cognitive mapping that incorporates many of Tolman's cognitive concepts. Deutsch assumed that when an animal explores a given environment, it learns that stimuli follow one another in a given sequence. Internal representations of the stimuli are linked together in the order they are encountered by the animal. Deutsch suggested that a given drive activates its goal representation, which in turn activates the linked representations of stimuli connected to it. When the animal is placed in the maze, it searches for stimuli stimulated by the goal representation. When a stimulus activated by the goal is perceived, the activation of lower stimuli in the chain is cut off and behavior is controlled by the stimulus closer to the goal. Deutsch's theory can account for latent learning in the following terms. When animals are exposed to the maze without being rewarded at the goal box, they learn about the connections between different places in the maze. When a reward is subsequently presented, the goal representation is activated thereby activating the representations of the stimuli connected to it.

Lieblich and Arbib (1982) addressed the question of how animals build a cognitive model of the world. Lieblich and Arbib posited that spatial representations take the form of a directed graph in which nodes represent a *recognizable situation* in the world. In Lieblich and Arbib's scheme, a node represents not only a place but also the motivational state of the animal. Consequently, a place in the world might be represented by more than one node if the animal has been there under different motivational states. Lieblich and Arbib postulated that each node in the world graph is labeled with *learned* vectors R that reflect the drive-reduction properties for multiple motivations of the place represented by that node. Based on the value of R, the animal moves to the node most likely to reduce its present drive.

Hampson (1990) analyzed maze navigation in terms of one stimulus–response model and two stimulus–stimulus models, all capable of assembling action sequences in order to reach a final goal. The stimulus–response system consists of (1) response operators trained by an evaluation function, (2) a shared memory that categorizes the input space, and (3) an evaluation function that achieves sequential credit assignment (i.e., the allocation of credit or blame for an eventual gain or loss) and is used to modify stimulus–response connections. The first stimulus–stimulus model searches backward from the goal looking for a sequence of states that connect the goal with the current state of the system. As the system moves from state to state, backward stimulus–stimulus links are formed. As states are activated by the goal by a process of spreading activation, all states become subgoals. Goal activation is propagated decrementally by arbitrarily setting the strength of all connections. With decremental propagation the output level of each state node indicates both whether the final goal can be reached from that state and how far it is.

Hampson combined stimulus–response and stimulus–stimulus mechanisms into one system. As before, the stimulus–response portion of the system (1) trains stimulus–response operators, (2) trains a shared memory of the input space, and (3) predicts future evaluations. The additional stimulus–stimulus portion of the system (1) produces a state-map of the transitions the system has encountered and (2) spreads the activation at the goal to subgoal states whose outputs reflect their distance to the goal. Hampson's (1990) second stimulus–stimulus model searches forward from the current state looking for a path to the goal. As the system moves from state to state, forward stimulus–stimulus links are formed. For forward search, the current state is activated and this activation is decrementally propagated. Alternative transitions leaving the current state are evaluated by the strength with which they stimulate the goal. By sequentially considering all alternative transitions out of the current state, the next state on the shortest path can be determined. In contrast to the forward-searching model, the backward-searching model does not require sequential scanning of alternative next states, directly providing the best transition.

Sutton (1991) reviewed different architectures that incorporate stimulus–response (which he calls state–action) associations. "Stimulus–response only" architectures establish correlations between the response made and the reward received. Stimulus–response associations are able to distinguish between clearly defined high and low rewards, but not when a low reward in one situation can be the highest reward attainable in another situation. "Reinforcement comparison" architectures learn to associate input stimuli with a baseline reward and compare the present reward with the current input baseline. Reinforcement comparison architectures are able to optimize immediate rewards but not total reward in the remote future. "Adaptive heuristic critic" architectures are able to optimize long-term reward by learning to predict cumulative future reward as a function of the input stimuli. "Q-learning" architectures predict future reward as a function of the input stimuli and the selected response. Finally, "Dyna" architectures incorporate internal models of the world as a replacement for the actual world. Sutton (1991) showed that Dyna architectures are able to find detours and shortcuts in a simple maze.

Klopf et al. (1993) presented a computational model of operant conditioning that uses stimulus–response associations. The model is a hierarchical network of control systems that learns by employing a drive-reinforcement mechanism proposed by Klopf (1988) that describes a wide range of classical conditioning phenomena. In multiple-T mazes, the model learns to chain responses that avoid punishment and lead to reward.

Schmajuk and Thieme (1992) presented a neural network that builds a cognitive map representing the adjacency between places. Based on this cognitive map and making use of a stimulus-approach principle, the network is able to describe correctly latent learning and detour behavior. In addition, the network can be applied to problem-solving paradigms such as the Tower of Hanoi puzzle (Winston, 1977).

BOX 11.1 Schmajuk and Thieme's (1992) maze-learning model

The fundamental properties of the model are:

1. The system is composed of (a) a goal-seeking system and (b) a cognitive map.
2. The goal-seeking system implements a "stimulus-approach" view of behavioral control by which animals approach appetitive stimuli and avoid aversive stimuli.
3. As the animal explores the maze under the control of the goal-seeking system, it builds a cognitive map of the external world.
4. The cognitive map is a topological map, that is, it represents only the adjacency, but not distances or directions, between places in the maze.
5. The cognitive map is updated whenever there is mismatch between real-time predictions and actual inputs (see Box 3.1). Real-time predictions are generated simultaneously with the occurrence of environmental events.
6. The cognitive map interacts with the goal-seeking system by generating fast-time predictions. These are produced in advance of what occurs in real time, when the information stored in the cognitive map is used to predict the remote future.

The neural network

Figure 11.1 represents a block diagram of the model showing the interaction between the goal-seeking system, a cognitive system (cognitive map), and the environment. The neural network is capable of guiding the animal's search for a specific goal in the environment with the assistance of the cognitive map. It is a real-time mechanism (see Box 1.4) that describes behavior as a moment-to-moment phenomenon. Box 11.1 summarizes the main properties of the model. Appendix 11.A presents a formal description of the network as a set of differential equations that depicts changes in the values of neural activities and connectivities as a function of time.

As the animal explores the environment under the control of the goal-seeking system, it builds a cognitive map of the external world. Once the cognitive map is available, search behavior can also be inhibited by environmental stimuli directly associated with the goal, or by stimuli that predict the goal through the cognitive map (the internal representation of the world). The cognitive map is updated whenever a mismatch between predicted and actual inputs is detected by a learning modulation system. Whereas changes in the cognitive map are regulated by real-time predictions, the cognitive map interacts with the goal-seeking system by generating fast-time predictions. The following sections explain (1) how the cognitive map is built and updated, and (2) how maze navigation is accomplished with the assistance of the cognitive map.

Cognitive mapping and maze learning

To the extent that the layout of a maze can be accurately portrayed in terms of the connections between its contiguous places, it seems reasonable to describe maze connectivity in terms of the associations between the representations of adjacent places. However, because associations between spatially adjacent places are insufficient to predict remote places, navigation to remote goals re-

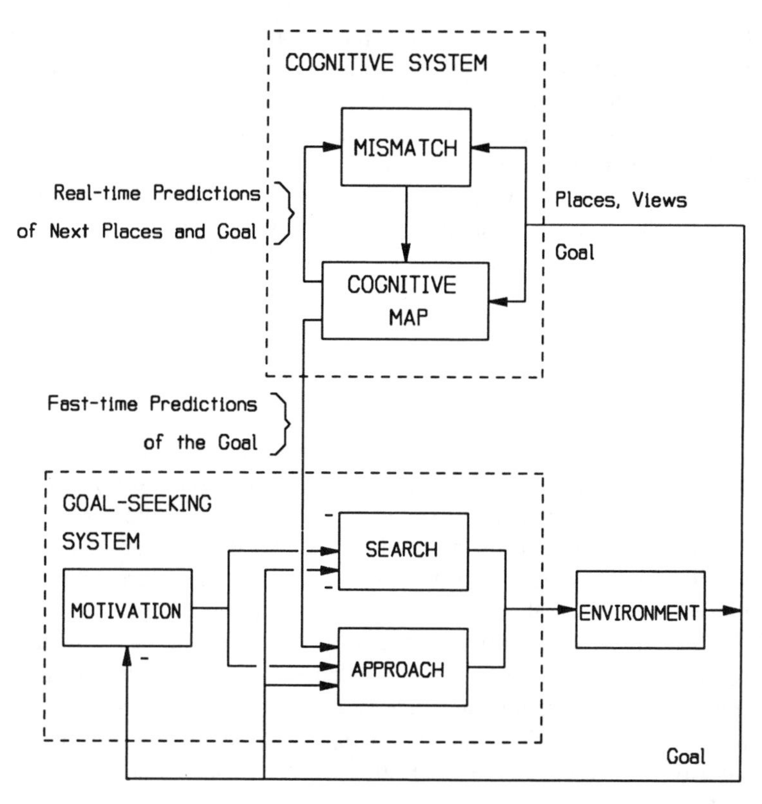

Figure 11.1. A block diagram of a system capable of maze navigation. Block diagram of the model showing the interaction between the goal-seeking system, cognitive map, and the environment. The goal-seeking system approaches appetitive stimuli defined by the motivational state. A given motivation elicits a search behavior that receives negative feedback from the appropriate goal. When the goal is encountered, it activates approach responses and inhibits search responses. As the animal explores the environment under the control of the goal-seeking system, it builds a cognitive map. Once the cognitive map is available, the search behavior also can be inhibited by stimuli that predict the goal when applied to the cognitive map. The cognitive map is updated whenever a mismatch between predicted and actual inputs is detected. Whereas changes in the cognitive map are regulated by real-time predictions, the cognitive map interacts with the goal-seeking system by generating fast-time predictions.

quires the building and reading of a cognitive map. This map should combine information about spatially adjacent places to infer by reasoning the connections to remote places.

Although both Gelperin's (1986) and Schmajuk's (1986, 1987) networks successfully describe inferential processes and cognitive mapping in classical conditioning (see Chapter 3), the networks cannot be applied to maze learning. Gelperin's approach cannot be applied to maze learning because autoassociative recursive networks generate associations between all inputs even when they have not been presented together. For instance, if place *A* predicts place *B* and place *B* predicts place *C*, when the animal is located in place *A*, the representation of place *B* is activated, which in turn activates the representation of place *C*. Consequently, because the representations of place *A* and place *C* are active simultaneously, they become associated. In the case of maze learning, this means that each place becomes associated with all other places, therefore making it impossible for the system to navigate through the maze. Schmajuk's

(1986, 1987) approach cannot be applied to maze learning either. Even though Schmajuk's network allows the generation of second-order predictions (place A predicts place B), the network cannot produce the higher-order predictions (place A predicts place B, place B predicts place C, . . . , place Z predicts the goal) needed to describe maze learning.

In sum, although several neural network models capable of cognitive mapping in classical conditioning have been proposed, they cannot be applied to maze learning. Therefore, this study presents a mechanistic, neural network approach to cognitive mapping capable of describing different maze-learning paradigms.

Building a cognitive map

Figure 11.2 shows a real-time network capable of cognitive mapping. Two types of inputs are considered: places and views. Places refer to discernible regions in the maze; views refer to the sights of places as seen from another place. For simplicity, the current implementation assumes that animals can only perceive views of places *adjacent* to the place where they are located. Although in the present version of the model we assume independence between places and views, a given place can be defined by the set of views available from that place. Therefore, views may be regarded as egocentric representations (depictions of spatial locations as perceived from the animal's vantage point), and places may be considered allocentric representations (portrayals of spatial locations relative to the positions of other discriminable locations) of the environment.

We assume that both place and view inputs are constant while the animal stays in the same place. Place i in a maze is assumed to give rise to a short-term memory (STM) trace x_i. Place trace x_i may become associated with the views of other places to form long-term associations $V_{i,j}$ between place i and view j. The strength of the modifiable synapses, indicated by open triangles in Figure 11.2, represents $V_{i,j}$ associations. Nonmodifiable synapses are indicated by arrows in Figure 11.2. When the animal is at place i and perceives view j, $V_{i,j}$ increases. When the animal is at place i and cannot perceive view j, $V_{i,j}$ decreases. Each time the animal enters place i, $V_{i,j}$ associations are activated to generate *real-time* predictions of views j to be seen from place i. That is, as long as the animal stays in place i, place i activates neurons p_j proportionally to $V_{i,j}$, and this activity represents the prediction that view j is available from place i (Equation 11.3 in Appendix 11.A). If there is a mismatch between the actual and the predicted view, $V_{i,j}$ is readjusted to reflect the maze configuration (Equation 11.4 in Appendix 11.A).

The cognitive map built by our network is a *topological map*, that is, it represents only the adjacency, but not distance or direction, of places. $V_{i,j}$ associations are the elementary internal learned representations of the topological connections in the external world.

Reading a cognitive map

As explained above, navigation through a maze requires a cognitive map. Cognitive maps should combine multiple associations $V_{i,j}$ to infer spatially remote

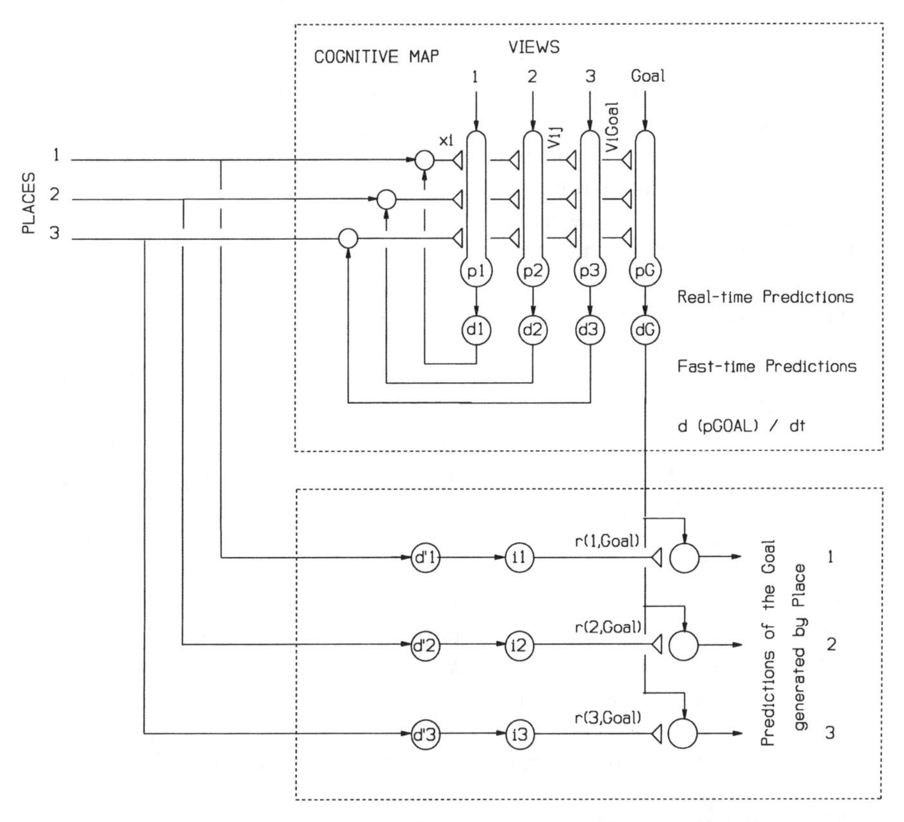

Figure 11.2. Cognitive network. Neural network that associates place representations with view representations. The first derivatives of the predictions of view j, p_j are fed back into the representation of place i, $x_{i=j}$. The first derivative of the prediction of goal, pGOAL, is stored in working memories r (h, goal). $V_{i,j}$: long-term memory of place-view associations; d_j and d_j': neurons computing first derivatives; i_j: neurons integrating the pulses generated by neurons d_j'; G: goal. Arrows represent fixed excitatory connections, open triangles variable excitatory connections.

goal locations. This can be achieved by recurrently reinjecting the signal representing view j as predicted by place i according to $V_{i,j}$, into the representation of place j. Place j now predicts view k according to $V_{j,k}$, and the signal representing view k is reinjected into the representation of place k. The process might continue until the representation of the goal is eventually activated.

Two major problems may hamper the recurrent signal reinjection process: (1) the extinction of previously stored $V_{i,j}$ associations and (2) the formation of spurious $V_{i,j}$ associations. This section explains how the network shown in Figure 11.2 avoids both problems and achieves cognitive mapping.

Suppose, for example, that a maze consists of place A connected to place B and place B connected to place C. The potential extinction of previously stored place–view associations can be explained in the following terms. Assume that the animal is located in place A. Then, place A activates a representation of place B through the $V_{A,B}$ association, and the representation of place B activates the representation of place C through the $V_{B,C}$ association. Because (the view of) place C is not available at place A, and hence is not activated, association

$V_{B,C}$ might extinguish according to Equation 11.4 in Appendix 11.A. This undesired result might be described as an extinction produced by the mismatch between the prediction that place C will be encountered in the future and the present absence of place C.

The potential acquisition of spurious place–view associations is described as follows. Assume again that the animal is located in place A. Place A activates through association $V_{A,B}$ the representation of place B, which in turn activates through association $V_{B,C}$ the representation of place C. Because the representation of place C is active together with the representation of place A, association $V_{A,C}$ might increase by Equation 11.4 in Appendix 11.A. This undesired result might be described as the acquisition of a spurious association between the predictions of (the view of) place C with the present location (place A) of the animal.

The network shown in Figure 11.2 allows the combination of multiple $V_{i,j}$ associations while avoiding the potential modification of previously stored place–view associations. In Figure 11.2, the output of cells p_j excites cells d_j, which generate an output proportional to the first derivative of p_j with respect to time, $d(p_j)/dt$. The output of cells d_j excites the corresponding place traces x_j. Since slow-changing signals weakly activate cells d_j, only fast-changing signals are reinjected into x_j. In other words, cells d_j allow fast-changing signals, but not slow-changing signals, to spread their activation over the network. Therefore, the network operates as a heteroassociative *nonrecurrent* network for slow-changing signals and as a *recurrent* network for fast-changing signals. These dual recurrent and nonrecurrent properties of the network are fundamental for cognitive mapping.

The extinction of previously stored place–view associations is now avoided for the following reasons. Suppose that the animal is steadily located in place A. Then, place A activates the $V_{A,B}$ synapse, but the representation of place B, x_B, is not activated because y_B does not activate d_B. Therefore, although view C is not available from place A, association $V_{B,C}$ is not changed. The acquisition of spurious place–view associations is also averted. Assume that the animal is steadily located in place A. Because the representation of view C is not active together with the representation of place A, x_A, the association $V_{A,C}$ does not change. In sum, undesirable changes in place–view associations are avoided because the network does not reinject slow-changing signals. In other words, the inclusion of d_j cells allows the network to function as a *nonrecurrent* heteroassociative network for slow-changing signals.

Like Tolman, we assume that before making a decision, the animal briefly examines all the *alternative next places h* linked to place i. These brief inspections (reflected in vicarious trial-and-error behavior, VTE) result in fast, short, and relatively weak activation pulses of the place inputs and their corresponding traces x_h. Place traces x_h activate cells p_j in proportion to their $V_{h,j}$ connections. Therefore, view cells p_j are activated by fast-changing signals proportional to $x_h V_{h,j}$, that are recurrently reinjected into x_j through cells d_j. Subsequently, x_j activates p_k, p_k activates x_k, and so forth, spreading the activation over the network. Because activation spreads decrementally (inversely proportional to the magnitude of a reinjection constant), the output of cell d_j reflects the distance of the alternative next place h to place j.

The output of cells d_j consists of a sequence of pulses that reflects the organization of the maze starting at the alternative place under examination (see Figure 11.6). This sequence of pulses might include the activation of the view of the goal. When present, this activation is inversely proportional to the number of reinjections needed to elicit it. In sum, cognitive mapping is achieved by combining multiple associations $V_{i,j}$ as the network recurrently reinjects fast-changing view representations into their corresponding place representations. In other words, the inclusion of d_j cells allows the network to function as a *recurrent* heteroassociative network for fast-changing signals.

Notice that in contrast to the *real-time predictions* of view j generated by the slow-changing place i signal, internal navigation generates *fast-time predictions* of views, that is, imaginary navigation proceeds at a faster pace than the real movement through the maze.

Goal-seeking system and maze navigation

As previously explained, before deciding which place it will enter, the animal sequentially evaluates all alternative next places h linked to its present place i. When the animal examines all the alternative next places, the output of cell d_{goal} is stored in a working memory $r_{h,\text{goal}}$. Working memories $r_{h,\text{goal}}$ are proportional to the prediction of the goal by the alternative next place h, and they decay at a relatively fast rate (Equation 11.5 in Appendix 11.A). By simultaneously comparing $r_{h,\text{goal}}$, the animal is able to decide which of the sequentially scanned places is the best predictor of the goal. We assume that working memories $r_{h,\text{goal}}$ are back to zero by the time the animal arrives at the next choice point.

Under the stimulus-approach view, the alternative next place that best leads to the goal becomes a *subgoal* and is subsequently approached. Because the model approaches the best predictor of the goal, the goal always has preeminence over its predictions. If the magnitude of the goal or its prediction is smaller than a certain minimum value, animals engage in random exploratory behavior. If the magnitudes of two predictions are identical, animals decide at random between the two stimuli generating the predictions. We assume that the total time and, consequently, the total number of reinjections allowed for the inspection of the connectivity of each alternative place are fixed. Maze navigation proceeds until reward is found or the animal is withdrawn from the maze.

To summarize, the recurrent and nonrecurrent properties of the neural network shown in Figure 11.2 allow the exploration of the connectivity of the maze without altering the associations formed between adjacent places and views. In other words, the network is capable of internally navigating the maze without modifying its cognitive map. The neural architecture of Figure 11.2 may be regarded as consisting of two functional overlapping subsystems. One functional subsystem maintains a topological map of the environment and generates real-time predictions for adjacent places. The other subsystem reads out the cognitive map by generating fast-time predictions of remote places. Fast-time predictions are used to guide the animal's behavior but, given their short duration, do not modify place–view associations. Real-time predictions are generated simultaneously with the occurrence of environmental events and are used to update place–view associations that reflect the connectivity of the external world. At a given choice point, animals briefly examine all alternative next places connected to their present place, thereby generating fast-time predictions of the goal. These sequentially generated fast-time predictions are stored in working memories that permit the simultaneous comparison of all the alternative next

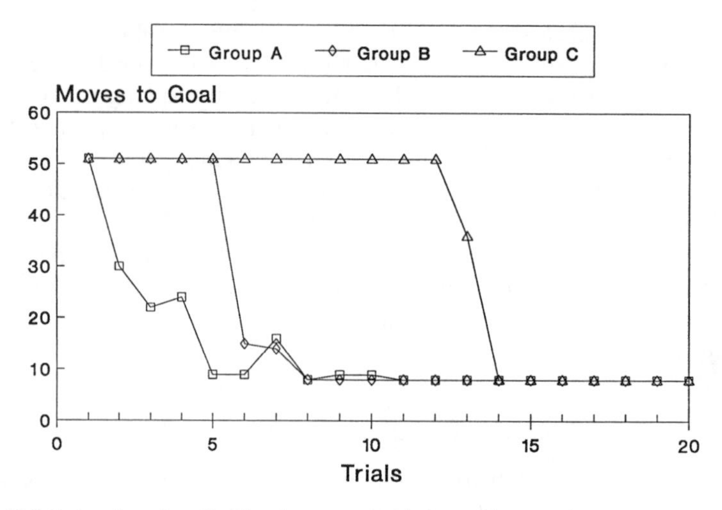

Figure 11.3. Latent learning. Simulated moves to reach the goal in a multiple-T maze for three different groups. Group *A* is rewarded at the goal box on the first trial, group *B* on the seventh trial, and group *C* on the thirteenth trial.

places. As a result of this comparison, a decision is made and the animal enters the place that best leads to the goal.

Computer simulations

This section presents computer simulations obtained with the network for two maze-learning paradigms, latent learning and detour learning, that according to Tolman (1932a, b) illustrate animals' capacity for reasoning. All simulations were carried out with identical parameter values. Parameter values used in the simulations are presented and justified in Appendix 11.B.

Latent learning

A diagram of the maze to study latent learning is shown in Figure 9.3 in Chapter 9. During the latent learning period, the model constructs a cognitive map of place–view associations. When no place–goal associations are available, the map is not utilized. When place–goal associations are established, the animal is able to determine the correct path, therefore showing faster improvement in performance than unexposed animals.

Figure 11.3 shows the number of moves to reach the goal as a function of trials for three different groups. Group *A* is rewarded at the goal box on the first trial, group *B* on the seventh trial, and group *C* on the thirteenth trial. According to Figure 11.3, the model shows latent learning because, when the reward is presented after a period of latent learning, animals with a preconstructed cognitive map (groups *B* and *C*) display rapid improvement in performance to the same level of group *A*. These results are in accordance with Blodgett's (1929) latent learning data.

Interestingly, because place–view associations increase with the number of exploratory movements and this number decreases as the animal is able to encounter the goal, group *C* shows more place–view associations than group *B*, and group *B* shows more place–view associations than group *A*. In Trial 20, the

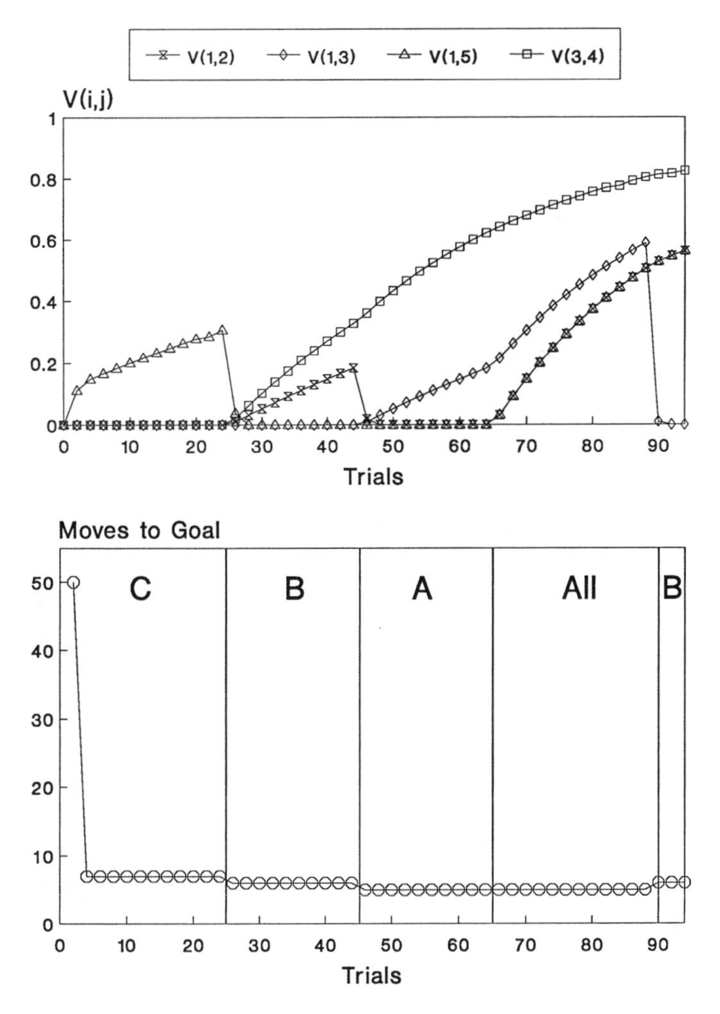

Figure 11.4. Short detour problem. Simulated moves to reach the goal and place–view associations as a function of trials for different maze configurations. Trials 1–24: The animal is forced through the longest path, path *C*. Trials 25–44: The animal is forced through path *B*. Trials 45–64: The animal is forced through path *A*. Trials 65–89: The animal is allowed to enter all pathways. Trials 90–95: Block ***a*** (see Fig. 9.4) is introduced after trial 90.

three groups show identical performance, solving the maze in the minimum number of movements, but they differ in the knowledge about the maze that each one has gathered.

Detour problem

A diagram of the maze employed to study a detour problem is shown in Figure 9.4 in Chapter 9. Path *A* is formed by places 1, 3, 4, 8, and goal. Path *B* is formed by places 1, 2, 3, 4, 8, and goal. Path *C* is formed by places 1, 5, 6, 7, 4, 8, and goal.

Figure 11.4 shows (1) the number of moves to the goal and (2) place–view associations $V(1, 2)$, $V(1, 3)$, $V(1, 5)$, and $V(3, 4)$ as a function of trials in which different maze configurations are adopted. In the simulation, the animal is successively forced to take paths *C*, *B*, and *A* for 25, 20, and 20 trials, respectively. When perfectly learned, path *A* is traversed in five moves, path *B* in six moves,

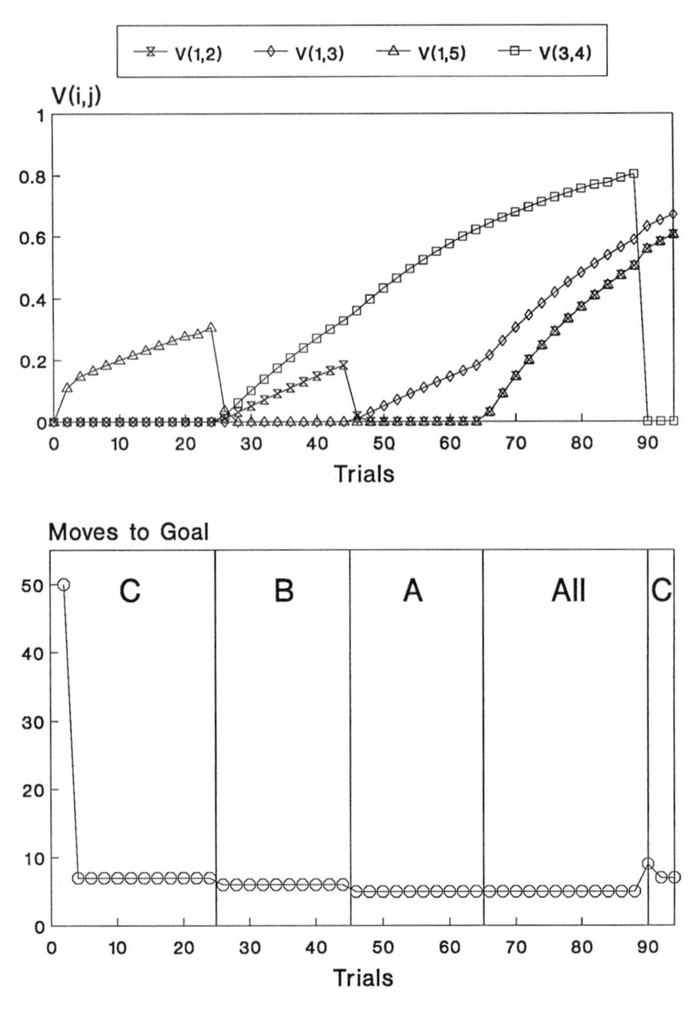

Figure 11.5. Long detour problem. Simulated moves to reach the goal and place–view associations as a function of trials for different maze configurations. Trials 1–24: The animal is forced through the longest path, path *C*. Trials 25–44: The animal is forced through path *B*. Trials 45–64: The animal is forced through path *A*. Trials 65–89: The animal is allowed to enter all pathways. Trials 90–95: Block ***b*** (see Fig. 9.4) is introduced after trial 90.

and path *C* in seven moves. After the forced phase, the simulated animal is free to choose from all three paths, choosing the shortest path, *A*. In trial 90, block ***a*** is introduced in path *A* between places 1 and 3 (see Fig. 9.4), and the animal immediately detours to the second shortest path *B*.

Figure 11.4 shows that, as the animal navigates through path *C*, association $V(1,5)$ increases. When the animal navigates through path *B*, associations $V(1,2)$ and $V(3,4)$ increase and association $V(1,5)$ decays to zero. When the animal navigates through path *A*, associations $V(1,3)$ and $V(3,4)$ increase and association $V(1,2)$ extinguishes. Finally, when all pathways are open, associations $V(1,2)$, $V(1,3)$, $V(1,5)$, and $V(3,4)$ increase. Figure 11.4 shows that the introduction of block ***a*** causes association $V(1,3)$ to decay to zero without altering the other associations, allowing the animal to detour to the second shortest path *B*.

Figure 11.5 shows (1) the number of moves to the goal and (2) place–view associations $V(1,2)$, $V(1,3)$, $V(1,5)$, and $V(3,4)$ as a function of trials in which

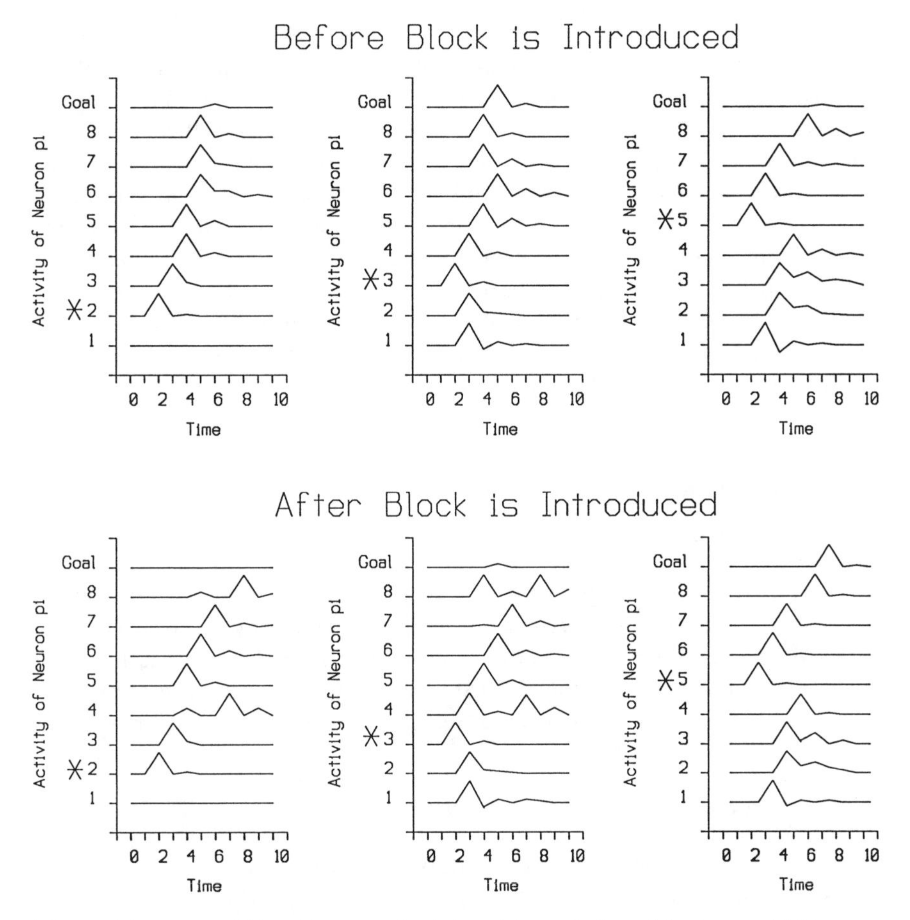

Figure 11.6. Detour problem. Real-time simulation of the predictions of the goal in trial 90 generated by glances of alternative next places 2, 3, and 5 from place 1 before and after introducing block ***b*** (see Fig. 9.4). *Left panels:* alternative 2. *Central panels:* alternative 3. *Right panels:* alternative 5. *Upper panels:* activity before introducing block ***b***. *Lower panels:* activity after introducing block ***b***. Notice that alternative 3 generates the strongest prediction of the goal before block ***b*** is introduced, and alternative 5 generates the strongest prediction of the goal after block ***b*** is introduced. In order to maximize display clarity, scales for each activity trace are different in every case except for the goal. Goal activity is scaled to its maximum value for alternative 3 in the upper panels and for alternative 5 in the lower panels.

different maze configurations are adopted. As in the previous simulation, the animal is successively forced to take paths *C*, *B*, and *A* for 25, 20, and 20 trials, respectively. After this forced phase, the simulated animal is free to choose from all three paths, choosing the shortest path, path *A*. Before the introduction of the block in Trial 90, variations in place–view associations are identical to those presented in Figure 11.5. In trial 90, block ***b*** is introduced between places 3 and 4, blocking paths *A* and *B* (see Fig. 9.4). Figure 11.5 shows that the introduction of block ***b*** causes association $V(3, 4)$ to decay to zero, and although association $V(1, 2)$ remains unchanged, the animal immediately detours to the longest path *C*. This simulation result is in agreement with Tolman and Honzik's data showing that rats, after backing out of path *A*, took the longest path *C*.

Internal navigation of the maze occurs in fast time, that is, it proceeds at a faster pace than the real movement through the maze. Figure 11.6 illustrates how fast-time predictions are generated by the network. The upper panel in Figure 11.6 shows a real-time simulation of the activity of neurons representing different maze views when the animal briefly glances from place 1 into places 2, 3, and 5, *before* block ***b*** is introduced (see Fig. 9.4). The upper panel in Figure 11.6 shows that place 3 evokes the strongest representation of the goal and, therefore, it will be entered by the animal. The lower panel in Figure 11.6 illustrates a real-time simulation of the activity of the nodes representing different maze views when the animal briefly glances from place 1 into places 2, 3, and 5, *after* finding block ***b***. The lower panel in Figure 11.6 shows that after block ***b*** is introduced, place 5 evokes the strongest representation of the goal and, therefore, it will be entered by the animal. Notice that the network generates the sequence in which different places in the maze are interconnected independently of the sequence in which these places were originally visited.

Discussion

Properties of the network

In accordance with Blodgett's (1929) data, the model shows latent learning because, when the reward is presented after a period of latency, rats with a preconstructed cognitive map display rapid improvement in performance to the same level as the group rewarded from the first trial. Also in agreement with Tolman and Honzik's (1930) results, the network correctly describes detour behavior in rats.

Cognitive mapping and maze navigation

The cognitive map built by our network is a *topological map,* that is, it represents only the adjacency, but not distance or direction, of places. $V_{i,j}$ associations are the elementary internal learned representations of the topological connections in the external world. Notice that, whereas Tolman proposed describing spatial maps in terms of S–R–S expectations, we assume a more radical cognitive posture, proposing that place–view associations, similar to classical CS–CS associations, suffice to describe behavior. Chapter 3 (Schmajuk, 1986, 1987, 1989; Schmajuk and Moore, 1988) offers a similar approach to describe classical conditioning paradigms such as sensory preconditioning. The present chapter shows how a comparable strategy can be applied to the description of latent learning and detour learning in mazes, thereby providing a unified approach to cognitive mapping in classical conditioning and spatial mapping.

Because the network functions as a heteroassociative nonrecurrent network for slow-changing signals, associations between adjacent places remain in "leak-proof" compartments. However, because the network also functions as a *recurrent* network for fast-changing signals, memories stored in the leak-proof compartments can be transiently combined into a cognitive map. Therefore, in contrast to Gelperin's (1986) autoassociative recurrent network, the network presented here shares the leak-proof properties of heteroassociative nonrecurrent networks (see Box 1.3) and, consequently, maintains a topological model of the world based on its experiences but not on its internal inferences. Also in contrast to Schmajuk's (1986, 1987, 1989) autoassociative nonrecurrent network, the network is capable of combining multiple predictions by recurrently reinjecting output information into its inputs.

Short-term, long-term, and working memories

The present model comprises three different types of memories: a trace short-term memory, an associative long-term memory, and an associative working memory. Trace memory x_i increases when place i is present and gradually decays back to zero when place i is absent. Short-term memory x_i does not represent associations between events. Associative long-term memory $V_{i,j}$ is used to generate the prediction that view j is available from place i. Long-term memory $V_{i,j}$ can be recalled by the presentation of place i and does not decay spontaneously. Finally, associative working memory $r_{h,\text{goal}}$ represents the prediction of the goal by the alternative next place h. Working memory $r_{h,\text{goal}}$ can be recalled by the presentation of place h and spontaneously decays at a relatively fast rate.

Fast-time and real-time predictions

A distinction between two types of predictions is introduced in this chapter: fast-time and real-time predictions. Fast-time predictions, that is, predictions produced in advance of what occurs in real time (Pribram and Isaacson, 1975), are generated at choice points when animals briefly examine all alternative next places connected to their present place. These sequentially generated fast-time predictions are stored in working memories that permit the simultaneous comparison of all the alternative next places. As a result of this comparison, a decision is made and the animal enters the place that best leads to the goal. Real-time predictions, that is, predictions of environmental events as moment-to-moment phenomena, are generated when place–view relationships are evaluated and, eventually, changed. When an animal enters place i, place–view associations $V_{i,j}$ are activated and the animal generates predictions of the views j that were available from that place in the past. These predictions are compared with the actual views and, if there is a mismatch, place–view associations $V_{i,j}$ are modified.

Comparison with other formal approaches to cognitive mapping

The main difference between our network and Deutsch's (1960) model is that, whereas our network generates forward predictions of the goal, Deutsch's model generates backward activation of the nodes. According to Sutherland and Mackintosh (1971), Deutsch's theory has several difficulties. First, the theory does not postulate mechanisms for learning the connections between stimulus representations. Second, the theory does not offer performance rules to guide the animal through the maze. Third, the theory wrongly predicts that, if a rat is placed close to the goal and then returned to the start of the maze, it will not run because lower excitation links are cut off by the animal's being close to the goal. Fourth, the theory cannot account for learning in a Skinner box where the animal has first to press a bar and then approach the food magazine. The theory predicts that the animal will directly approach the magazine without pressing the bar first. The network presented in this chapter remedies some of the problems in the Deutsch model. First, our theory postulates mechanisms for learning the connections between stimulus representations. Second, our theory offers performance rules to guide the animal through the maze. Third, our theory predicts that if a rat is placed close to the goal and then returned to the start of the maze, it will run because the cognitive map is not affected by the animal being close to the goal. However, like the Deutsch model, our theory predicts that the animal will directly approach the magazine in a Skinner box without pressing the bar first.

Unlike Lieblich and Arbib's world graph, the cognitive map implemented by our network simply represents connections between places, and the drive-reduction properties of a given place are computed on the basis of the information stored in the map. Although Hampson's (1990) approach shares aspects with the model introduced in the present study, our network differs from Hampson's in various respects. First, our system is implemented as a neural, biologically plausible model. Second, we specifically describe how the neural network builds (and uses) a cognitive map of the environment. Third, in contrast to Hampson's use of stimulus–response operators, our network adopts a purely stimulus-approach strategy. Hampson's use of stimulus–response operators might constitute an important feature to account for experimental data suggesting that, as maze learning is thoroughly mastered, animals switch from a stimulus-approach to a response-selection strategy (Mackintosh, 1965).

Choice-point behavior

Although the network presented in this chapter is capable of successful navigation through the environment, the issue of choice-point behavior remains unsolved. Bower (1959) detailed a mathematical model that describes the behavior of a subject at a choice point in a T maze. At the choice point, the subject might orient to the right or to the left; when the subject is oriented in one direction, it might move in that direction or reorient. Bower described this VTE behavior between discriminative stimuli in terms of a Markov process. Some states in the Markov process represent orientations to different stimuli, for example, right or left. Other states represent the subject's final approach to either stimulus. In representing choice-point behavior as a Markov process, Bower assumed that transition probabilities (1) depend only on the last state and (2) remain constant for the duration of the trial. Bower contrasted the model's predictions about (1) the probabilities of transitions between orienting responses and (2) the total number of transitions with the VTE behavior of rats in a T maze. Experimental data show that (1) animals increased the transitions toward the rewarded side and decreased the transitions to the nonrewarded side, and (2) animals decreased the total number of transitions. Bower found that the model could correctly describe most experimental results.

In contrast to Bower's (1959) choice-point model, our model assumes that VTE behavior does not change over trials. Duration of VTE behavior is always proportional to the number of alternative next places accessible from a given choice point. The experimental finding that decision time at choice points decreases over trials may imply that the magnitude of the prediction of the goal $r_{h,\text{goal}}$ might control decision time and VTE behavior. As $r_{h,\text{goal}}$ increases over trial, decision time at choice points would decrease.

Problem solving

Tolman (1932b, p. 177) suggested that the relations between initial and goal positions can be represented by a directed graph and called this graph a means-end

Table 11.1. *Simulations obtained with the Schmajuk and Thieme (1992) maze-learning model.*

Paradigm	Model
Water maze	Y
Latent learning in the water maze	Y
Detour problems in the water maze	Y
Velocity profile in a runway	
Appetitive-aversive interactions in a runway	
Multiple-T maze	Y
Latent learning in the multiple-T maze	Y
Detour problems in the multiple-T maze	Y

Note. Y: The model can simulate the experimental data.

field. Many years later, artificial intelligence theories described problem solving as the process of finding a path from an initial to a desired state through a directed graph (Winston, 1977).

Directed graphs can be transformed into trees by having terminating paths lead to previously visited nodes. Two basic methods of tree searching have been proposed, namely, depth-first and breadth-first searches. In the depth-first method, one alternative is selected and pursued at each decision point until forward movements are blocked. Then forward exploration begins again from the nearest decision point with unexplored alternatives. In a breadth-first method, all alternatives at each decision point are equally explored. According to Winston (1977, p. 92), when search trees are very large, both depth- and breadth-first techniques become inefficient. The inefficiency of a sequential approach to tree searching is avoided by our neural network. Our neural network combines depth- and breadth-first procedures: At each decision point, the network first explores *in depth* all alternatives and then selects the best one.

In order to ascertain the power of the network in tree-structured problem solving, we applied the network to the Tower of Hanoi task. This task begins with three pegs and three rings of increasing sizes placed on the left peg. The goal is to move the three rings from the left to the right peg. The rules to move the rings are (1) only one ring can be moved at a time, and (2) a larger ring cannot be placed on top of a smaller one. Following Hampson (1990), we represented the task in terms of the possible transitions between different ring arrangements in the three pegs. Simulations show that the network takes a few trials to learn to solve the problem in the minimal number of moves (seven).

Conclusion

The present chapter presents a real-time, biologically plausible neural network capable of describing cognitive mapping. Computer simulations show that the network can be successfully applied to the description of maze learning, latent learning, and detour behavior (see Table 11.1). In addition, Schmajuk et al. (1993) showed that the model can also describe learning in the Morris tank. Simulations also demonstrate that the network can be applied to problem-solving paradigms, such as the Tower of Hanoi puzzle.

Appendix 11.A

Formal description of the model

Changes in the representation of place *i*, x_i, are given by

$$dx_i/dt = -K_1x_i + K_2(1 - x_i)\,T_i \tag{11.1}$$

where K_1 is a decay constant and K_2 a rise constant. x_i is bound between 0 and 1.
T_i is given by

$$T_i = \text{Place}_i + K_3 f[d(p_i)/dt] \tag{11.2}$$

where place$_i$ is 1 if the animal is at place *i* and 0 otherwise. Place$_i$ is 0.1 if the animal glances into place *i* while at another position and 0 otherwise. K_3 is the reinjection constant. $f[d(p_i)/dt]$ is the *fast-time prediction* of place$_i$. $f[\,d(p_i)/dt] = d(p_i)/dt$ if $d(p_i)/dt > 0$, and $f[d(p_i)/dt] = 0$ otherwise. Therefore, the representation of place *i*, x_i, is active when the animal is at place *i* or place *i* is predicted by the network.

The *real-time prediction* of view *j*, p_j, is given by

$$p_j = \Sigma_j V_{i,j} x_i \tag{11.3}$$

where V_{ij} is the association between place *i* and view *j*.

Changes in association V_{ij} between place *i* and view *j* are given by

$$dV_{ij}/dt = K_4\, x_i(\text{view}_j - p_j) \tag{11.4}$$

According to Equation 11.4, $V_{i,j}$ increases when view *j* and x_i are present together and decreases when x_i is present in the absence of view *j*. Because $d(p_j)/dt$ is a short pulse generated when the animal glances into different places, x_i is essentially proportional to place$_i$ (see Equation 11.2), and therefore $V_{i,j}$ basically reflects the association between place$_i$ and view$_j$.

Before making a decision, the animal briefly examines all the alternative next places *h* linked to place *i*. The animal chooses the highest value $d(p_{\text{goal}})/dt$. This is accomplished by associating a STM of the pulse, generated by neurons i_i in Figure 11.2, with $d(p_{\text{goal}})/dt$:

$$d(r_{h,\text{goal}})/dt = -K_5 r_{h,\text{goal}} + K_6 i_h(1 - r_{h,\text{goal}})\, d(p_{h,\text{goal}})/dt \tag{11.5}$$

According to Equation 11.5, $r_{h,\text{goal}}$ increases when alternative *h* and $d(p_{h,\text{goal}})/dt$ are active together and decreases otherwise. $r_{h,\text{goal}}$ is bound between 0 and 1. If the largest $r_{h,\text{goal}}$ is less than a threshold value (K_7), we assume that the next place is randomly visited. If two $r_{h,\text{goal}}$ have identical values, the next place is randomly decided between those two places.

Appendix 11.B

Parameter values

Simulations assume that one time step is equivalent to 1 msec. Parameters used in all simulations were $K_1 = 0.99$, $K_2 = 0.25$, $K_3 = 1$, $K_4' = 10^{-3}$, $K_4'' = 2$, $K_5 = 10^{-5}$, $K_6 = 0.3$, and $K_7 = 10^{-5}$. Parameters K_1 and K_2 were selected to obtain a moderate rise and a rapid decay in place short-term memory x_i. Parameter K_3, which controls the rate of attenuation over repetitive reinjections in the network, was chosen to obtain an adequate signal at the neuron representing the goal. Greater values of K_3 would allow more reinjections without appreciable attenuation. Parameters K_4' and K_4'' were selected to achieve relatively slow acquisition and fast extinction of place–view associations $V_{i,j}$. A relatively slow acquisition rate is needed to describe latent learning and a fast extinction rate is required to generate a fast detour. When the animal enters a given place, it stays in that place for 30 time units before starting VTE behavior. Each place connected to the currently occupied place is briefly entered for one time unit. In order to avoid interference, 10 time units separate the examination of each alternative next place. The total time allowed for the analysis of the consequences of choosing a given path is 30 time units. K_5 and K_6 are selected to store values of different working memories $r_{h,\mathrm{goal}}$ for a length of time that allows the comparison and selection of the best predicting path.

IV Learning, cognition, and the brain

12 Learning, cognition, and the hippocampus: data and theories

Parts I–III describe how different neural networks can be applied to the behavioral aspects of classical, operant, and cognitive paradigms. Part IV shows that, in addition to describing behavior, when regarded as models of specific brain circuits, neural networks can provide theories that extend to the anatomical and physiological aspects of learning.

Modeling of brain systems has been acomplished by means of both top–down and bottom–up approaches. Top–down analysis consists first of describing a given behavior, then proposing possible neural mechanisms that might subserve that behavior, and finally mapping these mechanisms onto brain circuits. Bottom–up analysis consists of describing the functional properties of given brain circuits and then integrating this function with the control of the behavior under study. In general, top–down approaches are useful in a first approximation of the understanding of the neural basis of behavior. Bottom–up approaches are valuable when the physiological interactions within a given brain circuit are known and the relation between circuit functioning and overt behavior is to be established.

Grossberg (1974) adopted a top–down approach and suggested that, when specific psychological postulates are established, a minimal neural architecture can be derived that realizes these postulates. The next step is to analyze the psychological and neural capabilities of this network in order to determine what the network can and cannot do. This information suggests new psychological postulates that are needed to derive the next, more complex network. Iteration of this process presumably leads to neural architectures that closely resemble the actual anatomy of the brain.

Like Grossberg, Marr also supported a top–down approach to information processing in the brain. According to Marr (1982), before one can say that a system is completely understood, it should be analyzed at different levels: computational, the goal of the computation; representational, the representations of inputs and outputs; algorithmic, the transformation of the input into the output; and implementation, the brain implementation of the algorithm. To paraphrase Marr, learning algorithms are more readily grasped by understanding the nature of the problem being solved (e.g., generating predictions in the case of classical conditioning) than by examining the neural mechanism in which they are embodied. Understanding the computational level helps the understanding of the algorithmic level. Understanding the algorithmic level helps the understanding of the implementation level. Many studies have made use of a "top–down" approach to brain function: first portraying a neural network that describes a given behavior and then proposing a plausible mapping of the network onto the brain circuitry. The advantage of a top–down approach is that it can organize a large amount of *otherwise seemingly conflicting* data into the framework of a functional theory. However, an important limitation of top–down approaches is that they might disregard some aspects of the brain circuitry.

Bottom–up approaches consist of (1) describing the neuronal properties and synaptic interactions in a given brain network, (2) building a biologically *realistic,* quantitative model of the network, and (3) determining through computer simulations whether such empirically derived circuit can quantitatively account for the behavioral output of the actual circuit. In contrast to top–down approaches, "bottom–up" approaches have sometimes serious difficulties in their attempt to derive the behavioral meaning of complex brain circuits from functional, neurophysiological, and anatomical information.

According to Rumelhart and McClelland (1986), the connectionist approach combines top–down considerations about cognitive processes with bottom–up brain style, neurally inspired processing principles. Neural network approaches seem to permit the integration of top–down (providing a functional interpretation for components and connections) and bottom–up strategies (contributing descriptions of these physiological components and connections). Known properties of different neural circuits might be incorporated into top–down descriptions in order to characterize the functional properties of a given brain area. The resulting model might account for a wide range of data in normal and lesioned animals, and for the neural activity that accompanies behavior.

Chapter 6 introduced the SD network and showed that it describes a large number of classical conditioning data. The SD model is built with neural elements whose properties (described in terms of differential equations) are constrained by neurophysiological data: (1) real-time nonlinear differential equations are constrained by neurophysiological properties (membrane dynamics, active excitation and inhibition, and saturation properties), and (2) the mathematical rules for controlling changes in these learning sites are constrained by empirical, quantitative studies of such plasticity. Chapter 13 shows how nodes and connections in the SD network can be mapped onto regional cerebellar, cortical, and hippocampal circuits. In the brain-mapped model, (1) neural elements in the model represent specific neural populations in the brain, (2) the activity of neural elements in the model represents neural activity of specific neural populations, (3) connections between neural elements represent known anatomical connections, (4) learning sites in the model represent known locations of plasticity in the brain, and (5) different manipulations (e.g., "lesions") of neural elements in the model represent equivalent manipulations of specific neural populations in the brain. The outcome is a neurophysiologically plausible model capable of describing animal behavior.

The neurophysiology of classical conditioning of the nictitating membrane response

This section briefly reviews neurophysiological, anatomical, and behavioral data that will constrain the mapping of the SD model onto brain circuits related mainly to classical conditioning of the rabbit's nictitating response.

Hippocampal neuronal activity during classical conditioning

In the rabbit nictitating membrane (NM) preparation, hippocampal activity during classical conditioning is positively correlated with the topography of the conditioned response (CR) (Berger and Thompson, 1978a). Berger et al. (1983) found that CA1 and CA3 pyramidal cells are characterized by an increase in frequency of firing over conditioning trials, and by a within-trial pattern of discharge that models the NM response. Berger, Clark, and Thompson (1980) reported that the activity correlated with the CR is present also in the entorhinal cortex, but is amplified over trials in CA1 and CA3 hippocampal regions. During extinction, Berger and Thompson (1982) found that pyramidal cells in dorsal hippocampus decrease their frequency of firing correlated with behavioral extinction during the US period (defined during acquisition), but in advance of behavioral extinction during the CS period.

In addition to CR-related neural activity, other types of activity have also been recorded from hippocampal regions. For instance, Vinogradova (1975) found that neural activity in CA3 and CA1 pyramidal neurons, dentate gyrus, and entorhinal cortex of the rabbit was correlated with the presentation of sensory stimuli (tones and light). Activity in CA3 and CA1 showed habituation after repeated presentation of the stimuli. Berger et al. (1983) found that hippocampal neurons coding CS information do not respond to the US and are also segregated from those coding CR information. Wible et al. (1986) observed that, in a variety of learning tasks, hippocampal cells respond not only to individual stimuli, but to combinations of stimulus dimensions such as color, shape, and spatial location.

Berger et al. (1983) also reported that theta cells, presumably basket cells that provide recurrent inhibition to pyramidal neurons, respond during paired conditioning trials with a rhythmic 8-Hz bursting pattern. Weisz, Clark, and Thompson (1984) found that granule cells in the dentate gyrus exhibited a CS-evoked theta firing when rabbits were trained with a CS followed by a US, but not when they were trained with CS and US unpaired presentations.

In summary, pyramidal neural activity in CA3 and CA1 might be correlated either with the behavioral CR or with the presentation of simple and compound CSs. Granule cell activity in the dentate gyrus exhibits theta firing evoked by CS presentation.

Lateral septal neural activity during classical conditioning

As mentioned, Berger and Thompson (1978a) and Berger et al. (1983) found that the activity of CA1 and CA3 pyramidal cells is positively correlated with the topography of the CR. Because one important hippocampal output is mediated through CA3 axons that reach the lateral septum, it is not surprising that Berger and Thompson (1978b) and Salvatierra and Berry (1989) reported that neural activity in the lateral septum during acquisition of classical conditioning in the rabbit NM preparation is similar to that recorded in hippocampal pyramidal cells. These results suggest an excitatory CA3-lateral septum pathway. However, in contrast to this view, Vinogradova et al. (1980) reported that interruption of hippocampal connections to the lateral septum at the level of the septo-fimbrial nucleus increased spontaneous activity in the lateral septum.

In summary, hippocampal activity reaches the lateral septum through CA3 axons and neural activity in the lateral septum during acquisition of classical conditioning reflects the activity of hippocampal pyramidal cells.

Cerebellar involvement in classical conditioning

Experimental evidence using the rabbit NM and the rat eyeblink conditioning preparations suggests that cerebellar areas are essential for the acquisition and maintenance of classical conditioning (Skelton, 1988; Thompson, 1986). The association of the CS and the US would be mediated by plastic changes at the interpositus nucleus of the cerebellum and/or at the Purkinje cells of the hemispheric portion of cerebellar lobule VI. Sensory representations of the CS may reach the interpositus nucleus and the cerebellar cortex via mossy fibers from the pontine nuclei, and the US representation seems to reach the interpositus nucleus and the cerebellar cortex via climbing fibers from the dorsal accessory olive (McCormick, Steinmetz, and Thompson, 1985). CR-related activity originates in the cerebellar lobule VI and/or the interpositus nuclei, is relayed to the contralateral red nucleus, and reaches the contralateral accessory abducens nuclei where the NM response is controlled. The inferior olive is selectively active during *unexpected* somatic events (Andersson and Armstrong, 1987). Weiss, Houk, and Gibson (1990) showed that stimulation of the red nucleus inhibits activity in the dorsal accessory olive.

Lesions of the dentate and interpositus cerebellar nuclei ipsilateral to the trained eye cause abolition of both the behavioral CR and CR-related neural activity in CA1 (Clark et al. 1984; Sears and Steinmetz, 1990). Information about the behavioral response might be conveyed to the hippocampus via cerebellar-thalamic-cortical pathways (see Ito, 1984).

To summarize, cerebellar areas seem critical for the acquisition and maintenance of eyeblink and NM conditioning. Whereas the pontine nuclei seem to convey CS information, the dorsal accessory olive appears to carry US information to the cerebellar loci of learning. CR-related activity originates in the cerebellar lobule VI and/or the interpositus nuclei, is relayed to the contralateral red nucleus, and reaches the contralateral accessory abducens nuclei where the NM response is controlled. Lesions of cerebellar nuclei cause abolition of both the behavioral CR and the CR-related neural activity in pyramidal cells.

Hippocampo–cerebellar interactions: modulation of classical conditioning

Berger et al. (1980) proposed that hippocampal projections to the pontine nucleus modulate the activity of mossy fiber projections to cerebellar cortex and interpositus nucleus (Berger et al., 1986; Steinmetz, Logan, and Thompson, 1988). Two hippocampo-pontine pathways have been described: a hippocampal-retrosplenial cortex projection via the subiculum that reaches the ventral pons (Berger, Bassett, and Weikart, 1985; Berger et al., 1986; Semple-Rowland, Bassett, and Berger, 1981) and a cingulo-pontine projection (Weisendanger and Weisendanger, 1982; Wyss and Sripanidkulchai, 1984).

In addition to its modulation of the pontine nucleus, Schmajuk and Di Carlo (1992) suggested that the hippocampus also acts on the dorsal accessory olive, thereby modulating the activity of climbing fiber projections to cerebellar cortex and interpositus nucleus. The hippocampus might act on the dorsal accessory olive through a lateral septum-raphe nucleus-dorsal accessory olive pathway. This suggestion is supported by Vinogradova's (1975) data showing that CA3 stimulation has an excitatory influence on the raphe nuclei and Weiss and Pellet's (1982a, b) results demonstrating that the raphe inhibits the inferior olive. More specifically, raphe fibers reaching the inferior olive innervate the dorsal accessory nucleus (Wilklund, Björklund, and Sjölund, 1977), the area proposed to convey US information to cerebellar areas where CS–US associations are stored (McCormick et al., 1985).

To recapitulate, hippocampal output might modulate postulated CS (pontine nucleus) and US inputs (dorsal accessory olive) to cerebellar areas.

Hippocampo–reticular formation interactions: control of theta rhythm

Vinogradova (1975, p. 39) reported that stimulation of the reticular formation evokes theta rhythm in the medial septum, whereas hippocampal stimulation inhibits activity in the medial septum and leads to a decrease in theta rhythm. Vinogradova suggested that the hippocampus excites the raphe nuclei that, in turn, inhibit theta. Consistent with this view, Vertes (1982) suggested that (1) the medial raphe nucleus is responsible for inhibiting theta, and (2) the pontis oralis is responsible for exciting hippocampal theta, through the medial septum (but see Graeff, Quintero, and Gray, 1980).

Berger and Thompson (1978b) found that neuronal activity in the medial septum during acquisition of classically conditioned NM responses decreased over trials during both the CS and US periods. They suggested that the medial septum might provide the hippocampus with an "arousal" signal that decreases across trials. In agreement with Vinogradova's (1975) view that the hippocampus inhibits the medial septum, thereby decreasing hippocampal theta, Berger and Thompson (1978b) noted that medial septal and hippocampal pyramidal activity is negatively correlated. Also compatible with Vinogradova's view is the negative correlation between hippocampal theta cell activity and pyramidal cell activity reported in an odor-discrimination paradigm in rats (Eichenbaum et al., 1986).

In summary, hippocampal output excites the raphe, which in turn inhibits the medial septum, and thereby decreases hippocampal theta rhythm. Consistently, during acquisition of classical conditioning, as hippocampal activity increases, medial septum neuronal activity decreases.

Medial septal modulation of hippocampal activity

Axons of the medial septum project to the dentate gyrus, CA3, and to a lesser extent the CA1 region, via the dorsal fornix and fimbria. Medial septal input plays a major role in the generation of hippocampal theta rhythm and in controlling the responsiveness of pyramidal cells in CA1 and CA3 regions and of granule cells in the dentate gyrus.

Permanent and temporary inactivation of the medial septal input to the hippocampus results in decreased activity of pyramidal cells. Vinogradova (1975) reported that lesions of septol-hippocampal pathways decreased the number of neurons responsive to sensory inputs in CA3, but not in CA1, particularly those activated by multimodal stimuli. Recently, Mizumori et al. (1989) examined the effects of reversibly inactivating the medial septum (with lidocaine) on spontaneous hippocampal single-unit activity. They reported that septal inactivation reduced spontaneous firing rates in granule cells, hilar/CA3 complex spike cells, and CA1 theta cells, but not in CA1 complex spike cells.

In agreement with the facilitatory view of medial septal function suggested by the inactivation studies, Krnjevic, Ropert, and Casullo (1988) showed that medial septal stimulation can strongly depress tonic inhibition and inhibitory postsynaptic potentials evoked in CA1 and CA2/3 pyramidal cells by fimbrial stimulation. Bilkey and Goddard (1985) reported that stimulation of the medial septum, although unable to elicit a field potential of its own, facilitated the granule cell population spike evoked by medial perforant path stimulation.

In addition to inactivation and stimulation studies, data obtained from freely moving rats show that the excitability of CA1 pyramidal cells and dentate gyrus granule cells reaches a maximum during the positive phase of theta rhythm (Rudell, Fox, and Ranck, 1980).

It has been suggested that medial septal "modulation" of dentate granule cells, CA1 and CA3 pyramidal cells, is mediated through either (1) a GABA-ergic inhibition of inhibitory interneurons or (2) a cholinergic excitation of pyramidal and granule cells (Bilkey and Goddard, 1985; Krnjevic et al., 1988; Rudell et al., 1980; Stewart and Fox, 1990).

Besides modulating pyramidal and granule cell activity, the medial septum regulates the generation of long-term potentiation (LTP) in perforant path-dentate gyrus synapses. Robinson and Racine (1982) demonstrated associative LTP involving septal and entorhinal inputs to the dentate gyrus in the rat. Concurrent medial septal and perforant path activation resulted in stronger LTP of the perforant path-dentate gyrus synapses than could be produced by perforant path stimulation alone. Similarly, Robinson (1986) reported that LTP induced in the rat dentate gyrus is enhanced by coactivation of septal and entorhinal inputs. Functionally important is the observation that these perforant path-dentate gyrus synapses are endogenously potentiated during classical conditioning of the rabbit NM response (Weisz, Clark, and Thompson, 1984).

In summary, the medial septum modulates hippocampal CA1 and CA3 pyramidal cells and dentate granule cells. In addition, medial septal activity enhances LTP of perforant path-dentate gyrus synapses.

Hippocampo–cortical interactions

Hippocampo–cortical interactions have recently been summarized by Squire, Shimamura, and Amaral (1989). A prominent hippocampal input is the entorhinal cortex. The entorhinal cortex receives inputs from polysensory associational cortices. A major hippocampal output is comprised of CA1 axons that extend to the subiculum, which in turn excites cells in the entorhinal cortex. Cells in the entorhinal cortex that do not project to the dentate gyrus project to many of the associational cortical fields from which the entorhinal cortex receives input.

To summarize, the hippocampus receives information from and sends information back to different polysensory associational cortices through the entorhinal cortex.

Cortico–cerebellar interactions

Parietal association cortex and frontal motor cortex project to the cerebellum. Stimulation of the parietal cortex produces responses in mossy fibers that originate in the pontine nuclei, whereas stimulation of the motor cortex produces responses in motor areas of the cerebellar cortex. In turn, cerebellar lateral and interpositus nuclei project to parietal association cortex and motor cortex (Ito, 1984, p. 314).

In short, while the pontine nuclei receive information from neocortical areas, interpositus nuclei project back to association and motor cortices.

Effects of hippocampal manipulations on classical conditioning

Hippocampal lesion techniques

For a long time, aspiration lesions of the hippocampus were the technique of choice (for a description of the technique see Isaacson and Woodruff, 1975). Aspiration lesions primarily damage the dentate gyrus and hippocampus proper (CA3 and CA1 regions), but might also extend to the adjacent subiculum, presubiculum, overlying cortex, and sometimes, the entorhinal cortex. Jarrard (1986) showed that fibers of passage, such as rostral projections from subiculum and entorhinal cortex, are interrupted by the lesion. According to Jarrard and Davidson (1991), aspiration lesions are best described as lesions of the hippocampal formation (HFL).

In addition to aspiration techniques, kainic acid and/or colchicine have also been used to produce hippocampal lesions (Jarrard, 1983). According to Jarrard (1983; Jarrard and Davidson, 1991; Jarrard and Meldrum, 1990; Whishaw and Tomie, 1991), injections of kainic acid and/or colchicine in the hippocampus might cause secondary damage to subicular, parasubicular, and presubicular cortices, entorhinal cortex, amygdala, prepyriform nuclei, and deep layers of neocortex (occipital and temporal cortices). Therefore, if kainic acid and/or colchicine lesions extend to parahippocampal areas from which fibers of passage originate (such as subicular areas or entorhinal cortex), they may be functionally similar to aspiration lesions. In contrast to the lesions obtained with the aspiration and kainic acid techniques, multiple injections of small amounts of ibotenic acid seem to selectively damage the dentate gyrus and hippocampus proper (CA3 and CA1 regions), without interrupting fibers of passage (Jarrard, 1986; Jarrard and Meldrum, 1990). According to Jarrard and Davidson (1991), ibotenic acid lesions are best described as lesions of the hippocampus proper (HPL).

This section reviews the effects of HFL and HPL on different classical conditioning paradigms.

Acquisition

The effects of HFL on acquisition of classical conditioning have been studied under a wide variety of experimental parameters, including different combinations of CS and US durations, interstimulus intervals (ISI), and types of US.

Acquisition of delay conditioning of the rabbit NM response has been reported to be unaffected or facilitated by HFL. Schmaltz and Theios (1972) found that HFL rabbits showed faster acquisition using a 250-msec CS, 50-msec shock US, and 250-msec ISI. Solomon and Moore (1975) determined that HFL rabbits displayed normal acquisition using a 450-msec CS, 50-msec shock US, and 450-msec ISI. Berger and Orr (1983) found normal acquisition using a 850-msec CS, 100-msec air puff US, and 750-msec ISI. Port and Patterson (1984) determined that HFL rabbits showed shorter CR onset latency and normal acquisition rate using a 500-msec CS, 50-msec shock US, and 450-msec ISI. Port, Mikhail, and Patterson (1985) found shorter CR onset latency and faster acquisition using a 800-msec CS, 50-msec shock US, and 150-msec ISI. When the ISI was extended to 300 msec, they determined that HFL rabbits showed normal CR onset latency and normal acquisition rate. Finally, with a 600-msec ISI, they found that HFL rabbits showed normal CR onset latency and acquisition.

Acquisition of trace conditioning of the rabbit NM response has been reported to be impaired, unaffected, or facilitated by HFL. Port et al. (1986) reported that HFL rabbits showed longer CR onset latency and normal acquisition rate using a 250-msec CS, 50-msec shock US, and 750-msec ISI. When the shock US was replaced by a 100-msec air puff US, they determined that HFL rabbits showed shorter CR onset latency and normal acquisition rate. Solomon et al. (1986) reported shorter CR onset latencies and slower acquisition in HFL rabbits using a 250-msec CS, 100-msec air puff US, and 500-msec ISI. James, Hardiman, and Yeo (1987) reported normal acquisition rate and shorter onset latency using a 250-msec CS, 50-msec shock US, and 750-msec ISI. Recently, Moyer, Deyo, and Disterhoft (1990) found normal CR onset latency and a slightly faster acquisition rate using a 100-msec CS, 150-msec air puff US, and 300-msec ISI. However, they determined shorter CR onset latency and a deficit in acquisition when the ISI was 500 msec.

Extinction

Three studies describe the effect of HFL on extinction of the rabbit NM response. Berger and Orr (1983) found normal extinction using a 850-msec CS, 100-msec air puff US, and 750-msec ISI in an explicitly unpaired procedure with trials in which the CS was presented alone, alternating with trials in which the US was introduced alone. Moyer et al. (1990) determined impaired extinction using a 100-msec CS, 150-msec air puff US, and 300-msec ISI in a simple extinction procedure with CS-alone presentations. In an acquisition–extinction series of the NM response in rabbits, Schmaltz and Theios (1972) found that the first extinction (with CS-alone presentations) appeared to be unaffected by HFL.

Savings

Schmaltz and Theios (1972) studied the effect of exposing rabbits to successive acquisition and extinction series using a 250-msec CS, 50-msec shock US, and 50-msec ISI. They found that HFL rabbits show faster acquisition than normals in the first acquisition series. HFL rabbits did not differ from normals in the first extinction series, but in the following extinction series normal rabbits decreased the number of trials to reach criterion (showed savings), whereas HFL rabbits had increased the number of trials to the extinction criterion.

Blocking

Solomon (1977) found that HFL disrupted blocking of the rabbit NM response and Rickert et al. (1978) discovered impairment of blocking after HFL using a conditioned suppression paradigm in rats. In contrast to these findings, Garrud et al. (1984) determined that blocking was not affected by HFL using a conditioned suppression paradigm in rats.

Overshadowing

Rickert et al. (1979), using a conditioned suppression paradigm, and Schmajuk, Spear, and Isaacson (1983), using a simultaneous discrimination procedure, found that overshadowing is disrupted in HFL rats. In contrast to these findings, Garrud et al. (1984) and Solomon (1977) determined that overshadowing was not affected by HFL.

Discrimination acquisition and reversal

Buchanan and Powell (1982) examined the effect of HFL on acquisition and reversal of eyeblink discrimination in rabbits. They found that HFL slightly impaired acquisition of discrimination and severely disrupted its reversal by increasing responding to CS–. Berger and Orr (1983; Orr and Berger, 1985) contrasted HFL and control rabbits in a two-tone differential conditioning and reversal of the NM response. Although HFL did not affect initial differential conditioning, HFL rabbits were incapable of suppressing responding to CS–. Because HFL rabbits showed normal explicitly unpaired extinction, Berger and Orr (1983) suggested that the increased responding to the initially reinforced CS_1 did not simply reflect a deficiency in extinction of the CS_1–US association.

Similar results were reported by Weikart and Berger (1986) in a tone-light discrimination reversal of the rabbit NM response, suggesting that deficits in the two-tone reversal learning after HFL are not due to an increased within-modality generalization to the tone CSs serving as CS+ and CS–. Port, Romano, and Patterson (1986) found that HFL impaired the reversal learning of a stimulus duration discrimination paradigm using the rabbit NM preparation. In addition, Berger et al. (1986) determined that lesions of the retrosplenial cortex, which connects the hippocampus to the cerebellar region, produced deficits in reversal learning of the rabbit NM response.

Berger (1984) found that entorhinal cortex stimulation that produced LTP in the perforant path-granule cell synapses increased the rate of acquisition of a two-tone classical discrimination of the rabbit NM response. Robinson, Port, and Berger (1989) showed that kindling of the hippocampal perforant path-dentate projection (which induces LTP) facilitates discrimination acquisition, but impairs discrimination reversal of the rabbit NM response.

Feature-positive discrimination

Ross et al. (1984) studied the effect of HFL in a serial feature-positive discrimination using rats. They found that, when HFL preceded training, the lesions prevented the acquisition and retention of the conditional discrimination.

Loechner and Weisz (1987) studied simultaneous feature-positive discrimination in the rabbit NM response preparation. They did not find a significant difference between HFL rabbits and controls when a light was the nonreinforced component, because neither HFL rabbits nor controls showed conditioning to the light. However, when a tone was the nonreinforced component, HFL rabbits exhibited a high level of responding to both the compound and the component, thereby failing to show positive-feature discrimination.

Inhibitory conditioning

Solomon (1977) found that HFL rabbits yield normal conditioned inhibition of the NM response. Interestingly, although HFL did not impair conditioned inhibition, Micco and Schwartz (1972) determined that it impaired inhibitory conditioning to CS_2 obtained in a differential conditioning (discrimination acquisition) paradigm using rats.

Negative patterning

So far, the effect of HFL on the acquisition and retention of negative patterning has not been studied in classical conditioning. However, because of the relevance of this paradigm, we will refer to Rudy and Sutherland's (1989) study on HFL effects on negative patterning using rats in an operant discrimination task. Reference to this work is further justified by the view that discriminative stimuli become classically conditioned to the reinforcer during instrumental conditioning (see Mackintosh, 1983, p. 100).

Rudy and Sutherland (1989) found that, when HFL preceded training, the lesions prevented the acquisition of negative patterning. When HFL followed the acquisition of negative patterning, the lesions impaired the correct performance of negative patterning. Before the surgical procedure, both experimental and control groups acquired negative patterning and responded to the components in about 90 percent of the trials and to the compound in about 40 percent of the trials. After the surgical procedure, both control and HFL groups dramatically increased their responding to the compound: By the end of the first postlesion session (120 trials), the HFL and control groups responded to the compound 90 percent and 70 percent, respectively, of the trials. After four sessions, controls responded to the compound only in 30 percent of the trials, whereas HFL rats responded to the compound in 90 percent of the trials.

In contrast to Rudy and Sutherland's (1989) results, Davidson, McKernan, and Jarrard (1993) reported that kainic acid/colchicine lesions did not impair the acquisition of negative patterning.

Positive patterning

The effect of HFL on the acquisition and retention of positive patterning has not been studied in classical conditioning.

Combined negative and positive patterning

Whishaw and Tomie (1991) trained animals to perform a simultaneous negative and positive patterning task during an operant conditioning task. In this paradigm, a pair of small strings T_1 and a pair of large strings T_2 were combined with the presence O_1 or the absence O_2 of an almond odor in four arrangements: $([T_1\ O_1]+, [T_1\ O_2]-)$, $([T_1\ O_1]+, [T_2\ O_1]-)$, $([T_2\ O_2]+, [T_1\ O_2]-)$, and $([T_2\ O_2]+, [T_2\ O_1]-)$. Animals with kainic acid/colchicine lesions of the hippocampus did not show impairment in conditional discrimination, but they were impaired in the simultaneous negative and positive patterning task. When rats in the patterning task were tested in a place-learning task in a water maze, HFL animals were impaired.

Using ibotenic acid lesions, Gallagher and Holland (1992) trained animals to perform an operant simultaneous negative and positive patterning task. A panel light P, a tone T, and white noise N were presented in four combinations: PN–, N+, T–, and PT+. In contrast to animals with kainic acid/colchicine lesions in

Rudy and Sutherland's (1989) and Whishaw and Tomie's (1991) experiments, Gallagher and Holland (1992) found that animals with HPL were not impaired in either the positive or negative component of the simultaneous negative and positive patterning task. However, the same lesioned animals were impaired in the Morris tank task.

Generalization

Using the rabbit NM response preparation, Solomon and Moore (1975) reported that HFL rabbits showed increased stimulus generalization to tones of different frequencies. Generalization testing consisted of nonreinforced presentations of tones of different frequencies (400, 800, 1,200, 1,600, and 2,000 Hz) following conditioning to a 1,200-Hz tone.

Effects of cortical lesions on classical conditioning

This section reviews the effects of cortical lesions (CL) in several classical conditioning paradigms. It is important to note that CL refers to the extensive removal (around 80 percent) of the neocortex and not simply the aspiration of the region of the neocortex overlying the hippocampus, a procedure sometimes used to control for the effect of HFL. All results refer to classical conditioning obtained in the rabbit NM response preparation.

Acquisition

Acquisition of delay conditioning of the rabbit NM response has been reported to be slightly retarded or unaffected by bilateral CL (Oakley and Russell, 1972). Acquisition of trace conditioning of the rabbit NM response has been reported to be unaffected by CL (Yeo et al., 1984).

Extinction

Extinction of the rabbit NM response has been reported to be normal after CL (Oakley and Russell, 1972).

Blocking

Moore (personal communication, 1990) found that CL rabbits show normal blocking of the NM response.

Discrimination acquisition and reversal

Oakley and Russell (1975) found that CL do not affect the rate of discrimination acquisition and facilitate the rate of reversal of the rabbit NM response.

Inhibitory conditioning

Conditioned inhibition of the rabbit NM response is not affected by CL (Moore et al., 1980).

Effects of cerebellar lesions on classical conditioning

A considerable amount of data shows that lesions of several cerebellar areas permanently abolish the classically conditioned NM and eyeblink response in the rabbit (see Thompson, 1986, for a review). Although, according to Thompson, cerebellar lesions do not affect the generation of the US, Welsh and Harvey (1989) reported deficits in the responses to both CS and US. Recently, Skelton (1988) determined that bilateral lesions of the dentate-interpositus region of the cerebellum disrupted eyeblink conditioning in the rat.

The neurophysiology of spatial learning and cognitive mapping

This section briefly reviews some behavioral data relating spatial learning and cognitive mapping to the anatomy and neurophysiology of the hippocampus and cortical systems.

Neuronal activity during spatial learning

Different cells in the hippocampus, entorhinal cortex, and subiculum seem to code information about spatial navigation. As a rat moves through an environment, different hippocampal cells are maximally activated at different spatial locations (Best and Ranck, 1982; Breese, Hampson, and Deadwyler, 1989; Eichenbaum et al., 1986; McNaughton, Barnes, and O'Keefe, 1983; Miller and Best, 1980; O'Keefe, 1976; O'Keefe and Dostrovsky, 1971; O'Keefe and Speakman, 1987; Olton, Branch, and Best, 1978; Muller, Kubie, and Ranck, 1987; Muller and Kubie, 1987). As summarized by Breese et al. (1989), place fields are characterized by (1) a decrease in firing density with increasing distances from the center to the borders of the field, (2) borders where firing significantly increases or decreases, (3) greater increments in firing when entering than when exiting the field, and (4) directional and velocity firing biases. Hippocampal place cells receive information from neurons in the entorhinal cortex, which display a noisier but also place-related activity (Muller et al., 1991).

The fact that pyramidal cells have a discrete preferred firing location in a given environment may suggest that hippocampal activity could simply reflect the animal's spatial position. Some studies suggest that this is not the case. First, Ranck (1973) found cells in CA3 ("approach-consummate" cells) that are most active before and during consummatory behaviors. Second, Eichenbaum et al. (1986) reported that hippocampal cells do not simply reflect the spatial location of the animal, but the animal's orientation and approach toward significant cues and places. They trained rats to perform a sequence of behaviors that include approach to a stimulus-sampling port, orientation and approach toward a reward location, and water reward consumption. They found that the activity of hippocampal cells increased (1) after onset of odor-cue sampling when the cue preceded reward ("cue-sampling cells"), and (2) prior to arrival at either the odor-sampling port or reward cup ("goal-approach cells"). Cells exhibiting place-field firing were subgroups of cue-sampling and goal-approach cells. Third, Breese et al. (1989) communicated that, in some cases, place cell firing represents biologically significant environmental stimuli. They recorded complex spike cell activity during exploration for water delivered to cups located in various regions of an elevated platform. They found that selective delivery of water to a single location on the platform shifted the location of the place field of a given neuron to the location where the water was available.

Further results support the view that hippocampal activity reflects more than the animal's present spatial position. O'Keefe and Speakman (1987) showed that to the extent that an animal knows where it is, virtually the entire set of sensory cues can be removed without disruption of the place-field specificity, a result that indicates that the activity of place-field units is based on a memory mechanism rather than on the set of inputs active at a given time. More recently, Quirk, Muller, and Kubie (1990) reported that place firing is stable for long periods (up to 8 min) in the dark if the rat is placed in the experimental chamber with the lights on, but not if placed in the chamber with the lights off. In addition, Quirk et al. reported that when animals are placed in the chamber in initial darkness, place cells maintain their activity even when the lights are turned on.

In addition to place fields, other types of activity have also been recorded from hippocampal regions. For instance, Wible et al. (1986) observed that, in a variety of learning tasks, hippocampal cells respond not only to individual stimuli, but to combinations of stimulus dimensions such as color, shape, and spatial location.

Effects of hippocampal lesions on place and cue learning

Morris et al. (1982) found that HFL caused a profound impairment in spatial navigation in a place learning task in which rats had to search for a hidden escape platform in a tank filled with opaque water. However, HFL did not affect a cue learning task in which rats had to approach a visible escape platform. Confirming Morris et al.'s (1982) results, DiMattia and Kesner (1988) reported that HFL impair not only acquisition, but also retention of previously acquired Morris tank performance. Jarrard (1983) and Jarrard et al. (1984), using a radial-arm maze, and Okaichi (1987), using a T maze, also found that place but not cue learning was impaired by HFL.

Morris et al. (1990) and Gallagher and Holland (1992) determined that HPL result in impaired performance of the task. Interestingly, Morris et al., (1990) reported that, with overtraining, HPL animals eventually reached near-normal performance levels, by navigating in circles around the pool instead of taking a navigation course in the direction of the submerged platform. Schenk and Morris (1985) reported that radio-frequency lesions of the entorhinal cortex alone, as well as combined lesions of entorhinal cortex/subiculum, impaired acquisition of place discrimination in the water maze.

Effects of cortical lesions on place learning

Whishaw and Kolb (1984) reported that decortication abolishes place learning, but not cue learning, in rats. More specifically, lesions of parietal (DiMattia and Kesner, 1988), entorhinal (Schenk and Morris, 1985), frontal (Kolb, Sutherland, Whishaw, 1983; Sutherland, Kolb, and Whishaw, 1982), or cingulate (Sutherland, Whishaw, and Kolb, 1988) cortex, but not of temporal cortex (Kolb, Buhrman, and McDonald, 1989), impair place-learning acquisition. Lesions of parietal (DiMattia and Kesner, 1988) or cingulate (Sutherland et al., 1988) cortex also impair retention of place learning.

Theories of hippocampal function

It has been proposed that the hippocampus is involved in a wide variety of processes such as attention, spatial mapping, temporal mapping, working memory, declarative memory, recognition memory, comparisons, contextual retrieval, and configuration (see Schmajuk, 1984a, for a review). The present section briefly reviews some of the theories.

Attentional theories

Attentional theories of hippocampal function emphasize that the hippocampus participates in the control of the level of processing assigned to each stimulus, thereby controlling what information is to be stored in the brain. For instance, Grastyan et al. (1959) suggested that the hippocampus inhibits orienting responses to nonsignificant stimuli. Douglas and Pribram (1966) proposed that the hippocampus excludes from attention CSs that have been associated with nonreinforcement. Kimble (1968) proposed that the hippocampus enables the organism to uncouple its attention from one stimulus and shift it to new and more consequential environmental events. Building on this idea, Douglas (1972) suggested that the hippocampus is involved in correlating a CS with nonreinforcement, thereby reducing its attentional priority. Solomon and Moore (1975) introduced the closely related hypotheses that the hippocampus participates in "tuning out" CSs poorly associated with reinforcement. Moore (1979a) proposed a neuronal model to explain how the hippocampus might participate in tuning out CSs during conditioning of the rabbit's nictitating membrane (NM) response. Similarly, Solomon and Moore (1975) emphasized that the hippocampus is involved in the control of stimulus processing. In one way or another, attentional theories propose that the hippocampus regulates the attentional priority of CSs according to their association with reinforcement.

Moore and Stickney (1980) proposed the first mathematical attentional model of hippocampal function. This seminal model and other computational attentional theories of hippocampal function are described in Chapter 13.

Spatial map theory

Primarily supported by data showing that hippocampal lesions impair some types of spatial learning and that neuronal activity in the hippocampus is correlated with locations in space, O'Keefe and Nadel (1978) introduced the view of "the hippocampus as a cognitive map" and suggested that it is "a specific memory system dealing only with the representations of environments and not with other types of information" (O'Keefe, 1989, p. 226). O'Keefe, Nadel, and Wilner (1979) argued that classical conditioning involves spatial strategies because normal animals define the spatial context in which conditioning occurs.

According to O'Keefe and Nadel, spatial navigation can be performed using either a cognitive map system or a route system. Cognitive maps are defined as a set of connected places that represent the world in three-dimensional space and provide the animal with a large choice of possible paths between any two points in the environment. Routes are defined as a list of cues to be approached by any available behavior. The cognitive map system is built on curiosity (exploration), learns rapidly, and is insensitive to intertrial interval; the route system is trained by reward or punishment, learns incrementally, and is sensitive to intertrial interval. According to O'Keefe and Nadel (1978), the cognitive map system contains (1) a system that combines simple environmental stimuli into place representations (equated with the hippocampal dentate gyrus), (2) a system that represents the transitions between places (located in hippocampal CA3), and (3) a system that signals the presence of changes in the environment (located in hippocampal CA1). Although O'Keefe and Nadel suggested that spatial navigation in animals with an intact hippocampus is carried out with the help of the cognitive map system, they do not specify the interaction between cognitive and route systems in normal animals.

Computational spatial map theories of hippocampal function are described in Chapter 13.

Temporal map theory

Solomon (1979; Moore, 1979b) argued that the hippocampus would participate in the construction of a temporal map. This temporal map consists of the registration of the temporal sequence of events. A similar view was proposed by Rawlins (1985), who suggested that the hippocampus is needed to process stimuli to be associated with other stimuli or events that are temporally discontiguous. After HFL, animals are unable to form temporally discontiguous associations.

Working memory

Olton (1986) proposed that the hippocampus was involved in working memory. Whereas reference memory refers to the permanent aspects of an experiment, working memory refers to the variable aspects (spatial or temporal) from trial to trial in the experiment. According to Olton (1986), working memory is more flexible than reference memory, shows more interference than reference memory, and preserves information about the temporal order of stimuli.

Recognition memory

Gaffan (1972) suggested that the hippocampus is involved in recognition but not associative memory. Associative memory refers to the information that two stimuli had occurred simultaneously. Recognition memory refers to information about the familiarity or novelty of a stimulus. Impairments in recognition memory imply failures in detecting novel stimuli.

Hippocampus as a comparator

According to Smythies (1966), the hippocampus compares environmental information coming from the entorhinal cortex with internal information coming from the septum. If two similar patterns of inputs are received, pyramidal neurons fire and the information is stored in the temporal cortex. Vinogradova

(1975) argued that the CA3 region would be involved in evaluating the novelty of a signal coming from the reticulo-septal circuit, as compared to its counterpart in the cortical input. When novelty is detected, an orienting response is elicited. According to Gray (1982), if actual and predicted events are the same, then behavior is maintained; if there is a mismatch, the hippocampus inhibits the current behavior and attention is increased.

Declarative memory

Cohen and Squire (1980; Cohen and Eichenbaum, 1993) suggested that the hippocampus is responsible for the storage of new information about the world (declarative memory), but not for the acquisition of new perceptual-motor skills (procedural memory).

Contextual retrieval

Hirsh (1974) proposed that the hippocampus is responsible for contextual retrieval. It is defined as "the retrieval of one item of stored information initiated by a cue which refers to the retrieved information." Two types of information storage are possible, on-line and off-line. In on-line storage, learning is gradual and presentation of a CS results in a response R. In off-line storage, learning is abrupt and memory retrieval is controlled by the contextual cues.

Configural theories

According to Wickelgren (1979), the organization of various stimuli into an associative group (chunking) is the basis of configuring in conditioning. Wickelgren (1979) proposed that the hippocampus plays a critical role in cortical chunking. The hippocampus partially activates free, as opposed to bound, cortical neurons. Bound neurons reduce their connections to the hippocampus, consolidating memory by protecting the neurons from hippocampal input. HFL produce amnesia due to a disruption in chunking.

Rudy and Sutherland (1989) proposed that the hippocampus participates in the acquisition and storage of configural associations. According to Rudy and Sutherland (1989), although two memory systems (simple and configural) subserve learning, only the configural association system depends critically on the integrity of the hippocampal formation. They assume that (1) memory is simultaneously stored in both the simple and configural systems and (2) if the configural association has a greater predictive accuracy than a simple association involving one of the relevant elements, then the simple association's output is suppressed. Rudy and Sutherland claim that the processing afforded by the hippocampal configural association system may be the basis for declarative memory, as described by Cohen and Squire (1980).

Chapter 13 describes some computational configural theories of hippocampal function.

13 Hippocampal modulation of learning and cognition

The present chapter describes a brain mapping of the SD model presented in Chapter 6 under the behavioral and neurophysiological constraints presented in Chapter 12. In the case of hippocampal and cortical areas, the mapping refers in general to classical conditioning as obtained in different preparations. In the case of cerebellar areas, however, this mapping alludes specifically to the classically conditioned eyeblink or nictitating membrane (NM) response.

The mapping of different layers in the SD network onto various brain regions, as advanced in this chapter, is indicated by labels in parentheses in Figure 13.1. Figure 13.2 displays an equivalent mapping of the network onto a schematic diagram of brain interconnections. Figures 13.1 and 13.2 show that output units map onto cerebellar circuits, hidden units map onto cortical circuits, and circuits controlling output and hidden-unit errors map onto the hippocampus. Box 13.1 delineates the basic assumptions. Based on the brain-mapped SD network, this chapter provides computer simulations of the effects of hippocampal and cortical lesions on classical conditioning.

Mapping of the SD model onto the brain circuitry

The hippocampus and the computation of error signals

Considering that (1) classical conditioning is still possible after lesions of the hippocampal formation (HFL), (2) induction of hippocampal long-term potentiation (LTP) does not cause classical conditioning but only facilitates discrimination acquisition, and (3) the activity of pyramidal cells in the hippocampus *reflects* the temporal topography of the CR, Schmajuk (1984a, 1989; Schmajuk and Moore, 1985, 1988) suggested that the hippocampus is involved in the computation of the aggregate prediction of ongoing events B. Aggregate prediction, $B = \Sigma_i CS_i VS_i + \Sigma_j CN_j VN_j$, represents the prediction of the intensity of the US based on all the stimuli present at a given time. Figure 13.2 shows that aggregate prediction B is used to determine the output error signal, EO = (US–B), that controls the association of different CSs and CNs with the US in the cerebellum. Notice that, because the CR is also proportional to $\Sigma_i CS_i VS_i + \Sigma_j CN_j VN_j$, B is an efference copy of the CR.

Because classical conditioning paradigms that require stimulus configuration seem to be impaired by HFL, it is conjectured that the computation of the error signals employed by hidden units EH_j also takes place in the hippocampal formation. As shown in Figure 13.1, information about the $CN_j VN_j$ activity and information about the output error are combined in the hippocampus to generate hidden-unit error signals EH_j.

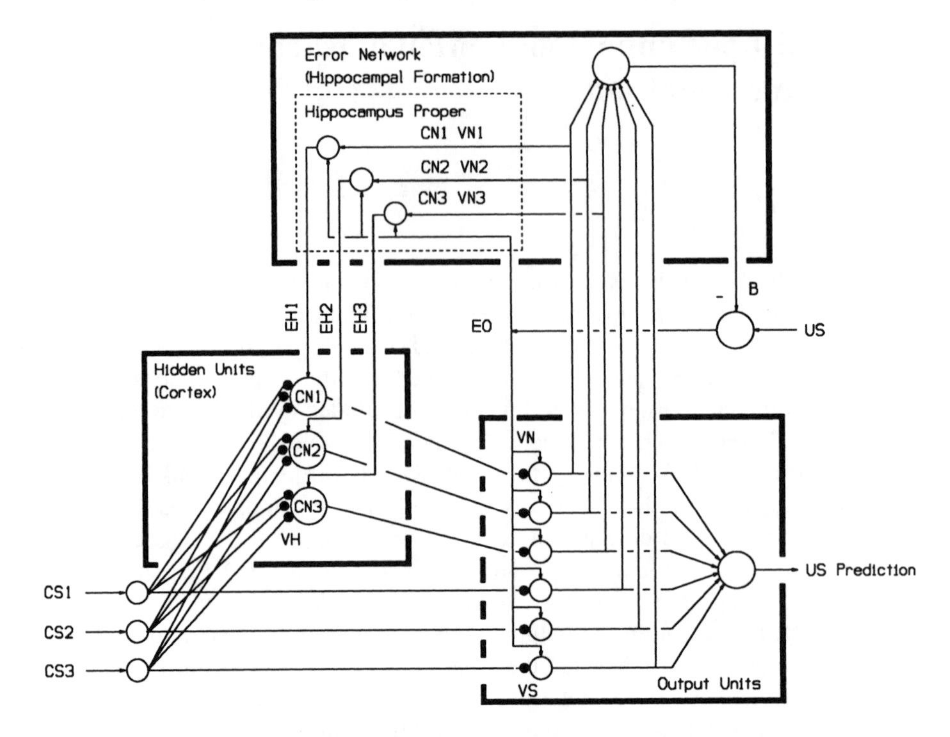

Figure 13.1. Diagram of the SD model (Adapted from Schmajuk and DiCarlo, 1992). CS: conditioned stimulus; CN; configural stimulus; VS: CS–US associations; VN: CN–US associations; VH: CS–CN associations; US: unconditioned stimulus; *B:* aggregate prediction; EH: error signal for hidden units; EO: error signal for output units; CR: conditioned response. Arrows represent fixed synapses, solid circles variable synapses. Anatomical areas indicated in parentheses refer to the mapping of different nodes in the network onto various brain regions.

Hippocampal neural activity during classical conditioning

Figure 13.2 shows that neural activity proportional to CS_iVS_i and CN_jVN_j reaches the hippocampus through an interpositus nucleus-thalamic-entorhinal cortex pathway. This information is (1) summed to generate the aggregate prediction B and (2) combined with the medial septal input to compute the error signal for hidden units, $EH_j = f(\theta CN_jVN_j)$ (see Appendix 6.A in Chapter 6). Because each hidden unit receives its own error signal, a measure of hippocampal neural activity reflecting the computation of all error signals is given by $\Sigma_j \theta CN_jVN_j$. Therefore, according to the model, total hippocampal neural activity is proportional to $B + \Sigma_j \theta CN_jVN_j$.

Lateral septal neural activity during classical conditioning

Because neural activity in the lateral septum during acquisition of classical conditioning is similar to that recorded in hippocampal pyramidal cells, Figure 13.2 shows an excitatory hippocampal output (from CA3) to the lateral septum, proportional to $B = \Sigma_i CS_iVS_i + \Sigma_j CN_jVN_j$.

Cerebellar involvement in classical conditioning

Considering that classical conditioning of the eyeblink or nictitating membrane response is impaired after cerebellar lesions, Figure 13.2 assumes that CS_i–US

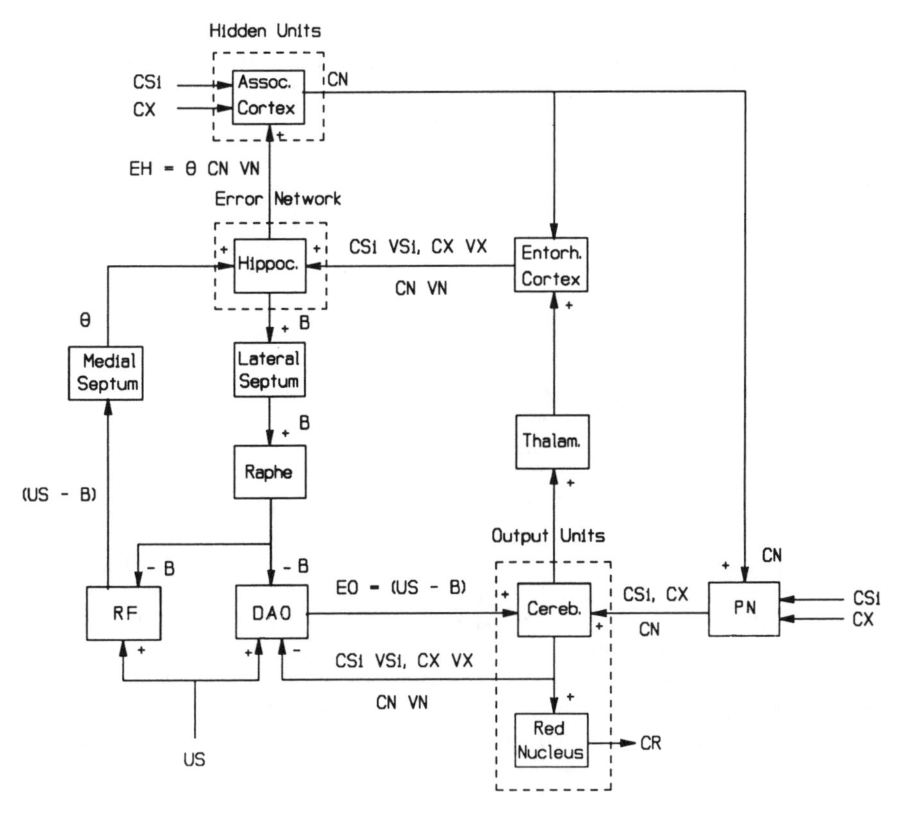

Figure 13.2. Mapping of the network onto a schematic diagram of cortical, hippocampal, and cerebellar interconnections. CS_i: conditioned stimulus; CX: context; US: unconditioned stimulus; CN: configural stimulus; VS_i: CS_i–US associations; VX: context–US associations; VN: CN–US associations; CR: conditioned response; *B:* aggregate prediction; θ: theta rhythm; EO = (US – *B*): output error; EH = θ CN VN: hidden-unit error. Assoc. cortex: association cortex; entorh. cortex: entorhinal cortex; hippoc.: hippocampus; RF: reticular formation; DAO: dorsal accessory olive; cereb.: cerebellum; PN: pontine nucleus; thalam: thalamus.

and CN_j–US associations are stored in cerebellar areas, thereby controlling the generation of CRs. Simple stimuli CS_i and configural stimuli CN_i reach the pontine nuclei. In the cerebellum, these sensory representations are associated with the US representation conveyed by the dorsal accessory olive. The cerebellar output reflects the magnitude of CS_iVS_i and CN_jVN_j and generates a CR by acting on the red nucleus.

The red nucleus is also responsible for an inhibitory gating of olivary responsiveness to the US. We assume that each CS and CN controls its own associations with the US by inhibiting the dorsal accessory olive in proportion to their individual associations with the US. This *local* cerebellar control of CS_i–US and CN_j–US associations is assumed to be independent of the associations accrued by all other CSs and CNs with the US.

To the extent that (1) the activity of pyramidal cells in the hippocampus *reflects* the temporal topography of the CR and (2) this neural activity *depends* on the integrity of cerebellar circuits, Figure 13.2 shows that copies of CS_iVS_i and CN_jVN_j are relayed from the interpositus nucleus to the thalamus, entorhinal cortex, and the hippocampus.

BOX 13.1 Mapping of Schmajuk and Dicarlo's (1992) configural model onto the brain circuitry

Schmajuk and DiCarlo's (1992) model is mapped onto the brain circuitry according to the following principles:

1. The hippocampus is involved in the computation of (1) the aggregate prediction of ongoing events B and (2) the error signals employed by hidden units EH_j.
2. Hippocampal neural activity is proportional to (1) the aggregate prediction B and (2) the sum of all error signals $\Sigma_j CN_j VN_j EO$.
3. Lateral septal activity is proportional to B.
4. The aggregate prediction B modulates the formation of CS_i–US and CN_j–US associations in cerebellar areas.
5. CS_i–US and CN_j–US associations are stored in cerebellar areas, thereby controlling the generation of CRs.
6. Activity proportional to $CS_i VS_i$ and $CN_j VN_j$ is relayed from the cerebellum to the entorhinal cortex and the hippocampus.
7. Medial septal theta represents the mismatch between the predicted and actual magnitude of the US, $\theta = |US - B|$.
8. Medial septal activity has a "modulatory" effect on CA1 and CA3 pyramidal cells. The output of the pyramidal cells is proportional to $CN_j VN_j \theta$ and equivalent to the error signal for the cortical hidden units, $EH_j = CN_j VN_j EO$.
9. The output of CA1 pyramidal neurons, proportional to $CN_j VN_j \theta$, is conveyed via the entorhinal cortex to association cortical areas, where they regulate learning in cortical hidden units, thereby modulating stimulus configuration.
10. Association cortex projections to the pontine nuclei are responsible for conveying CN_j information to the cerebellum, where they become associated with the US.

Hippocampo–cerebellar interactions: modulation of classical conditioning

According to the circuit shown in Figure 13.2, the hippocampus inhibits (through the lateral septum and raphe nuclei) the dorsal accessory olive, which conveys the US input to cerebellar areas. By inhibiting the dorsal accessory olive, the hippocampus controls all CS_i–US and CN_j–US associations. Because hippocampal inhibition is proportional to B, $B = \Sigma_i CS_i VS_i + \Sigma_j CN_j VN_j$, this *global* control signal reflects the associations accrued by all other CSs and CNs with the US. As $CS_i VS_i$, $CN_j VN_j$, and B increase during acquisition, the US becomes less unexpected and the dorsal accessory olive decreases its activity, thereby preventing further increments in CS_i–US and CN_j–US cerebellar associations. Details of the interaction between local and global inhibitory mechanisms are presented in Equation 6.6 in Appendix 6.A in Chapter 6.

Reticulo–hippocampal interactions: control of theta rhythm

In Figure 13.2, medial septal theta represents the mismatch between the predicted and actual magnitude of the US, $\theta = |US–B|$. Appendix 13.A shows that two theta populations may exist: one active when the US is underpredicted ($\theta_p = US - B$), and another one active when the US is overpredicted ($\theta_n = B - US$). Schmajuk and Moore (1988; Schmajuk, 1989) showed that medial septal activ-

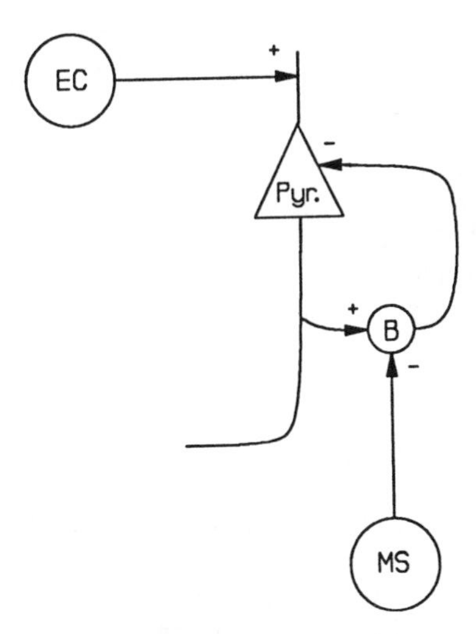

Figure 13.3. Medial septal modulation of inhibitory feedback of pyramidal cells. One aspect of the hippocampal intrinsic circuitry showing how medial septal inputs inhibit the recurrent inhibition provided by basket cells. EC: entorhinal cortex; pyr.: pyramidal cell; *B:* basket cell; MS: medial septum.

ity can be adequately described by the sum of all mismatches between predicted and actual environmental events, including CSs and the US. Because the present version of the SD model does not include CS–CS associations and, therefore, CS predictions, the only mismatch considered here is $|US - B|$.

Medial septal modulation of hippocampal activity

According to the data presented in Chapter 12, medial septal activity has a "modulatory" effect on dentate granule cells, CA1 and CA3 pyramidal cells. This modulation might be mediated by a GABAergic inhibition of inhibitory basket cell interneurons.

Figure 13.3 shows a diagram of a neuronal arrangement in which a medial septal input inhibits the recurrent inhibition provided by a basket cell. Appendix 13.A demonstrates that the output of the pyramidal cells is proportional to $CN_jVN_j\theta$ and equivalent to the error signal for the cortical hidden units, $EH_j = f(CN_jVN_jEO)$ (see Equation 6.8 in Appendix 6.A). This output is conveyed via the entorhinal cortex to association cortical areas, where it regulates learning in cortical hidden units, thereby modulating the stimulus configuration.

In sum, the modulatory effect of the medial septum on the hippocampal pyramidal cells allows the model shown in Figures 13.1 and 13.2 to implement a biogically plausible version of backpropagation.

Hippocampo–cortical interactions

According to Figure 13.2, stimulus configuration takes place in association cortex. The output of CA1 pyramidal neurons, proportional to θCN_jVN_j, is conveyed via the entorhinal cortex to association cortical areas, where it regulates learning in cortical hidden units, thereby modulating the stimulus configuration.

Figure 13.2 also shows cortical areas projecting to the hippocampus through the entorhinal cortex. In a version of the SD model that included CS–CS asso-

BOX 13.2 Effects of hippocampal, cortical, cerebellar, and septal lesions

1. *Lesions of the hippocampal formation (HFL).* Aspiration and colchicine-kainic acid lesions of the HF are simulated by assuming that the aggregate prediction B and the hidden-unit error signals EH_j equal zero.
2. *Lesions of the hippocampus proper (HPL).* The effect of lesioning the HP with ibotenic acid is described by assuming that hidden-unit error signals EH_j equal zero.
3. *Cortical lesions (CL).* The effect of CL is simulated by removing all hidden units.
4. *Cerebellar lesions.* The effect of cerebellar lesions is described by eliminating all output units.
5. *Medial septal lesions.* The effect of medial septal lesions is described by assuming $\theta = |US - B|$ equals zero and, therefore, EH_j is zero.
6. *Lateral septal lesions.* Lateral septal lesions are equivalent to the elimination of hippocampal inhibition (proportional to the aggregate prediction B) on the dorsal accessory olive.

ciations, these projections would provide information about CS–CS associations to the hippocampus in order to compute the aggregate prediction of a given CS.

Cortico–cerebellar interactions

In Figure 13.2, association cortex projections to the pontine nuclei are responsible for conveying CN_j information to the cerebellum, where CN_js become associated with the US.

Effects of hippocampal, cortical, cerebellar, and septal lesions

Based on the mapping of nodes and connections in the SD model onto the brain circuitry presented in Figures 13.1 and 13.2, the effect of hippocampal, cortical, cerebellar, and septal lesions can be described by removing or disconnecting the appropriate blocks in the diagram. Box 13.2 summarizes the effects of the different lesions in terms of the variables in the model.

Lesions of the hippocampal formation (HFL)

Schmajuk and Blair (1993) suggested that the effect of aspiration and colchicine/kainic acid lesions of the hippocampal formation can be simulated by removing or disconnecting the "hippocampal formation" block in Figure 13.1. This manipulation produces important changes in the performance of the model. First, the aggregate prediction of the US, $B = \Sigma_i CS_i VS_i + \Sigma_j CN_j VN_j$, is no longer computed. B controls competition among CSs and CNs to gain association with the US. In the absence of B, CS_i–US and CN_j–US associations are independent of one another, yielding impairment in paradigms such as blocking, conditioned inhibition, and negative patterning. Removal of the hippocampal formation block also prevents the computation of hidden-unit error signals EH_j and, therefore, no *new* configurations are formed in the association cortex. However, previously stored configurations are not modified. Because HFL removes inhibition of the dorsal accessory olive, the SD model predicts that HFL animals will show increased neural activity in the dorsal accessory olive. Notwithstanding the lack of (1) aggregate prediction signals and (2) hidden-unit error signals, HFL animals are still capable of classical conditioning by storing VS_i and VN_j associations in cerebellar areas.

In summary, we assume that aspiration and kainic acid/colchicine lesions of the HFL affect the computation of both B and EH_j in the network shown in Figure 13.1. Animals are absolutely impaired at solving configural problems in the absence of competition (no B) that follows HFL. A formal description of the effects of HFL is presented in Appendix 13.B.

Lesions of the hippocampus proper (HPL)

The effect of lesioning the hippocampus proper with ibotenic acid can be described by removing or disconnecting the block labeled "hippocampus proper" in Figure 13.1 (Schmajuk and Blair, 1993). Removal of this block prevents the computation of hidden-unit error signals EH_j, so that no new stimulus configurations are formed in the association cortex. It is important to notice that, since previously stored configurations are not disrupted following lesions of the hippocampus proper, these lesions should not impair tasks that do not require changes in cortical unit configurations. By contrast, tasks that require retraining of cortical hidden units should be impaired.

In summary, we assume that localized lesions obtained with ibotenic acid are confined to the hippocampus proper and affect EH_j in the network shown in Figure 13.1. Surprisingly, animals might still be able to solve configural problems in the absence of configuration (no EH_j) that follows ibotenic acid lesions. A formal description of the effects of HPL is presented in Appendix 13.B.

Cortical lesions

The effect of cortical lesions (CL) can be described by removing the block labeled "cortex" in Figure 13.1. Removal of this block eliminates all hidden units, destroying old information stored in the association cortex and preventing the formation of new configural stimuli CN_j. Cortical lesions impair tasks that require configuration, but preserve tasks that only require competition among stimuli. However, the hippocampus is still available for the computation of the aggregate prediction B, and CL animals are still capable of classical conditioning by storing VS_i and VN_j associations in cerebellar areas. A formal description of the effects of CL is presented in Appendix 13.B.

Cerebellar lesions

According to the circuit shown in Figure 13.1, cerebellar lesions destroy previously stored (and preclude the formation of new) CS_i–US and CN_j–US associations. A formal description of the effects of cerebellar lesions is presented in Appendix 13.B.

Medial septal lesions

Because the medial septum provides the output error value included in the error signals used to train cortical units, medial septal lesions preclude the computation of hidden-unit error signals EH_j and, therefore, no *new* configurations are formed in the association cortex. This idea is compatible with Berry and Thompson's (1979) data showing that lesions of the medial septum produce retardation of acquisition of classical conditioning.

Lateral septal lesions

According to Figure 13.2, lateral septal lesions are equivalent to the elimination of hippocampal inhibition (proportional to the aggregate prediction *B*) on the dorsal accessory olive. In general, this prediction is supported by the finding that simultaneous medial and lateral septal lesions and HFL have similar effects in many learning paradigms (for a review, see Gray and McNaughton, 1983). However, the model predicts different effects of medial and lateral septum lesions. Whereas lateral septal lesions suppress the broadcast of aggregate prediction signals to the dorsal accessory olive, consequently altering the normal formation of CS_i–US and CN_j–US associations, medial septal lesions impair the transmission of output error signals to the hippocampus, thereby impairing the formation of configural cortical associations.

Computer simulations

This section demonstrates that the SD model can describe a broad range of experimental results from the animal learning literature, and also generate novel predictions concerning the effects of different lesion manipulations. Simulations are presented for normal animals and for animals with lesions of the hippocampus proper (HPL), hippocampal formation (HFL), and cortical regions (CL). In all simulations, CR amplitudes are expressed as a percentage of the network output for normal animals, after 10 simulated acquisition trials. All simulations were carried out with identical parameter values, as shown in Appendix 13.C (see Schmajuk and Blair, 1993).

Acquisition of delay and trace conditioning

Figure 13.4 shows real-time simulations in trials 1, 4, 8, 12, 16, and 20 in a delay conditioning paradigm with a 200-msec CS, 50-msec US, and 150-msec ISI for normal, HFL, and CL cases. In the normal case, as CR amplitude increases over trials, output weights VS_i and VN_j, and hidden weights VH_{ij} may increase or decrease over trials.

In the HFL case, CR amplitude and VS_i change at a faster rate than in the normal case. Hidden weights VH_{ij} do not change in the HFL case. In the HFL case, some CN_j–US learning occurs due to the initial random weights on the hidden units. This learning, however, is limited because the absence of the hippocampus precludes the formation of strong configural stimuli CN_j. In spite of the lack of strong CN_j–US associations, the absence of the aggregate prediction of the US in the HFL case allows more CS_i's and CN_j's to establish associations with the US, thereby attaining faster learning than in the normal case. This result is in

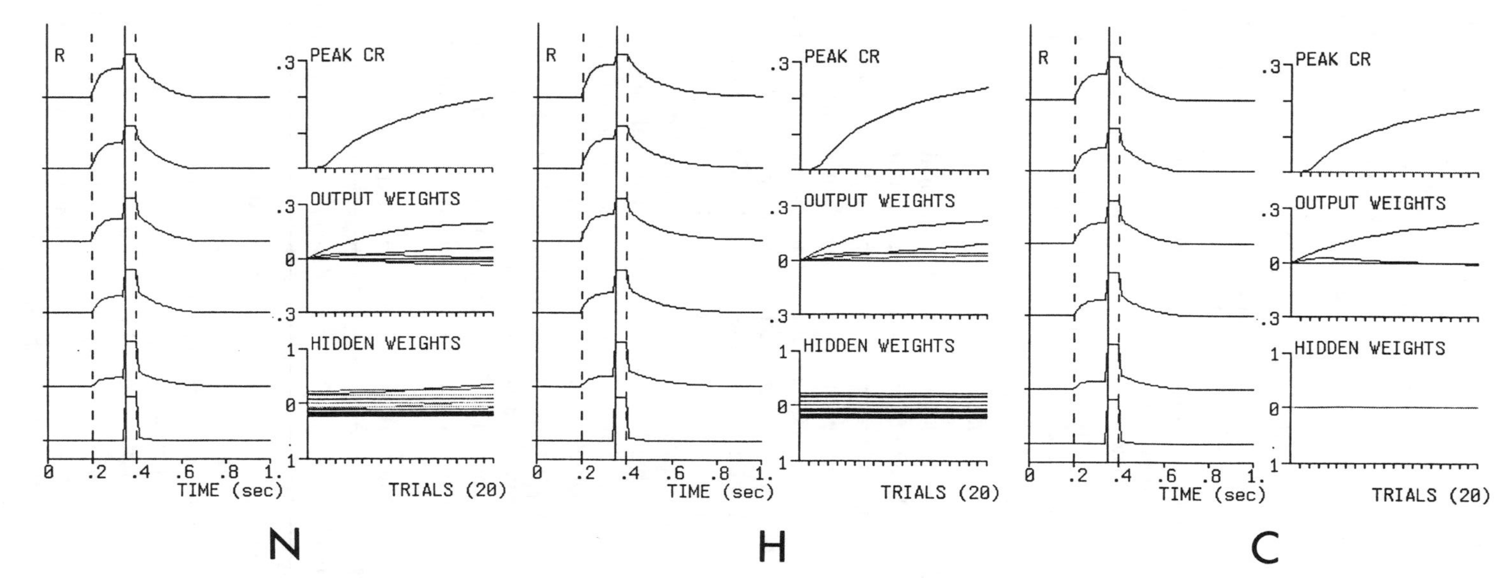

Figure 13.4. Effect of cortical (CL) and hippocampal formation lesions (HFL) on acquisition of classical conditioning. *N*: normal case; *H*: HFL case; *C*: CL case. *Left panels.* Real-time simulated conditioned and unconditioned response in trials 1, 4, 8, 12, 16, and 20. Vertical dashed lines indicate CS onset and offset, vertical solid line indicates US onset. Trial 1 is represented at the bottom of the panel. *Right panels.* Peak CR: peak CR as a function of trials; output weights: average VSs and VNs as a function of trials; hidden weights: average VHs as a function of trials.

agreement with Schmaltz and Theios's (1972) and Port, Mikhail, and Patterson's (1985) data.

No hidden weights VH_{ij} are present in the CL case. Because CN_j–US associations are lacking in the CL case, CL animals display a slightly slower rate of acquisition than the normal case. This result is in agreement with Oakley and Russell's (1972) data.

Simulated results for trace conditioning with a 200-msec CS, 50-msec US, and 0-, 200- , 400- , 600-, and 800-msec ISIs show that, in agreement with some empirical data (James, Hurdiman, and Yeo, 1987; Moyer, Deyo, and Disterhoft, 1990; Port et al., 1986), trace conditioning is not impaired (and may be facilitated) in HFL animals. In agreement with Yeo et al. (1984), trace conditioning is not impaired in the CL case. A more elaborate description of the interaction between CS duration and ISI in determining HFL effects on conditioning acquisition is offered by an alternative model (Schmajuk and DiCarlo, 1991a, b).

Acquisition and extinction series

Figure 13.5a shows the simulated number of trials to acquisition in acquisition series for normal, HPL, HFL, and CL cases. Five acquisition–extinction series were simulated by alternating acquisition and extinction of delay conditioning. Each acquisition phase continued until a maximum CR amplitude of 0.18 was reached, and extinction lasted until CR amplitude was less than 0.02, or a maximum of 40 extinction trials had been conducted. Consistent with Schmaltz and Theios's findings (1972), Figure 13.5a shows that HFL animals display faster acquisition than normals in the first acquisition series. In normal animals, reacquisition improves over the series because (1) after the initial acquisition phase, VS_i associations start at a higher level, closer to the behavioral threshold, and (2) during later acquisition phases, hidden units increase their responding to context–CS configurations, which cause relearning to occur at a faster rate. In the HPL, HFL, and CL cases, the second acquisition series shows some savings because, as in the normal case, VS_i associations start at a higher level after the initial acquisition phase. However, the HPL, HFL, and CL cases show less overall savings than the normal case, because hidden units do not increase their responding to the context and the CS.

Figure 13.5b shows the simulated number of trials to extinction in extinction series for normal, HPL, HFL, and CL cases. In the normal case, the model predicts faster extinction over series. These results are in agreement with the acquisition–extinction series of the Schmaltz and Theios (1972) study showing that normal animals decreased the number of trials to extinction criterion. Also in agreement with Schmaltz and Theios (1972) results, simulations show an increasing impairment in extinction in HFL animals. The model predicts that HPL and CL animals extinguish at approximately the same rate over series.

In the normal case, extinction becomes faster as the hidden-unit output increases over successive series. Hidden-unit output does not increase in the HPL, HFL, or CL cases, and therefore extinction is not facilitated. In the HPL and CL cases, extinction occurs at approximately the same rate over series. In the HFL case, the absence of the aggregate prediction B causes extinction to be retarded over series. Figure 13.5b shows that in later series, the HFL case fails to reach extinction criterion within 40 trials.

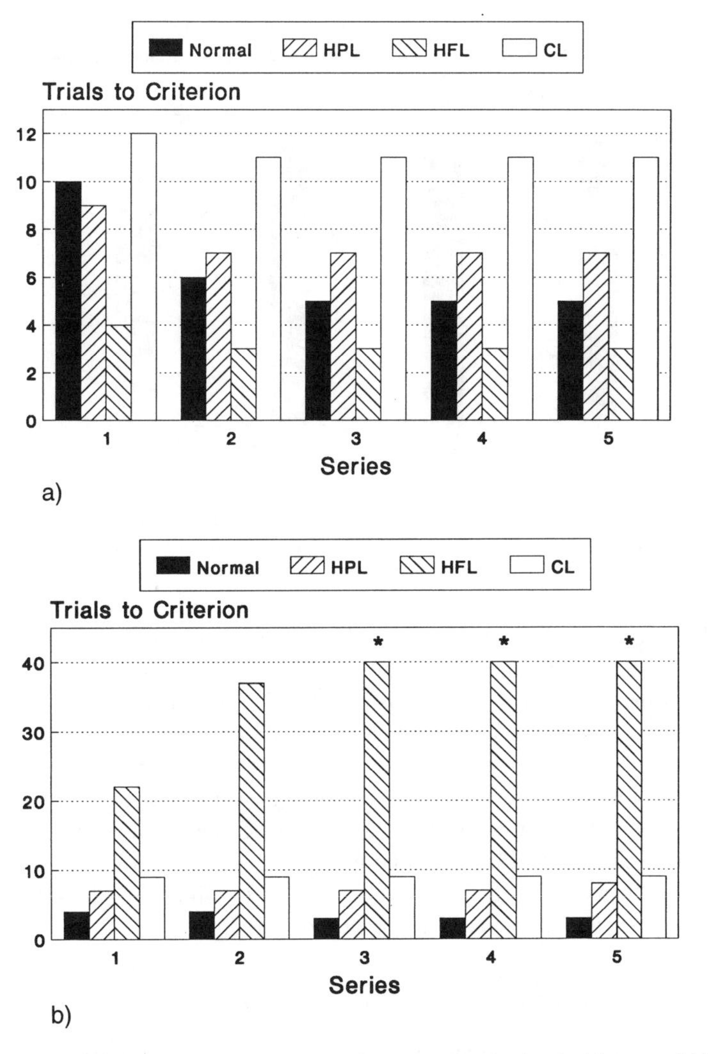

Figure 13.5. Acquisition and extinction series. (a) *Acquisition:* simulated trials to acquisition criterion (0.18) for normal, HPL, HFL, and CL cases in five series of acquisition trials. (b) *Extinction:* simulated trials to extinction criterion (0.02) following acquisition for normal, HPL, HFL, and CL cases. The asterisks indicate that the HFL case failed to achieve reversal within 40 trials.

Overshadowing and blocking

Figure 13.6 shows simulated peak CR amplitude evoked by CS_2 in acquisition, overshadowing, and blocking for normal, HPL, HFL, and CL groups. Acquisition consisted of CS_2 presentations paired together with the US during 10 trials. Overshadowing consisted of CS_1 and CS_2 presentations paired together with the US during 10 trials. Blocking consisted of 10 CS_1 acquisition trials followed by 10 CS_1–CS_2 acquisition trials.

Figure 13.6 indicates that simulated normal, HPL, and CL cases exhibit overshadowing. Consistent with Rickert et al.'s (1978) and Schmajuk, Spear, and Isaacson's findings (1983), but not Garrud et al.'s (1984), the model predicts that overshadowing is disrupted by HFL. In addition, Figure 13.6 shows that the

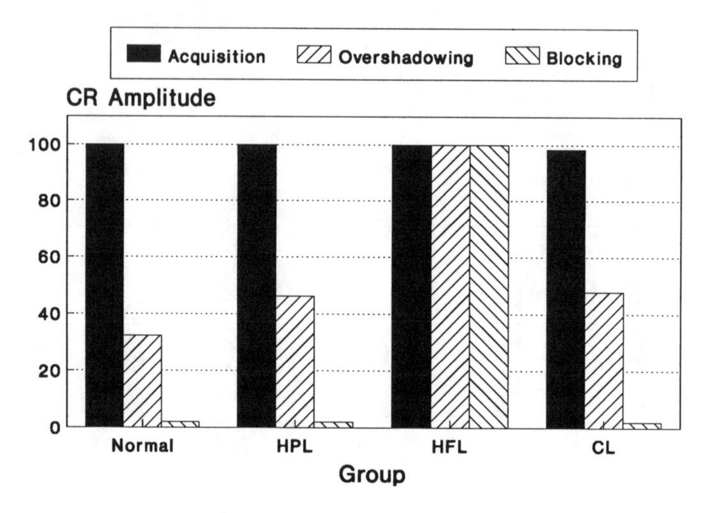

Figure 13.6. Blocking and overshadowing. Peak CR amplitude for normal HPL, HFL, and CL cases evoked by CS_2 after 10 reinforced CS_1–CS_2 trials, following 10 reinforced CS_1 trials in the case of blocking and 10 CS_1–CS_2 reinforced trials in the case of overshadowing.

model exhibits blocking in normal, HPL, and CL cases. In agreement with Rickert et al. (1979) and Solomon (1977), but not Garrud et al. (1984), the model predicts that HFL eliminate blocking. The CL result for blocking is in agreement with other experimental data (Moore, personal communication, 1990).

According to the SD model, HFL preclude the computation of the aggregate prediction *B*, thereby making the associations accrued by CS_2 independent of the associations accrued by CS_1, and producing deficits in overshadowing and blocking. Because the aggregate prediction *B* is unaffected in the HPL and CL cases, blocking and overshadowing are unimpaired.

Discrimination acquisition and reversal

Figure 13.7 shows the simulated number of trials to criterion during discrimination acquisition for normal, HPL, HFL, and CL groups. During discrimination acquisition, reinforced CS_1–US trials were alternated with nonreinforced CS_2 trials, until the response to CS_1 was at least twice that to CS_2. Figure 13.7 shows similar discrimination acquisition for simulated normal, HPL, HFL, and CL cases. These results agree with those obtained by Jarrard and Davidson (1991) for HPL; by Berger and Orr (1983), Berger et al. (1986), Port, Romano, and Patterson (1986), and Weikart and Berger (1986) for HFL; and by Oakley and Russell (1975) for CL. Also, in agreement with Micco and Schwartz (1972), simulations show that CS_2 acquires inhibitory association with the US during discrimination acquisition in normal but not HFL animals.

Figure 13.7 shows simulations of the number of trials to criterion during discrimination reversal. During reversal, the original nonreinforced CS_2 was reinforced and alternated with nonreinforced CS_1 trials, until the response to CS_2 was at least twice that to CS_1. Figure 13.7 indicates a strongly impaired discrimination reversal in the HFL case (asterisk indicates that the HFL case failed to reach criterion in 20 trials). This result is in agreement with results obtained by Berger and Orr (1983), Berger et al. (1986), Buchanan and Powell (1980), Port

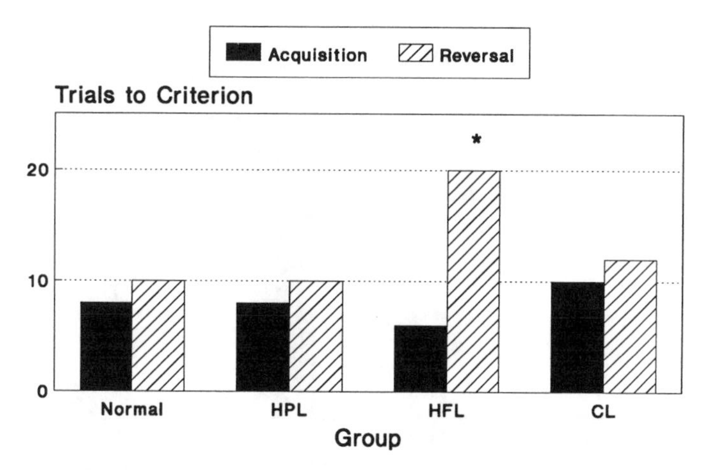

Figure 13.7. Discrimination acquisition and reversal. Trials to criterion for normal, HPL, HFL, and CL cases. Criterion was reached when the CR generated by the reinforced stimulus was at least twice as large as the CR generated by the unreinforced stimulus. The asterisk indicates that the HFL case failed to achieve reversal within 20 trials.

et al. (1986), and Weikart and Berger (1986) that illustrate a large impairment due to an increased responding to CS_1. Figure 13.7 shows that discrimination reversal is unimpaired in the HPL and CL groups. Schmajuk and DiCarlo (1992) found that, in simulations using a weaker intensity for CS_1 and CS_2, the CL group was impaired at discrimination reversal.

According to the model, discrimination acquisition is slightly facilitated by HFL because the absence of the aggregate prediction *B* allows more CSs and CNs to establish associations with the US. In the HPL and CL cases, the aggregate prediction is not disturbed and, therefore, discrimination acquisition and reversal are minimally affected. During discrimination reversal in normal animals, because the context is alternately reinforced in the presence of CS_2 but not reinforced in the presence of CS_1, the association of the context with the US becomes inhibitory, thereby contributing to a decreased responding to CS_1 and facilitating reversal. Discrimination reversal is impaired in HFL animals because, in the absence of the aggregate prediction, the association of the context with the US becomes excitatory (instead of inhibitory as in the normal case), thereby contributing to increased responding to CS_1 and hindering reversal.

Conditioned inhibition

Figure 13.8 shows simulated peak CR amplitude for normal, HPL, HFL, and CL cases evoked by CS_1 and CS_1–CS_2 after 20 reinforced CS_1 trials, alternated with 20 nonreinforced CS_1–CS_2 trials. In contrast to experimental results showing that conditioned inhibition is not affected by HFL (Kaye and Pearce, 1987a, b; Solomon, 1977), simulations do not display conditioned inhibition in the HFL case. The absence of inhibitory conditioning predicted by the SD model for the HFL case is consistent, however, with data showing that differential conditioning is impaired by HFL (Micco and Schwartz, 1972). According to the SD model, inhibitory conditioning is impaired in HFL animals because the aggregate prediction is needed in order to generate an inhibitory association between CS_2 and the US (see Schmajuk and DiCarlo, 1992). In agreement

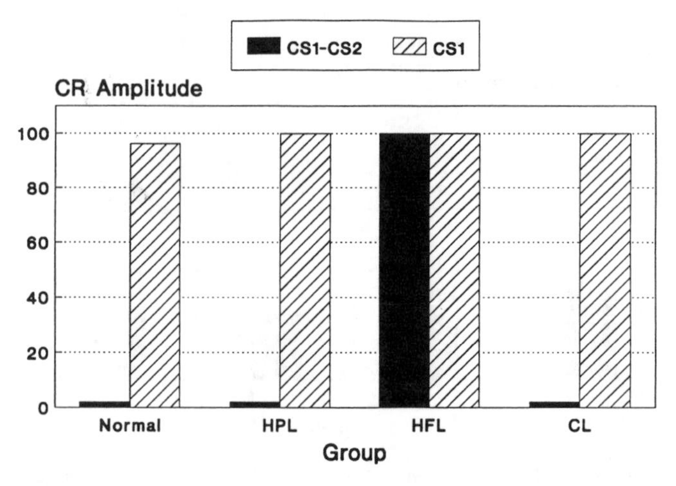

Figure 13.8. Conditioned inhibition. Peak CR amplitude for normal, HPL, HFL, and CL cases evoked by CS_1 and CS_1–CS_2 after 20 reinforced CS_1 trials, alternated with 20 nonreinforced CS_1–CS_2 trials.

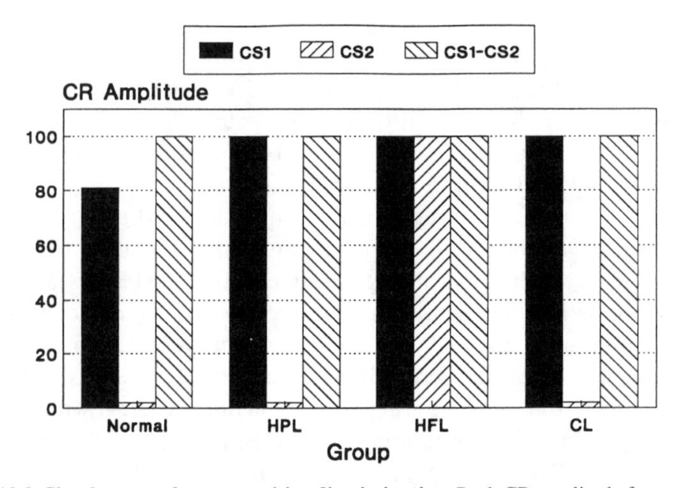

Figure 13.9. Simultaneous feature-positive discrimination. Peak CR amplitude for normal, HPL, HFL, and CL cases evoked by CS_1, CS_2, and CS_1–CS_2 after 75 reinforced CS_1–CS_2 trials, alternated with 75 nonreinforced CS_2 trials.

with Moore et al.'s findings (1980), simulations display normal conditioned inhibition in the CL case. No data are available for conditioned inhibition in the HPL case.

Simultaneous feature-positive discrimination

Figure 13.9 displays results for normal, HPL, HFL, and CL cases in simulations of simultaneous feature-positive discrimination. It shows simulated peak CR amplitude evoked by CS_1, CS_2, and CS_1–CS_2 after 75 reinforced CS_1–CS_2 trials, alternated with 75 nonreinforced CS_2 trials. Figure 13.9 demonstrates that normal, HPL, and CL cases exhibit feature-positive discrimination because the responses to CS_1–CS_2 and CS_1 is large, and the response to CS_2 is almost negligible. The HFL case does not show feature-positive discrimination because it strongly responds to all stimuli. These results are in agreement with Loechner and Weisz's (1987) data showing that acquisition of feature-positive discrimi-

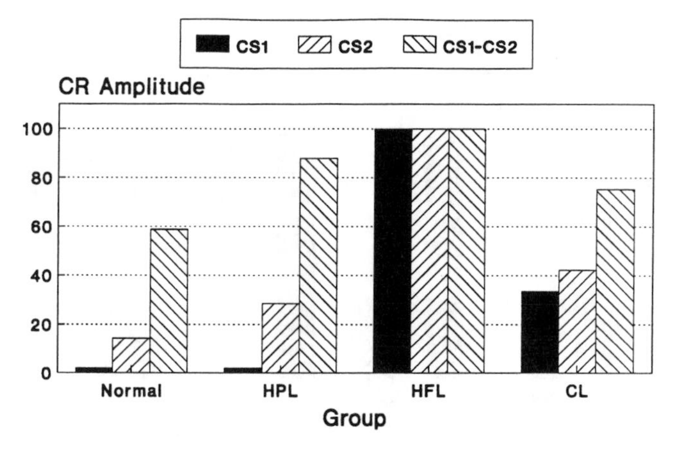

Figure 13.10. Serial feature-positive discrimination. Peak CR amplitude for normal, HPL, HFL, and CL cases evoked by CS_1, CS_2, and serial presentation of CS_1–CS_2 after 75 reinforced serial CS_1–CS_2 trials, alternated with 75 nonreinforced CS_2 trials.

nation is impaired in HFL animals. No data are available for CL effects on feature-positive discrimination.

Serial feature-positive discrimination

Figure 13.10 shows simulated results for normal, HPL, HFL, and CL cases in simulations of serial feature-positive discrimination. It indicates peak CR amplitude evoked by CS_1, CS_2, and CS_1–CS_2 after 75 reinforced CS_1–CS_2 trials, alternated with 75 nonreinforced CS_2 trials. It also shows that only the normal and HPL cases exhibit conditional discrimination because the response to CS_1–CS_2 is large, and the responses to CS_1 and CS_2 are small. The HFL and CL cases do not show conditional discrimination because they strongly respond to all stimuli. These results are in agreement with Ross et al.'s (1984) data showing that acquisition of conditional discrimination is impaired in HFL animals, and with Jarrard and Davidson's (1991) observation that conditional discrimination is impaired in HFL but not HPL animals.

Negative patterning

Figure 13.11 shows the effect of simulated HPL, HFL, and CL on negative patterning acquisition. Figure 13.11A illustrates peak CR amplitude evoked by CS_1, CS_2, and CS_1–CS_2 after 100 reinforced CS_1 trials, alternated with 100 reinforced CS_2 trials and 100 nonreinforced CS_1–CS_2 trials. It also indicates that although simulated normal and HPL animals exhibit negative patterning, HFL and CL animals do not. This result is in agreement with Rudy and Sutherland's (1989) results for HFL animals and Gallagher and Holland's (1992) data for HPL animals.

Figure 13.11B displays the effect of simulated HPL, HFL, and CL on negative patterning retention. Negative patterning was simulated as in Figure 13.11A; then simulated "lesions" were induced, followed by five reinforced CS_1 trials, alternated with five reinforced CS_2 trials, and five nonreinforced CS_1–CS_2 trials. Figure 13.11B shows postlesion peak CR amplitude evoked by CS_1, CS_2, and CS_1–CS_2 for the normal, HPL, HFL, and CL cases. According to

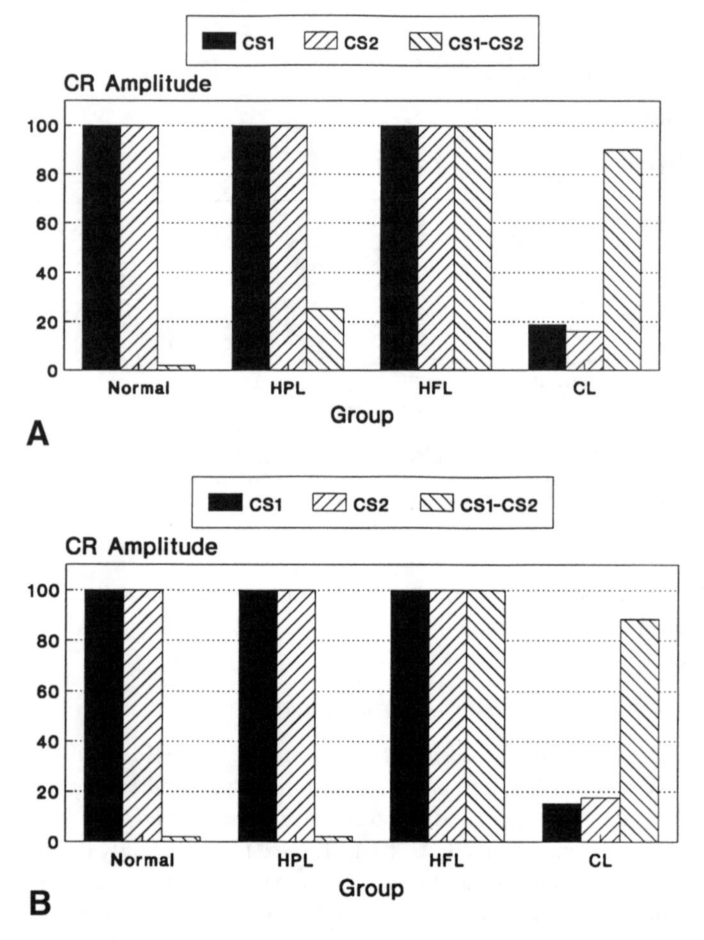

Figure 13.11. Negative patterning acquisition and retention. (A) *Acquisition:* peak CR amplitude for normal, HPL, HFL, and CL cases evoked by CS_1, CS_2, and CS_1–CS_2 after 100 reinforced CS_1 trials, alternated with 100 reinforced CS_2 trials and 100 nonreinforced CS_1–CS_2 trials. (B) *Retention:* peak CR amplitude for normal, HPL, HFL, and CL cases evoked by CS_1, CS_2, and CS_1–CS_2 after negative patterning acquisition followed by five reinforced CS_1 trials, alternated with five reinforced CS_2 trials and five nonreinforced CS_1–CS_2 trials

Figure 13.11B, CL and HFL eliminate the previously learned negative patterning. These results are in agreement with Rudy and Sutherland's (1989) results. Rudy and Sutherland (1989) found that both control and HFL animals lose some negative patterning after the surgical procedure, but control animals recover to prelesion levels and HFL animals worsen their responding, within 120 trials. No data are available concerning the effects of HPL on retention of negative patterning.

Figure 13.11 shows that the normal and HPL cases succeed at acquisition and retention of negative patterning, whereas the HFL and CL cases fail. The normal case solves the problem by training some hidden units to represent the CS_1–CS_2 compound, and then forming inhibitory associations between these hidden units and the network output. In the HPL case, negative patterning is achieved because, even though hidden units cannot be trained, hidden weights VH_{ij} are initialized to small random values, so that some hidden units may be preferentially activated when the CS_1–CS_2 compound is presented. Although the HFL case also has randomly initialized hidden units that may respond to the

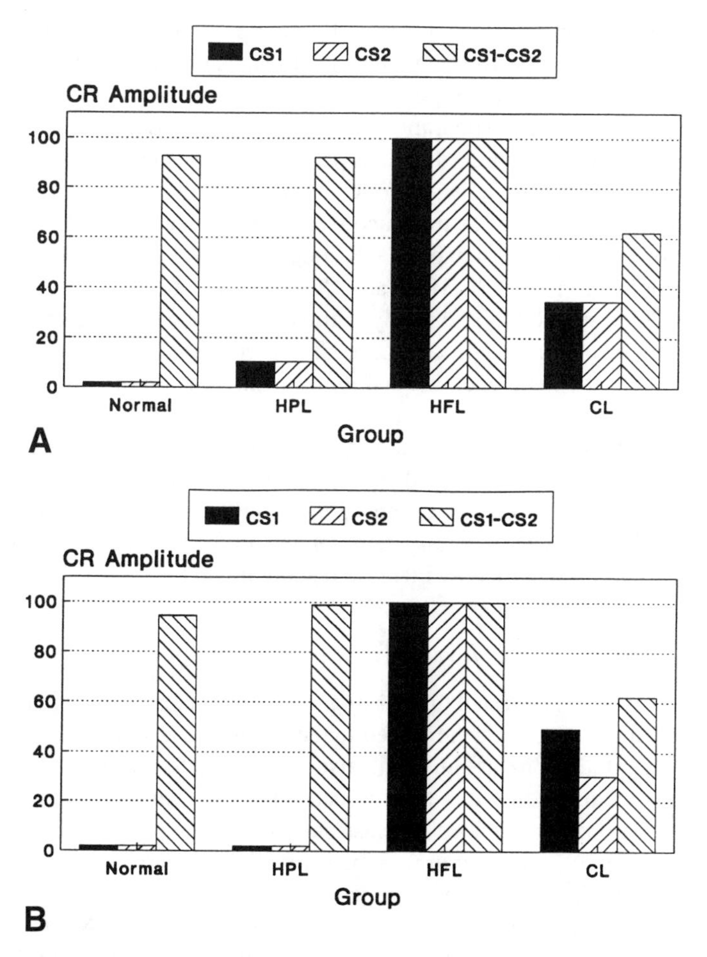

Figure 13.12. Positive patterning acquisition and retention. (A) *Acquisition:* peak CR amplitude for normal, HPL, HFL, and CL cases evoked by CS_1, CS_2, and CS_1–CS_2 after 60 nonreinforced CS_1 trials, alternated with 60 nonreinforced CS_2 trials and 60 reinforced CS_1–CS_2 trials. (B) *Retention:* peak CR amplitude for normal, HPL, HFL, and CL cases evoked by CS_1, CS_2, and CS_1–CS_2 after positive patterning acquisition followed by five reinforced CS_1 trials, alternated with five reinforced CS_2 trials and five nonreinforced CS_1–CS_2 trials.

CS_1–CS_2 compound, it cannot compute the aggregate prediction *B* and, therefore, is unable to suppress responding to the CS_1–CS_2 compound. The CL case has no hidden units and, therefore, cannot perform the stimulus configuration necessary for acquisition and retention of negative patterning.

Positive patterning

Figure 13.12A shows the effect of simulated HPL, HFL, and CL on positive patterning acquisition. It illustrates peak CR amplitude evoked by CS_1, CS_2, and CS_1–CS_2 after 60 reinforced CS_1 trials, alternated with 60 reinforced CS_2 trials and 60 nonreinforced CS_1–CS_2 trials. It also shows that normal and HPL animals exhibit positive patterning because the response to CS_1–CS_2 is larger than the sum of the responses to its components, CS_1 and CS_2. The HFL case does not show positive patterning, because the animal fails to suppress responding to the components CS_1 and CS_2. The CL case also fails to learn the task, because the responses to the components are more than half of the response to the compound. In agreement with Gallagher and Holland's findings (1992), the HPL case is not impaired at performing the task. No data are available for the effects of HFL or CL on acquisition of positive patterning.

Figure 13.12B displays simulations of the effects of CL and HFL on the retention of positive patterning. Positive patterning was simulated as in Figure 13.12A; then lesions were induced, followed by five reinforced CS_1 trials, alternated with five reinforced CS_2 trials and five nonreinforced CS_1–CS_2 trials. Figure 13.12B shows postlesion peak CR amplitude evoked by CS_1, CS_2, and CS_1–CS_2 for the normal, HPL, HFL, and CL cases. According to Figure 13.12B, CL and HFL eliminate the previously learned positive patterning. No data are available for the effects of HPL, HFL, or CL on retention of positive patterning.

According to the SD model, acquisition of positive patterning is impaired in CL and HFL animals because stimulus configuration is absent in both cases. The CL case does not show retention of positive patterning because the previously formed CN_j's are obliterated by the lesion. The HFL case does not show retention of positive patterning because, although the previously formed CN_j's are not changed after HFL, the system cannot limit the formation of CS_i–US simple associations and thereby increase responding to the components.

Simultaneous negative and positive patterning

Because it has been suggested that, during instrumental learning, discriminative stimuli become classically conditioned to the reinforcer (see Mackintosh, 1983, p. 100), our simulations assume that discriminative stimuli are analogous to CSs in a classical conditioning task. Because peak CR amplitude and CR percentage are strongly correlated in classical conditioning (e.g., Schmajuk and Christiansen, 1990), peak output was interpreted as CR percentage in the simulations. The simulated stimuli were a panel light P, a tone T, and white noise N. Prior to discrimination training, simulated animals received 40 N+ and 40 PT+ trials to simulate the pretraining phase of Gallagher and Holland's (1992) experiment. Simultaneous negative and positive patterning training was conducted by repeatedly presenting each of the four stimulus patterns (N+, T–, PT+, PN–) in random order. Four complete cycles through all four stimulus patterns were regarded as analogous to one daily training session in Gallagher and Holland's (1992) experiment. The ratio between the number of pretraining (80) and training (320) pattern presentations in the simulations was made identical to that between the number of pretraining (300) and training (1,200) pattern presentations in the experiment.

Figure 13.13 shows simulation results for normal and HPL animals in the simultaneous negative and positive patterning task after 20 training sessions.

Figure 13.13 shows that, as in Gallagher and Holland's (1992) experiment, the HPL case is facilitated at performing the task. These results can be explained in terms of the mechanisms used by the SD model to solve negative and positive patterning. As in other backpropagation procedures, the hidden-unit layer in the SD model is initialized with random input–hidden-unit weights VH_{ij} (see Figure 13.1). Through training, these initial input–hidden-unit weights are modified to generate the configural stimuli needed to solve the task. Although the model does not train input–hidden-unit weights in the HPL case, the network can still solve a simultaneous negative and positive patterning task if the untrained hidden units provide stimulus configurations that allow the network to solve the problem. The probability that randomly initialized hidden weights

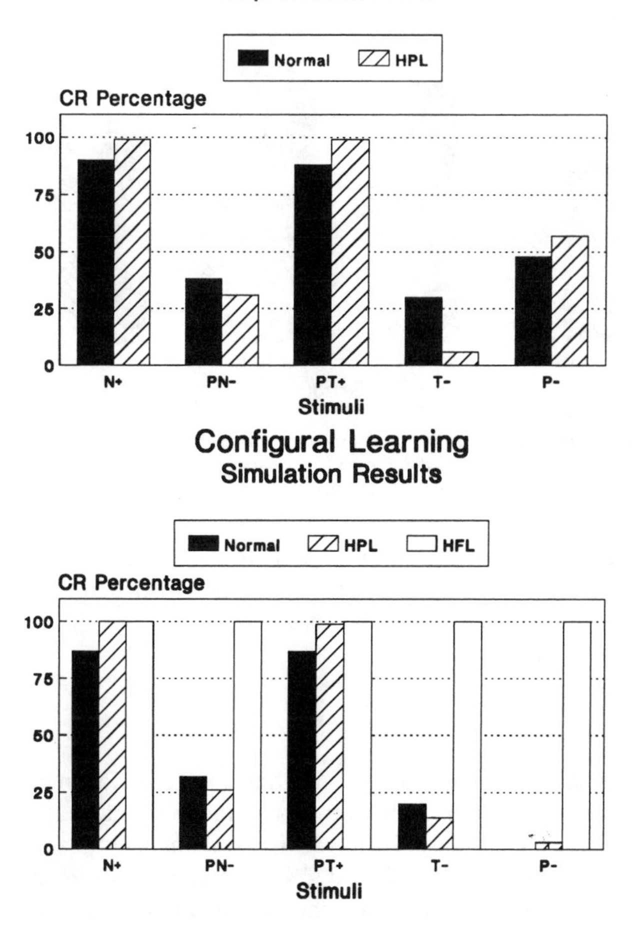

Figure 13.13. Combined negative and positive patterning. CR percentage after 20 training sessions for normal versus hippocampal animals in the simultaneous negative and positive patterning task. *Upper panel:* experimental data from Gallagher and Holland (1992). *Lower panel:* simulated results for normal, hippocampus proper (HPL), and hippocampal formation (HFL) cases.

will be sufficient to solve the task increases with the number of units in the hidden layer. In the case of negative patterning, a hidden unit that responds specifically to the combined presentation of both CSs is sufficient to solve the problem.

Furthermore, the HPL case solves the discrimination task faster than the normal case because the mutually dependent output and hidden-layer weights need time to attain their correct, final, and stable values in the normal case. In contrast, in the HPL case, hidden-layer weights do not change, and output weights can reach their correct values in a shorter period of time. In other words, the more flexible normal case needs some more time to reach a stable discrimination than the more rigid HPL case.

Figure 13.13 also shows simulations of simultaneous negative and positive patterning after aspiration or kainic acid/colchicine (HFL) lesions. In agreement with Rudy and Sutherland's (1989) and Whishaw and Tomie's (1991) results (but not with Davidson, McKernan, and Jarrard's, 1993) and in contrast to the

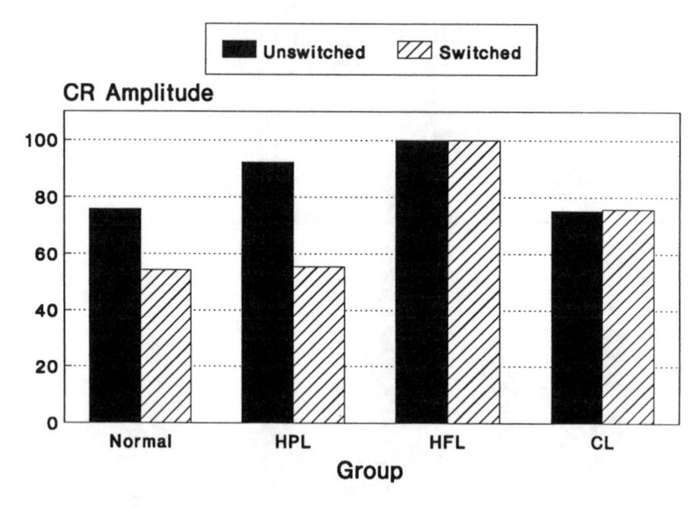

Figure 13.14. Contextual effects. Peak CR amplitude for normal, HPL, HFL, and CL cases evoked by CS_1 in context CX_1 (unswitched) and context CX_2 (switched) following 10 reinforced trials in context CX_1.

HPL case, simulated HFL impair negative and positive patterning. Because the model cannot generate inhibitory associations in the absence of the aggregate prediction *B*, it is impossible for the network to solve negative or positive patterning problems after HFL.

Contextual effects

In a context-switching paradigm, animals are trained to criterion in one context and tested in a new context that differs from the training context along visual, olfactory, and tactile dimensions. Figure 13.14 shows simulated peak CR amplitude evoked by CS_1 in contexts CX_1 and CX_2 after 10 CS_1 reinforced trials in context CX_1. In agreement with Penick and Solomon (1991), normal animals show reduced responding when the context is switched, whereas HFL animals respond similarly in both contexts. The model predicts that HPL animals should show reduced responding after context switching, and that CL animals should show similar responding in both contexts.

In the normal case, the model creates a configural stimulus out of the CS and the training context CX_1. This configural stimulus competes and gains some association with the US, along with the direct association between the CS and the US. When the context is switched, the prediction of the US by the configural stimulus is eliminated, and only the direct CS–US association remains, resulting in a smaller response. The HFL case responds maximally in both contexts because, in the absence of the aggregate prediction *B*, the CS does not compete with configural stimuli and, therefore, the direct CS–US association becomes strong enough to generate a maximum CR. The CL case responds equally in both cases because, without hidden units, there are no configural stimuli to compete against the CS for association with the US.

Hippocampal pyramidal cell activity during classical conditioning

As described before, the activity of some pyramidal cells is proportional to CS_iVS_i and CN_jVN_j values used to compute the aggregate prediction, $B =$

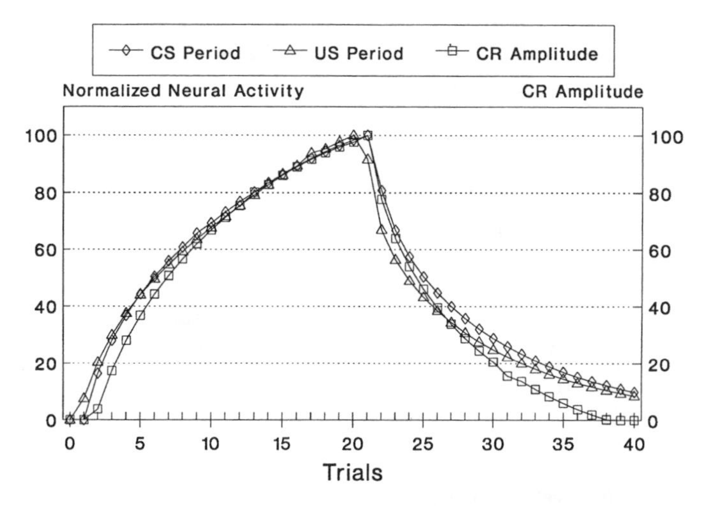

Figure 13.15. Hippocampal pyramidal activity during acquisition and extinction of classical conditioning. Simulated peak CR amplitude and hippocampal pyramidal activity during CS and US periods for normal animals over 20 acquisition and 20 extinction trials. CR amplitude and neural activity are expressed as a percentage of their maximum values. Hippocampal neural activity is assumed to be proportional to the aggregate prediction of the US, B, plus the sum of error signals for the hidden units $\theta\Sigma_j CN_j VN_j$.

$\Sigma_i CS_i VS_i + \Sigma_j CN_j VN_j$, whereas the activity of other pyramidal cells is proportional to the product $\theta CN_j VN_j$. Therefore, pyramidal cell activity is computed as $B + \theta\Sigma_j CN_j VN_j$. Because both the CR and pyramidal activity are proportional to $\Sigma_i CS_i VS_i + \Sigma_j CN_j VN_j$, in agreement with empirical results (Berger and Thompson, 1978a; Berger et al., 1983), pyramidal activity is positively correlated with the topography of the CR.

Figure 13.15 shows simulated CR amplitude and hippocampal neural activity during the CS and US periods for normal animals over 20 acquisition and 20 extinction trials. CR amplitude and neural activity are expressed as a percentage of their maximum values. Figure 13.15 indicates that, in agreement with experimental data (Berger and Thompson, 1978a, 1982; Berger et al., 1983), changes in hippocampal activity during the US period precede both behavioral acquisition and extinction. Changes in pyramidal activity precede behavioral acquisition during the US period because a behavioral threshold should be exceeded to generate a CR (see Equation 6.9 in Appendix 6.A). Changes in pyramidal activity precede extinction during the US period because the sum of error terms for the hidden units, $\theta\Sigma_j CN_j VN_j$, which contributes to determining pyramidal activity, decreases faster than $\Sigma_i CS_i VS_i + \Sigma_j CN_j VN_j$, which sustains both pyramidal activity and the CR.

However, in partial disagreement with empirical data (Berger and Thompson, 1982), simulated changes in pyramidal activity during the CS period precede behavioral changes during acquisition but not during extinction. During extinction, pyramidal activity and behavior decrease at the same time because θ does not decrease during the CS period.

Medial septal neural activity during classical conditioning

In the model, medial septal neural activity is proportional to the absolute difference between actual and predicted US values, $|US - B|$. Because the medial sep-

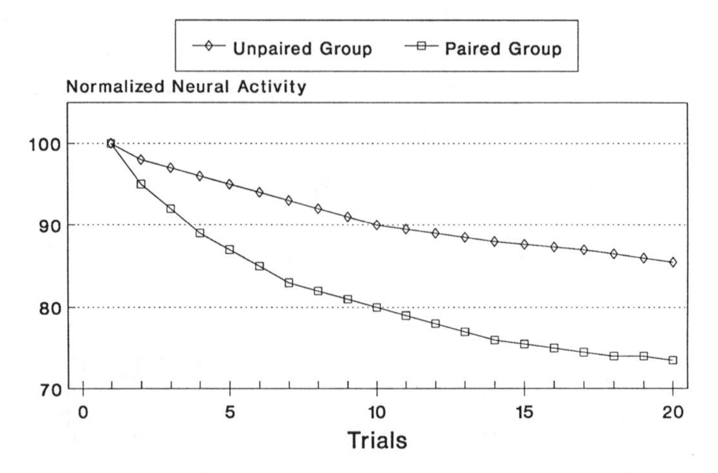

Figure 13.16. Medial septum activity during acquisition of classical conditioning. Simulated medial septum neural activity during the US period for normal animals over 20 paired and unpaired trials. Neural activity is expressed as a percentage of its maximum value at the beginning of acquisition. Medial septum neural activity is proportional to the absolute difference between actual and predicted US values.

tum controls the generation of theta, hippocampal theta rhythm is also proportional to this value. In agreement with Berger and Thompson (1978a), because hippocampal activity is proportional to B, medial septal and hippocampal activity are negatively correlated during acquisition of classical conditioning.

Figure 13.16 shows simulated medial septum neural activity during the US period for normal animals over 20 paired and unpaired trials. Paired and unpaired cases show reductions in $|\text{US} - B|$ because the US is predicted by the CS in the paired case and by the sound of the air puff in the second case. Because the US is better predicted by the CS and the sound of the air puff than by the air puff alone, the final value of $|\text{US} - B|$ during the US period is smaller in the paired case than in the unpaired case. These simulation results are in agreement with experimental data (Berger and Thompson, 1978b). Because the present version of the SD model does not include CS–CS associations, it cannot capture the decrease found in medial septal activity (and θ) during the CS period, a phenomenon that is well described by the S–P–H model (Schmajuk and Moore, 1988).

Place learning

With the SD model, place learning in the Morris water maze is simulated by exposing the network to different points in the tank and consistently rewarding it at the location where the platform is located. At each location, the network's input is presented with visual angles, $\Omega_i(d_i) = \arctan(O_i/d_i)$, that define a spatial location. These visual angles accrue a *direct* association with the output and become *configured* in hidden units, which become associated with the network output. Direct associations VS_i between visual angles Ω_i and the US acquire relatively large positive values, so that direct connections excite the output of the network. In the hidden layer, units are trained to exhibit output at various locations in the pool other than that of the platform. Because hidden units accrue an inhibitory association VN_j with the output of the network, network output decreases at locations other than the platform location. After training, the output of the system becomes maximally active at the spatial location where the hidden platform is encountered and displays decremental generalization at other locations. As in Chapter 10 (see Schmajuk, 1990; Schmajuk and Blair, 1993) we assume that animals navigate to the location of the platform by following the gradient of the network's output from any novel start point in the periphery of the tank. Interestingly, because the aggregate prediction B is essentially identical to the output of the network (see Equation 6.9 in Appendix 6.A) and B is assumed to be computed in the hippocampus, the model correctly predicts that some hippocampal neurons should exhibit place-specific activity at the location of the US (Breese, Hampson, and Deadwyler, 1989; Eichenbaum et al., 1986).

In sum, the SD model achieves stimulus configuration by adjusting the random initial values of input–hidden-unit associations VH_{ij}. In the case of classical conditioning, CS_i input–hidden-unit j associations VH_{ij} are modified so that hidden unit j becomes active under different combinations of input CS. In the case of place learning, visual angle i–hidden-unit j associations VH_{ij} are modified so that hidden unit j becomes active at precise distances from different spatial landmarks. Schmajuk and DiCarlo (1992, Appendix E) showed that the percentage of cases in which a negative patterning problem is solved by the network increases with an increasing number of hidden units, and the number of trials needed to reach this criterion decreases with an increasing number of hidden units. Whereas the number of hidden units needed to reliably solve negative patterning varies between five and ten, 20 hidden units were needed in order to attain reliable place learning.

Figure 13.17 shows the network's prediction of the location of the hidden platform after 20 simulated swimming trials for normal, HPL, HFL, and CL cases. The large circle represents the boundary of the Morris tank, four spatial landmarks are represented by solid boxes, and the asterisk indicates the location of the hidden platform. The magnitude of the network's prediction of the location of the platform at each point is represented by the sizes of the small circles.

Figure 13.17 shows that simulated normal animals exhibit place learning by displaying a peak output near the location of the hidden platform and decremental generalization at surrounding locations. For the normal case, the aggregate prediction B of the location of the platform is identical to the output of the network. Therefore, because the SD model assumes that B is computed in the hippocampus, the normal case in Figure 13.17 can be interpreted as hippocampal place-specific activity at the location of the US (see Breese et al., 1989; Eichenbaum et al., 1986).

In the normal case, the system usually solves place learning as follows. Associations VS_i between visual angles Ω_i and the US acquire relatively large positive values, so that direct connections excite the output of the network. In the hidden layer, units are trained to exhibit output at various locations in the pool other than that of the platform. Because hidden units accrue an inhibitory association VN_j with the output of the network, network output decreases at locations other than the platform location. Therefore, after training, the network output will show a peak at the location of the platform and steep generalization decrement at other points in the pool. Because the location of the platform is clearly discriminated, normal animals are capable of finding their way to the goal from any novel start point in the periphery of the tank.

Figure 13.17 also shows that, in agreement with Gallagher and Holland (1992) and Morris et al. (1990), the HPL case exhibits poor place discrimination, with no well-defined peak at the location of the platform and excessive

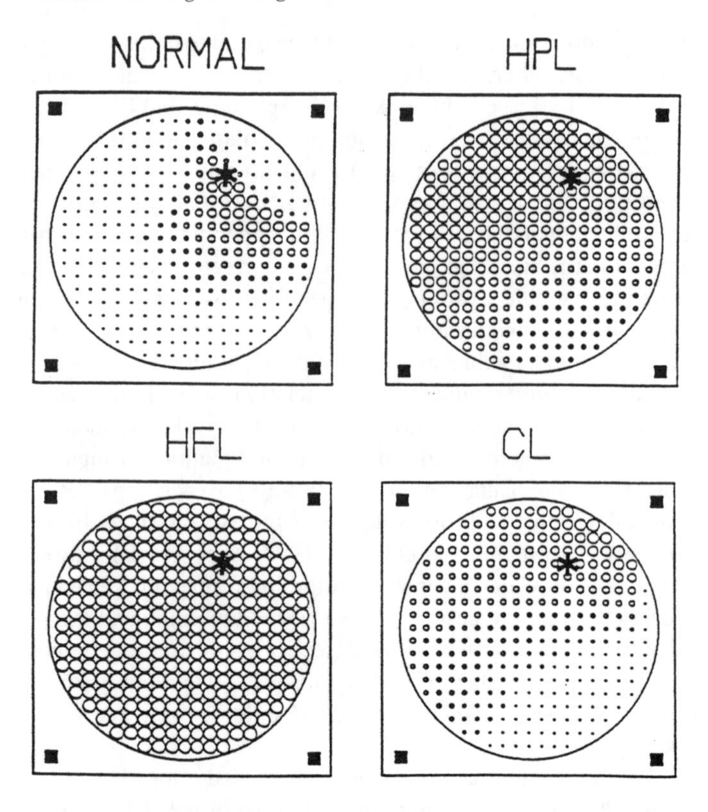

Figure 13.17. Place-learning acquisition. Simulated prediction of the location of the hidden platform at different points in the Morris water maze after 20 trials for normal, HPL, HFL, and CL cases. The large circle represents the boundary of the Morris tank, four spatial landmarks are represented by solid boxes, and the asterisk indicates the location of the hidden platform. The magnitude of the network's prediction of the location of the platform at each point in the pool is represented by the sizes of the small circles.

generalization to other areas of the pool. Place-learning acquisition is even more dramatically impaired by HFL, with no peak at the location of the platform and a flat generalization surface at all positions in the pool. Under the assumption that animals navigate toward the platform by following the gradient of the network's output (see Chapter 10; Schmajuk, 1990; Schmajuk and Blair, 1993), HFL animals are more impaired than HPL animals at finding the goal. These results are similar to Morris et al.'s (1990) data showing that animals with ibotenate HPL were impaired in the Morris tank, but less impaired than animals with more extensive aspiration lesions of the hippocampus (Morris et al., 1982) or animals with entorhinal cortex/subiculum lesions (Schenk and Morris, 1985).

In the HPL case, the system might solve place learning as follows. As in the normal case, associations VS_i between visual angles Ω_i and the US might acquire relatively large positive values, so that direct connections excite the output of the network. In the hidden layer, units cannot be specifically trained to exhibit output at locations other than that of the platform but they are active, instead, at places defined by their initial random weights. As in the normal case, hidden units also accrue inhibitory associations VN_j with the output of the network and, therefore, network output decreases at locations other than the platform location. Consequently, the system output peaks, with varying degrees of accuracy depending on the initial hidden-unit weight values, at the location of the submerged platform and shows a broad generalization to other points in the pool. Because the location of the platform is poorly discriminated, HPL animals are impaired at finding their way to the goal.

In the HFL case, the absence of the aggregate prediction causes associations VS_i and VN_j to acquire large positive values, so that direct and indirect connections excite the output of the network. Therefore, the output of the system saturates over the entire surface of the pool. Because the location of the platform is not discriminated, HFL animals cannot find their way to the goal. Whereas the impairment shown by the HPL case in place learning can be described in terms of (1) the error in the prediction of the precise location of the platform and (2) increased generalization, impairment shown by the HFL case in place learning can be described in terms of the extensive generalization over the entire surface of the pool.

Figure 13.17 also shows that the CL case exhibits poor place discrimination, with a peak in the vicinity of the landmark nearest to the platform, and not at the location of the platform. Experimental data show that lesions of parietal (DiMattia and Kesner, 1988), entorhinal (Schenk and Morris, 1985), frontal (Sutherland, Kolb, and Whishaw, 1982), or cingulate (Sutherland, Whishaw, and Kolb, 1988) cortex, but not temporal cortex (Kolb, Buhrman, and McDonald, 1989), impair place-learning acquisition. These results suggest that cortical areas, other than temporal cortex, might be involved in the stimulus configuration during spatial learning.

The model predicts that retention of place learning should be unaffected by HPL, because hidden-unit weights do not need to be adjusted once they have been trained. Morris et al.'s (1990) data regarding HPL effects on place-learning retention are inconclusive. By contrast, the model predicts that HFL should impair place-learning retention because, in the absence of competition, the system cannot maintain proper connection strengths in the output layer. This prediction is consistent with the findings of DiMattia and Kesner (1988) showing that HFL impair retention of the Morris tank task. Finally, the model predicts that cortical lesions should impair retention of place learning, because stimulus configuration in hidden units is essential for place discrimination. Experimental data, showing that lesions of parietal (DiMattia and Kesner, 1988) or cingulate (Sutherland et al., 1988) cortex impair retention of place learning, suggest that these regions might be involved in storing or retrieving spatial configural stimuli.

Conclusion

Simulation results

The evaluation of the brain-mapped model is a critical component in the present chapter. The model has been evaluated at anatomical, behavioral, and neural levels. At the anatomical level, the interconnections among neural elements in the model reflect the neuroanatomical data, that is, links in the neural architecture of the model have equivalents in pathways connecting different brain regions and in the intrinsic circuit of the hippocampus. At the behavioral level, computer-simulated results were contrasted with experimental data for normal

Table 13.1. *Simulations of HFL, HPL, and CL cases obtained with the Schmajuk and DiCarlo (1992) model compared with experimental results in different learning paradigms.*

	HFL		HPL		CL	
Paradigm	Data	Model	Data	Model	Data	Model
Delay conditioning	+, 0	+	?	+	0	0
Trace conditioning	+, 0, –	+	?	0	0	0
Extinction	0, –	–	?	–	0	0
Explicitly unpaired extinction	0	– [a]	?	–	?	–
Acquisition series	–	–	?	–	?	–
Extinction series	–	–	?	–	?	–
Blocking	0, –	–	?	0	0	0
Overshadowing	0, –	–	?	0	?	0
Discrimination						
Acquisition	0	0	0	0	0	0
Reversal	–	–	?	0	+	0[a]
Conditioned inhibition	0	– [a]	?	0	0	0
Simultaneous feature–positive discrimination						
Acquisition	–	–	0	0	0	0
Retention	–	–	0	0	0	0
Simultaneous feature–positive discrimination	–	–	0	0	?	–
Differential conditioning	–	–	?	0	?	0
Negative patterning						
Acquisition	–	–	0	0	?	–
Retention	–	–	?	0	?	–
Positive patterning						
Acquisition	?	–	0	0	?	–
Retention	?	–	?	0	?	–
Context switching	–	–	?	0	?	–
Place learning						
Acquisition	–	–	–	–	0, –	–
Retention	–	–	0	0	–	–

Note. – : Deficit, + : Facilitation, 0 : No effect, ? : no available data, [a] : the model fails to describe accurately the experimental data.

animals. By evaluating the model in the large variety of learning paradigms listed in Table 13.1 (using the same parameter values), we expect it to reflect properties of the real nervous system and to reveal more than a fortuitous selection of parameter values. At the neural level, the performance of neural elements in the model was compared with data about the activity of different neural populations. This comparison provides insights into the biological plausibility of the computational processes occurring in the model.

Further insights into the model's performance come from evaluating its success after "lesioning" (removing elements) from the network. Table 13.1 summarizes the results of the simulation experiments for HPL, HFL, and CL cases. It shows that the model is able to describe the effects of HPL, HFL, and CL on acquisition and extinction of delay and trace conditioning, acquisition and extinction series, explicitly unpaired extinction, blocking, overshadowing, discrimination acquisition and reversal, feature-positive discrimination, conditional discrimination, conditioned inhibition, differential conditioning, contextual effects, acquisition and retention of positive patterning, and acquisition and retention of negative patterning. Importantly, the model describes the effect of HPL, HFL, and CL on the acquisition and retention of place learning. Table 13.1 indicates that the model has difficulty simulating the effect of HFL on explicitly unpaired extinction and conditioned inhibition. Also as illustrated in

Table 13.2. *Simulations of neural activity obtained with the Schmajuk and DiCarlo (1992) model compared with experimental results in classical conditioning and place learning.*

Brain Region	Paradigm	Data	Model
Hippocampal pyramidal cells	Acquisition CS period	Increases Precedes behavior	Increases Precedes behavior
	Extinction CS period	Decreases Precedes behavior	Decreases Succeeds behavior[a]
	Acquisition US period	Increases Precedes behavior	Increases Precedes behavior
	Extinction US period	Decreases Precedes behavior	Decreases Precedes behavior
	Place learning	Maximal activity at US location	Maximal activity at US location
Lateral septum	Acquisition	Increases	Increases
Medial septum	Acquisition	Decreases	Decreases
Dorsal accessory olive	Acquisition	Decreases	Decreases

Note. [a] the model fails to describe accurately the experimental data.

Table 13.1, the model makes numerous novel predictions about the effects of HPL in many learning paradigms.

The model is also able to describe the effect of CL on delay conditioning, trace conditioning, extinction, conditioned inhibition, and discrimination acquisition. However, the model has difficulties describing the effects of CL on discrimination reversal.

In addition to simulating behavioral effects of brain lesions, the model also describes neural activity in normal animals. Table 13.2 summarizes the descriptions of neural activity provided by the brain-mapped SD model in different brain regions. It shows that the model correctly describes the neural activity of hippocampal pyramidal cells, medial septum, and lateral septum during the acquisition and extinction of classical conditioning. Only pyramidal cell activity during the CS period in the course of extinction was not correctly described. In addition, the model also describes the activity of pyramidal cells during place learning.

A nonunitary computational theory of hippocampal function

Most theories of hippocampal function are characterized by (1) assigning a unitary psychological function to the hippocampal formation and (2) proposing that different hippocampal regions collaborate in the same psychological function (Schmajuk, 1984a). In contrast, the brain-mapped SD model is distinguished by (1) assigning several computational functions to the hippocampal formation and (2) proposing that different hippocampal regions might process incoming information in different ways.

According to Figure 13.1, the hippocampal formation receives information about (1) the associations of simple and configural stimuli with the US and (2) the mismatch between the actual and predicted US intensity. Based on these inputs, the hippocampal formation computes and broadcasts (1) a signal proportional to the sum of the associations of simple and configural stimuli with the US (the aggregate prediction of the US) to cerebellar areas in order to control the associations formed with the US, and (2) error signals to the association cortex in order to modulate the stimulus configuration. At a more molecular level, the computation of aggregate prediction signals might be carried out in the entorhinal cortex, whereas the computation of error signals might be accomplished in the CA3/CA1 regions. In sum, the brain-mapped SD network strictly defines hippocampal function as the *multiple* transformations that occur between the hippocampal inputs and outputs, rather than as a unitary psychological function.

Computational approaches to classical conditioning and hippocampal function

Moore and Stickney (M–S) (1980) proposed the first mathematical attentional model of hippocampal function. The model is based on Mackintosh's (1975) attentional theory of conditioning. In applying their model to hippocampal function, Moore and Stickney (1980) proposed that HFL prevent associability from decreasing when it otherwise would. In this formal sense, therefore, HFL prevent irrelevant CSs from being "tuned out." Simulation studies (Moore and Stickney, 1980, 1982) show that the model describes the behavior of HFL animals in latent inhibition, blocking, and spatial learning. Schmajuk and Moore (1985) presented a revised version of the M–S model that corrects deficiencies in earlier versions in describing extinction, partial reinforcement, and reacquisition following extinction. Schmajuk and Moore (1985) also showed that the model can describe, in addition, the behavior of HFL animals in delay and trace conditioning, conditioned inhibition, extinction, and overshadowing.

Schmajuk and Moore (1985, 1989) analyzed the effects of various hippocampal manipulations on the classically conditioned NM response in an elaborated rendering of the Moore and Stickney (1980, 1982) model, called the M–S–S model. Under the "tuning out" hypothesis suggested by Moore and Stickney (1980), the M–S–S model correctly describes the experimental effects of HFL on delay conditioning, conditioning under optimal ISI, conditioned inhibition, extinction, latent inhibition, blocking, and mutual overshadowing. The model, however, is inconsistent with experimental findings describing the effects of HFL on trace conditioning with shock as the US under short and long ISIs, trace conditioning with air puff as the US under long ISIs, discrimination reversal, and sensory preconditioning. Schmajuk (1986) suggested that LTP facilitates the "tuning in" of good predictors. Under the tuning in hypothesis, the M–S–S model is unable to describe the effects of hippocampal induction of LTP in the acquisition of classical discrimination. Under the assumption that hippocampal neuronal activity is proportional to the magnitude of CS–US associations, the M–S–S model correctly describes hippocampal neuronal activity during the acquisition and extinction of classical conditioning.

Schmajuk (1984b) described the effect of hippocampal lesions in terms of Pearce and Hall's (P–H) (1980) attentional model. As mentioned in Chapter 2, according to the P–H model, when a CS is followed by a US, a CS–US association (a prediction of the US by the CS) is formed. Changes in CS–US associations are controlled by the absolute difference between the US intensity and the "aggregate prediction" of the US, computed on all CSs present at a given time. When the actual and predicted US are equal, no changes occur in CS–US associations. Schmajuk (1984b) hypothesized that the hippocampus computes the aggregate prediction used to control CS–US associations.

Schmajuk (1986, 1989; Schmajuk and Moore, 1985, 1988) introduced a real-time version of Pearce–Hall's (1980) attentional model, designated the S–P–H model. By applying Schmajuk's (1984) aggregate prediction hypothesis, the model is capable of generating multiple predictions. First, the aggregate prediction hypothesis assumes that hippocampal lesions imply impairments in the computation of the aggregate prediction. Under this view, the S–P–H model correctly describes the effect of hippocampal lesions on delay conditioning, conditioning with short, optimal, and long ISI with a shock as US, conditioning with long ISI and air puff as the US, extinction, latent inhibition, generalization, blocking, overshadowing, discrimination reversal, and sensory preconditioning. However, under the aggregate prediction hypothesis, the S–P–H model has difficulty describing the effect of hippocampal lesions on conditioned inhibition and mutual overshadowing. Second, the aggregate prediction hypothesis assumes that long-term potentiation induction increases the integration of multiple predictions into the aggregate prediction by way of increasing CS–CS associations. Unfortunately, under the aggregate prediction hypothesis, the S–P–H model has difficulty describing the effects of long-term potentiation on discrimination acquisition. Third, the aggregate prediction hypothesis assumes that neural activity in the hippocampus is proportional to the instantaneous value of the aggregate prediction. Under the aggregate prediction hypothesis, the S–P–H model correctly describes neural activity in hippocampus during the acquisition but not extinction of delay conditioning.

Schmajuk (1990) also tested the aggregate prediction hypothesis in spatial tasks. He presented a real-time neural network capable of describing temporal discrimination and spatial learning in a unified fashion. In this network, aggregate predictions, triggered by the CSs present at a given time, forecast *what* event is going to occur, *when* in time, and *where* in space. Schmajuk (1990) showed that the network correctly describes the activity of hippocampal pyramidal neurons and the effect of hippocampal lesions in spatial learning.

Schmajuk and DiCarlo (1991a, b) described hippocampal participation in classical conditioning in terms of Grossberg's (1975) attentional theory presented in Chapter 4. According to Grossberg's (1975) attentional theory, pairing of a CS with a US causes both an association of the sensory representation of the CS with the US (conditioned reinforcement learning) and an association of the drive representation of the US with the sensory representation of CS (incentive motivation learning). Sensory representations compete among themselves for a limited-capacity STM activation that is reflected in a LTM storage. Schmajuk and DiCarlo (1991a, b) proposed that, in the context of Grossberg's (1975) model, the hippocampus controls self-excitation and competition among sensory representations and stores incentive motivation associations, thereby regulating the contents of a limited-capacity short-term memory. Based on this hypothesis, termed the "STM regulation" hypothesis, the model predicts that HFL impair phenomena that depend on competition for STM, such as blocking and overshadowing, and phenomena that depend on the duration of STM, such as trace conditioning with long interstimulus intervals. In the case of a CS presented in isolation, the model specifies the conditions under which acquisition or extinction paradigms are impaired, unaffected, or even facilitated. Therefore, the Grossberg model replaces the notion that trace conditioning or extinction is or is not affected by HFL, by the concept that STM is more or less changed under different experimental parameters. The model also predicts that hippocampal LTP or kindling facilitate the acquisition of classical discrimination by increasing the stored values of incentive motivation associations. In addition, the model describes hippocampal neural population activity as proportional to the strength of CS–US associations. In sum, under the STM regulation hypothesis, Grossberg's model provides (1) a correct description of the effect of LTP induction on discrimination acquisition (but not on reversal), (2) novel descriptions of LTP induction and blockade effects in many classical conditioning paradigms, and (3) a description of the interaction between ISI and CS durations during the acquisition of classical conditioning.

Whereas the brain-mapped SD model describes how the hippocampus regulates cerebellar learning by interacting with the US representation at the dorsal accessory olive, the M–S–S and the Grossberg (1975) models describe how the hippocampus regulates cerebellar learning by interacting with the CS representation at the pontine nucleus. Schmajuk and DiCarlo (1992) suggested that, rather than being mutually exclusive, both approaches might be complementary. A neural network incorporating both features would be able to (1) describe a wider range of paradigms and (2) yield improved descriptions of hippocampal, cortical, and cerebellar manipulations. Whereas hippocampal control of the US representation in the dorsal accessory olive is needed to explain paradigms such as conditioned inhibition, negative patterning, and positive patterning, hippocampal regulation of the CS representation in the pontine nucleus is necessary to describe paradigms such as latent inhibition, sensory preconditioning, ISI effects, and the effects of LTP induction in discrimination acquisition.

The SD model (Schmajuk and DiCarlo, 1992) and the STM memory regulation hypothesis (Schmajuk and DiCarlo, 1991a, b) may also be combined to yield an improved functional description of the intrinsic hippocampal circuits. Schmajuk and DiCarlo (1992) suggested that the hippocampus inhibits the US input to cerebellar areas in proportion to the magnitude of the aggregate prediction of the US (as defined in the S–P–H and SD models) and excites the CS input to cerebellar areas in proportion to the magnitude of the association of the CS with the US [as defined in the M–S–S and Grossberg (1975) models]. Schmajuk and DiCarlo (1991a, b) mapped Grossberg's neural network onto the intrinsic circuit of the hippocampus according to the STM regulation hypothesis. This mapping is *constrained* by the three-dimensional organization of the hippocampus, the distribution and properties of hippocampal LTP, and the firing characteristics of hippocampal CA3, CA1, and granule neurons. According to this mapping, one class of CA3 pyramidal cell population receives CS representations from the entorhinal cortex and helps to sustain these sensory representation activities through a positive feedback loop. A second class of CA3 pyramidal cell population associates (in the form of LTP) CR-related inputs with CS representations from the entorhinal cortex and constitutes part of an incentive motivation loop. A third class of CA1 pyramidal cell population is assumed to add CS information from different rostrocaudal levels of the hippocampus and constitutes a competition loop. Hippocampal outputs regulate the CS representation at the pontine nuclei.

The intrinsic mapping suggested by the STM regulation hypothesis (Schmajuk and DiCarlo, 1991a, b) ignores important hippocampal components and connections, such as septal inputs and basket cells, and overlooks LTP in other than perforant path-CA3 synapses. The SD network contains elements that might complete these aspects of the intrinsic mapping. First, the SD model describes the interaction between the medial septal activation of basket cells and entorhinal activation of pyramidal cells. Second, it provides a functional interpretation for the information stored as LTP of perforant path-dentate gyrus synapses. LTP of perforant path-dentate gyrus synapses increases with the simultaneous activation of the perforant path and medial septal inputs to the dentate. Whereas the perforant path input conveys CS sensory representations, the medial septum input provides a signal proportional to the *novelty* of environmental events (theta rhythm). Therefore, LTP of the perforant path-dentate gyrus synapses might code the novelty value of a given CS. This association, stored in the form of LTP, enhances the STM of those CSs active at the time when novel events occur, thereby modulating their rate of conditioning.

Gluck and Myers (1993) presented a computational theory of the function of the hippocampal region and offered computer simulations for normal and hippocampal lesioned animals in a wide range of conditioning paradigms. According to Gluck and Myers (1993), the hippocampal region develops new stimulus representations that enhance the discriminability of differentially predictive cues while compressing the representation of redundant cues. Hippocampal representations are assumed to recode sensory representations in cortical and cerebellar regions. The authors assume three three-layer networks that work in parallel. One of the networks represents the hippocampal region, another one the cortex, and a third one the cerebellum. The output and hidden layers of the hippocampal network are trained to associate stimulus inputs with those same stimulus inputs and the unconditioned stimulus US. The output layers of both cortical and cerebellar networks are also trained by US. However, hidden units of both cortical and cerebellar networks are trained by those of the hippocampal network. According to Gluck and Myers, lesions of the hippocampal region eliminate training of the cortical and cerebellar hidden units and, therefore, cortical and cerebellar regions use previously established fixed representations.

Neurophysiological evidence suggests that the distinction between Schmajuk and DiCarlo's (1992) assumption (hippocampal lesions cause the error signals for cortical hidden units to be zero) and Gluck and Myers's (1993) assumption (hippocampal lesions cause the teaching signal for cortical hidden units to be zero) is an important one. McCormick, Steinmetz, and Thompson (1985) reported that lesions of the rostromedial portions of the inferior olive, which presumably serves as a pathway for information from the US to reach the cerebellum, cause a previously learned classically conditioned response to extinguish even with US presentations. That is, disruption of the US teaching signal for the cerebellum causes the extinction of previously acquired associations. Similarly, in Gluck and Myers's (1993) model, disruption of the hippocampal teaching signal for cortical hidden units should cause the extinction of previously acquired associations. Therefore, because hippocampal lesions do not produce devastating retrograde amnesia, hippocampal output probably does not act as a teacher for the cortex.

Computational approaches to spatial and cognitive mapping and hippocampal function

Several models have been proposed to describe cognitive mapping in terms of hippocampal function. For instance, Schmajuk and Segura (1980) described a model in which S–R–S associations are stored in the hippocampus. In the same vein, McNaughton (1989) described how a cognitive map might be implemented in the hippocampus. He proposed that conditional associations between places and movements are established during learning about an environment. These conditional associations are stored in a Hebb–Marr net (McNaughton and Nadel, 1990) that includes principal neurons whose outputs represent a desired place. Principal neurons receive input from (1) *Y* inputs connected through powerful "detonator synapses," (2) *X* inputs connected via Hebb synapses, and (3) an inhibitory interneuron that divides the activation of the principal neuron by a term proportional to the number of inputs *X* that are active. After training, a given pattern *X* activates the pattern *Y* with which it has been previously presented. McNaughton (1989) made use of a Hebb–Marr net in order to describe how the present position of the animal could be combined with the animal's movement to define the next position. The system is conferred additional flexibility by assuming that the links between successive local views are not fixed movements but the output of a system that calculates movement equivalency, which allows the rat to calculate novel routes to a goal (McNaughton, 1989). McNaughton, Leonard, and Chen (1989) proposed that parietal pathways may convey representations of conjunctions of motion and location to the hippocampus where they are associated with the next location. Interestingly, when the next location is reinjected into the current location input, chains of local views can be recalled using only information about the starting point and the sequence of movements. This mechanism would explain O'Keefe and Speakman's (1987) and Quirk, Muller, and Kusie's (1990) results showing that hippocampal place cells fire in their correct location in the darkness, provided that the animal is informed about the starting position.

O'Keefe (1989) criticized McNaughton's (1989) approach on the grounds that (1) it requires an excessive amount of memory to store local views and movements, (2) it does not explain how rats take shortcuts or detours, and (3) it does not explain how the rat reaches the goal from a new starting point. Muller et al. (1991) suggested three additional difficulties with McNaugton's model. First, they claim that place cells are not local-view cells, that is, their firing is independent of the direction in which the animal is entering the field. Second, Muller et al. claim that, because the number of active cells in entorhinal cortex is independent of the number of external stimuli, theta cells do not receive the right kind of signal to perform the division operation on the principal cells. Third, Muller et al. contend that place-cell firing involves more nonlinearities than simple pattern completion. In addition, although McNaughton's scheme predicts the next location based on the present place and movement, the system does not define the movement to make when the animal intends to go from one location to another.

Interestingly, local-view cells are not unique to the hippocampal formation. Spatially and directionally dependent cells are found also in the entorhinal cortex (Muller et al., 1991) and posterior parietal cortex (McNaughton et al., 1989). However, McNaughton et al. (1989) indicate that, given the large number of necessary associations between local views and movements, sparse coding (such as that present in the hippocampal formation) is advantageous for economical storage of information.

O'Keefe (1989) offered a computational model for a Cartesian navigational system that formally implements the cognitive map view of hippocampal function. This Cartesian mapping system consists of (1) a space that stores the coordinates of the stimuli experienced in the environment, (2) a mechanism that transforms the information about the sensory array, (3) a mechanism that manipulates maps in order to bring them in correspondence with the sensory array, and (4) a mechanism that compares the representation of current and desired locations and calculates the translation required to move between them. Based on the distance and angle to the cues in the environment, the mapping system computes (1) the geometrical centroid of the set of cues and (2) the slope of the distribution of the cues. The centroid is used as the origin and the slope as the reference for angles in a polar space framework. Place cells code some aspect of the centroid vector information. O'Keefe suggested that spatial vectors representing the location of cues can be coded as theta-frequency sinusoids with amplitudes coding the distance from the origin to the cue and phases coding the angle from the slope to the vector pointing to the cue.

O'Keefe's (1989) theory suggests that each granule and pyramidal cell acts as an oscillator that sums its sinusoidal inputs. Upper layers of the entorhinal cortex contain place-coded neurons and provide the information to the hippocampal mapping system. The direct projection from the perforant path to CA3 provides the basis for the calculation of the centroid. CA1 pyramids receive inputs from CA3 and the entorhinal cortex and use them to compute the movements necessary to navigate from one location to another.

According to O'Keefe, the model generates several predictions. For instance, the model predicts that the uniform compression or enlargement of the cue configuration will not change the centroid of a previously established representation and, therefore, will not elicit exploration. The model calculates an expected new location as the addition of the vector representing the current location and a movement vector. O'Keefe proposes that the movement vector is provided by the amplitude of the theta rhythm, but as he acknowledges, the frequency, but not the amplitude, of theta is proportional to the distance or speed of the movement. In contrast to McNaughton's (1989) model, which can only calculate expected new locations, O'Keefe's model can also calculate the movement required to shift from a current location to a desired location, thereby controlling the animal's motor behavior.

In contrast to O'Keefe's (1989) and McNaughton's (1989) theories, the present chapter suggests that the hippocampus is not specifically involved in spatial cognitive mapping but, instead, in the computation of variables that are fundamental to the performance of a variety of behaviors, including spatial tasks. According to the SD model, HFL and HPL preclude the correct configuration of places (see Figure 13.17) and, therefore, animals are prevented from building spatial and cognitive maps (see Figure 11.2; also Schmajuk, Thieme, and Blair, 1993).

Appendix 13.A

Medial septal modulation of hippocampal activity

Entorhinal cortex input excites pyramidal and granule cells according to

$$d(\text{pyr}_i) = -K_{10}\text{pyr}_i + K_{11}\text{ent}_i - K_{12}\text{bas}_i\text{pyr}_i \qquad (13.1)$$

where pyr_i is the activity of pyramidal (or granule) cell i, ent_i is the excitatory input from the entorhinal cortex, $\text{ent}_i = a_i V_i$, and bas_i represents the inhibitory activity of basket cell i.

The activity of basket cell i is given by

$$d(\text{bas}_i) = -K_{13}\text{bas}_i + K_{14}\text{pyr}_i - K_{15}\theta\text{bas}_i \qquad (13.2)$$

where θ represents the activity of the medial septum ($\theta = |\text{US} - B|$), which inhibits basket cell i.

If we combine Equations 13.1 and 13.2, the asymptotic output of the pyramidal (or granule) cell i is given by

$$\text{pyr}_i \propto \sqrt{\theta\ \text{ent}_i} \propto \sqrt{|\text{US} - B|\,\text{an}_i \text{VN}_i} \tag{13.3}$$

Equation 13.3 is positive when US > B and negative when US < B. Because the output of the same neuron cannot turn from excitatory to inhibitory, we assume that one neural population codes $\sqrt{(\text{US} - B)\ \text{an}_i\text{VN}_i}$ when $\theta > 0$ and a different one $\sqrt{(B - \text{US})\ \text{an}_i\text{VN}_i}$ when $\theta < 0$. The activity of a pyramidal cell yielded by Equation 13.3 is well approximated by the hidden-unit error function EH_j generated by Equation 6.8 in Appendix 6.A. In both cases, the output of the pyramidal cells, assumed to control cortical learning, is proportional to the product of the theta rhythm and the association of the hidden unit with the US.

Appendix 13.B

A formal description of the effects of hippocampal, cortical, and cerebellar lesions

A formal description of the SD model for normal animals is presented in Appendix 6.A in Chapter 6.

HFL effects

Input–output associations

After HFL, changes in input–output associations VS_i are given by

$$d(\text{VS}_i)/dt = K_6 \text{as}_i (1 - |\text{VS}_i|)\ \text{EO}_i \tag{13.4}$$

The output error EO_i in Equation 13.4 is given by

$$\text{EO}_i = \text{US} - a_i \text{VS}_i \tag{13.5}$$

where the aggregate prediction B_i is no longer present.

Hidden unit-output associations

In agreement with the aggregate prediction hypothesis, after HFL, changes in hidden-unit–output associations VN_j are given by

$$d(\text{VN}_j)/dt = K_6 \text{an}_j (1 - |\text{VN}_j|)\ \text{EO}_j \tag{13.6}$$

The output error EO_j in Equation 13.6 is given by

$$\text{EO}_j = \text{US} - a_j \text{VN}_j \tag{13.7}$$

HFL and HPL effects

Input–hidden-unit associations

After HFL, changes in input–hidden-unit associations VH_{ij} are given by

$$d(\text{VH}_{ij})/dt = K_7 X_i (1 - |\text{VH}_{ij}|)\ \text{EO}_j \tag{13.8}$$

The hidden-unit error EH_j is given by

$$\mathrm{EH}_j = 0 \qquad (13.9)$$

and, therefore, $d(\mathrm{VH}_{ij})/dt = 0$.

CL effects

After CL, VH_{ij} associations are zero.

$$\mathrm{VH}_{i,j} = 0 \qquad (13.10)$$

Cerebellar lesion effects

After cerebellar lesions, VS_i and VN_j associations are zero.

$$\mathrm{VS}_i = 0, \mathrm{VN}_j = 0 \qquad (13.11)$$

Appendix 13.C

Simulation parameters

For normal, HPL, HFL, and CL cases, simulations of classical conditioning and place learning were conducted with identical parameter values. In order to describe place learning, parameters used in Chapter 6 were slightly modified. Parameter values were $K_1 = 0.065$, $\alpha = 4.5$, $K_2 = 0.08$, $K_3 = 2$, $K_4 = 2.5$, $K_5 = 1.5$, $K_6 = 0.005$, $K_7 = 1.12$, $K_8 = 5$, $K_9 = 0$, $K_{10} = 0.5$, $\beta_1 = 0.5$, and $n = 1.5$. Also, $\mathrm{CS}_i = 2$ for all discriminative stimuli, $O_i = 3$ arbitrary units of length for all spatial landmarks L_i, and d_i varying between 1 and 10 arbitrary units of length for all positions in the Morris tank. The initial values of VS_i and VN_j were 0. Input–hidden-unit association weights VH_{ij} were randomly assigned using a uniform distribution ranging between $\pm$ 0.25 in all simulations. Because 20 hidden units are needed in order to attain reliable place learning, this number was used in all simulations.

Conclusion

14 The character of the psychological law

"When you are dealing with psychological matters things can't be defined so precisely". Yes, but then you cannot claim to know anything about it.

—Richard Feynman in *The Character of the Physical Law,* 1967.

In his series of lectures on the character of the physical law, Feynman (1967) explained that laws in physics are established by first generating hypotheses, then computing the logical consequences of the hypotheses, and finally comparing the result of the computations with experimental data. In the behavioral sciences, often the computation of the logical consequences of the basic hypotheses seems to be inadequate. Although numerous intriguing ideas have been proposed regarding mechanisms of learning and cognition, it is sometimes difficult or impossible to determine their logical consequences in some experimental situations. Hinzman (1991) suggested that the reason for the recent popularity of explicit theories is linked to the acceptance of the unreliability of intuitive reasoning for relating hypotheses and experimental results. Neural networks not only allow one to establish the logical consequences of the basic principles being tested, but they do it in a way compatible with brain-style computations. In other words, because neural networks are tools (1) sufficiently precise for deriving the logical conclusions of elemental assumptions and (2) appropriately constrained by physiological principles, they provide the attractive possibility of establishing connections between brain and behavior.

Some of the features of neural network theorizing are:

Integration of large amounts of data. In the absence of such models, there is only the "bloomin' buzzin' confusion" of the experimental data. In general, neural networks integrate, condense, compress, and provide a precise account of complex and intricate behavioral data.

Real-time descriptions of behavior and neural activity. Real-time neural networks (see Box 1.4), such as those presented in this book, provide accurate descriptions of behavior and simultaneous neural activity as moment-to-moment correlated phenomena. In this case, computer simulations can be compared with real-time experimental data to test the validity of the theory.

Real-time paradigms. Furthermore, real-time nets permit the precise description of the temporal arrangements used in different experimental designs, such as those regarding the effect of CS or US duration, ISIs and ITIs. These descriptions provide exact experimental parameter values instead of vague verbal labels (e.g., trace conditioning).

Complex paradigms. In addition, neural neworks easily apprehend the intricacies of complex behavioral paradigms difficult to grasp without the help of mathematical formalisms and computer simulations. Complex behavioral interactions can be described in terms of nonlinear dynamic equations that are easily simulated on computers. These simulations reliably display the authentic power

BOX 14.1 Some general properties of real-time neural network theories

1. Integrate and condense large amounts of data.
2. Provide accurate descriptions of behavior and associated neural activity in real time.
3. Computer simulations can be compared with experimental data to test the theory.
4. Continuous variables permit the precise description of experimental designs.
5. Provide accurate descriptions of complex behavioral paradigms.
6. Expose the simple principles that sometimes underlie complex networks.
7. Show emergent properties.
8. Provide natural links between different levels of description: computational, representational, algorithmic, and biological.

of the model, beyond the limitations of human intuition when large numbers of variables and their interactions are involved. At the same time, the computer simulations are unforgiving in exposing the weaknesses of the model beyond human wishful conclusions.

Simple principles, complex networks. Although the hope of discovering simple principles favors simple formal systems (Hinzman, 1991), that might not always be possible. As shown in the book, even relatively simple principles might necessitate a rather complex machinery to be implemented. Whereas simple psychological principles might be derived from the adaptive value of behavior, their neural network implementation might result in relatively complex, but physiologically plausible, mechanisms.

Emergent properties. Although some important properties of the network are the consequence of some basic principles, equally important *emergent properties* result from the physiological principles that constrain the design of neural networks. For example, in neural networks both reading in and reading out of CS–CS and CS–US associations are sometimes controlled by the magnitude of the internal representations of the CSs. Consequently, the model presented in Chapter 5 is able to describe both storage and retrieval processes in terms of a single biologically inspired mechanism.

Link between different levels of description. Neural nets seem to provide the appropiate language to link Marr's (1982) different levels of description: computational (the goal of the computation), representation (what are the representations of inputs and outputs), algorithmic (what is the algorithm transforming the input into the output), and biological (the brain implementation of the algorithm).

Box 14.1 summarizes the above-mentioned aspects of neural network theorizing. In many respects, neural network models seem to lead theoretical psychology into a realm so far reserved for the more mature natural sciences.

Neural network theories of animal learning

This book describes several neural network theories of animal learning and cognition. Starting at the simple assumption that psychological associations are represented by the strength of neural synaptic connections, mechanistic descriptions of complex cognitive behaviors are provided. Some of the psychological

Table 14.1. *Some general principles in animal learning theories.*

Concept	Application	Chapter
I. CLASSICAL CONDITIONING		
1. Real-time predictions	Delay and trace conditioning	3
2. Stimulus selection		
Best, earliest predictors	Blocking, overshadowing	3, 5, 6
3. Inference generation		
Fast-time predictions	Cognitive mapping, planning	11
	Sensory preconditioning	3
4. Selective attention		
Predictors of the US	Latent inhibition, blocking	
	Overshadowing	4
Predictors of novelty	Latent inhibition	5
5. Storage and retrieval processes	Attenuation of latent inhibition	5
6. Stimulus configuration		
Complex internal representations of simple external events	Negative and positive patterning	6
	Occasion setting	6
7. Timing	CR timing	7
II. OPERANT CONDITIONING		
8. Response-selection principle	Escape, avoidance	8
	Operant conditioning	
	Animal communication	
III. COGNITION		
9. Stimulus-approach principle		
Goal-seeking mechanism	Spatial navigation	10
	Cognitive mapping	11
	Some cases of avoidance	8
10. Fast-time predictions	Cognitive mapping	11

functions addressed at different points in the book and the mechanisms that implement them (see Boxes 1.1–1.4) are presented in this section and summarized in Table 14.1.

Part I describes different neural network theories of classical conditioning:

1. Real-time predictions. Chapter 2 introduces the idea that $B_k = \Sigma_i V_{i,k}\tau_i$ represents the aggregate prediction that different CS_i's will be followed by CS_k. Following Hebb (1949), changes in synaptic strength $V_{i,k}$ might be described by $\Delta V_{i,k} = f(CS_i)\, f(CS_k)$, where $f(CS_i)$ represents a function of the presynaptic activity and $f(CS_k)$ a function of the postsynaptic activity. Different $f(CS_i)$ and $f(CS_k)$ functions have been proposed. Learning rules for $V_{i,k}$ either assume variations in the effectiveness of CS_i, $f(CS_i)$, $f(CS_k)$, or both.

2. Stimulus selection. Chapter 3 introduces the delta rule (see Box 1.1). According to the simple delta rule, CS_i–US associations are changed until the difference between the US intensity and the "aggregate prediction" of the US computed on all CSs present at a given moment $(US - \Sigma_j V_{j,US} CS_j)$ is zero. The term $(US - \Sigma_j V_{j,US} CS_j)$ can be interpreted as the effectiveness of the US to become associated with the CS. Changes in CS–US associations are given by $\Delta V_{i,US} \sim CS_i\,(US - \Sigma_j V_{j,US} CS_j)$. Use of the delta rule results in the associative selection of the earliest and most reliable predictors of the US.

3. Inference generation. Chapter 3 also introduces a recurrent autoassociative network (see Box 1.3) capable of integrating multiple predictions into larger units through a process called inference. One simple example of inference formation is sensory preconditioning. Dickinson (1980) suggested that knowledge can be represented in declarative or procedural form. Whereas in the declarative form, knowledge is represented as a description of the relationships between events (knowing that), in the procedural form, knowledge is represented as the prescription of what should be done in a given situation (knowing how). Examples of declarative knowledge are classical CS–CS associations (CS_1 precedes CS_2) or CS–US associations (CS_2 precedes the US). Examples of procedural knowledge are operant S–R associations (if S is present, then do R). Dickinson indicates that declarative, but not procedural, knowledge can be integrated through inference rules. Changes in CS–CS and CS–US associations are given by a delta rule, $\Delta V_{i,k} \sim CS_i(CS_k - \Sigma_j V_{j,k} CS_j)$.

4. Selective attention. Chapter 4 introduces an attentional theory that assumes the effectiveness of CS_i to form CS_i–US associations (associability) depends on the magnitude of the "internal representation" of CS_i. In neural network terms, selective attention may be interpreted as the modulation of the CS representation that activates the presynaptic neuronal population involved in associative learning. Changes in CS–US associations are defined by the rule $\Delta V_{i,\mathrm{US}} = f(CS_i, V_{i,\mathrm{US}})\, f(\mathrm{US})$. In the attentional theory introduced in Chapter 4, $f(CS_i, V_{i,\mathrm{US}})$ is modulated by both (1) CS_i–US associations ($V_{i,\mathrm{US}}$) and (2) the competition with other CS representations for a limited-capacity short-term memory. As in the case of the delta rule, the use of this rule also results in the attentional selection of the earliest and most reliable predictors of the US.

Selective attention is also implemented in Chapter 5 by a network that extends the cognitive model described in Chapter 3 in order to account for latent inhibition. Chapter 5 combines variations in the effectiveness of both CS_i and CS_k (or the US). Whereas $f(CS_i)$ is modulated by the total environmental novelty, $f(\mathrm{US})$ is modulated by a delta rule. Novelty is defined as the sum of the mismatches of all predicted and observed events at a given time. Changes in CS–CS and CS–US associations are given by $\Delta V_{i,k} = f(CS_i, \text{novelty})\, [CS_k - \Sigma_j V_{j,k} f(CS_j, \text{novelty})]$.

5. Storage and retrieval processes. Chapter 5 introduces a model that combines both storage and retrieval processes, two different aspects of a single neural mechanism. Whereas storage of CS–CS and CS–US associations can be manipulated by controlling novelty at the time of training, retrieval of CS–CS and CS–US can be manipulated by controlling novelty at the time of testing. In the case of CS–US associations, $CR = \Sigma_j V_{j,k} f(CS_j, \text{novelty})$.

6. Stimulus configuration. Chapter 6 presents a network that employs a "generalized" delta rule (see Box 1.2) to train a layer of hidden units that combine simple CSs into configural stimuli, CN. According to the generalized delta rule, CS_i–US and CN_h–US associations change according to $\Delta V_{i,\mathrm{US}} = f(CS_i)\,(\mathrm{US} - \Sigma_j V_{j,\mathrm{US}} CS_j - \Sigma_h V_{h,\mathrm{US}} CN_h)$ or $\Delta V_{h,\mathrm{US}} = f(CN_h)\,(\mathrm{US} - \Sigma_j V_{j,\mathrm{US}} CS_j - \Sigma_h V_{h,\mathrm{US}} CN_h)$.

7. Timing. Chapter 7 introduces a network that describes timing by assuming that the presentation of a CS generates multiple traces, each one peaking at a different time after CS onset. Traces more closely overlapping with the US acquire stronger trace–US associations and generate a CR peak at the time of US presentation.

Part II describes neural networks of operant conditioning (primarily avoidance), introduces the concept of response-selection mechanisms, and applies the network to the description of animal communication.

8. Response selection. Chapter 8 introduces a network that describes operant learning by combining classical and operant systems. Changes in S–R associations are controlled by mismatches in the classical conditioning system. Dickinson (1980) proposed that operant S–R associations are an example of procedural knowledge (if S is present, then do R). Classical CS–US and R–US associations are defined by a delta rule, $\Delta V_{i,j} \sim X_i\,(\mathrm{US} - \Sigma_j V_{j,\mathrm{US}} X_j)$, where X_i represents CS_i or R_i. In the case of appetitive learning, changes in CS–R associations are given by $\Delta V_{i,j} \sim \mathrm{CS}_i \mathrm{R}_j(\mathrm{US} - \Sigma_j\, V_{j,\mathrm{US}} X_j)$. In the case of aversive learning (avoidance), changes in CS–R associations are given by $\Delta V_{i,j} \sim \mathrm{CS}_i \mathrm{R}_j(\Sigma_j V_{j,\mathrm{US}} X_j - \mathrm{US})$.

Chapter 8 also shows how two neural networks can be combined to describe animal communication. Two simulated animals communicate with each other through calls, that is, responses that can be used as warning stimuli by other simulated animals but do not have any effect on the environment.

Part III describes neural networks of animal cognition and introduces the notions of goal-seeking mechanisms and cognitive systems. Cognitive networks display remarkable adaptability and adopt alternative behavioral strategies that are independent of any specific set of responses. In general, behavior described by cognitive nets (Part III) is more flexible than that described by operant nets (Part II), and behavior described by operant nets more flexible than that described by classical networks (Part I).

9. Stimulus-approach principle. Chapters 10 and 11 present neural networks that incorporate goal-seeking mechanisms to approach appetitive stimuli during spatial navigation. Goal-seeking mechanisms approach appetitive stimuli defined by the animal's motivational state. A given motivation elicits a search behavior that receives negative feedback from the appropriate goal. When the goal is encountered, it activates approach responses and inhibits search responses. As the animal explores the environment under the control of the goal-seeking system, it builds a spatial map (Chapter 10) or a cognitive map (Chapter 11). Once the spatial (cognitive) map is available, the search behavior can also be inhibited by stimuli that predict the goal when applied to the spatial (cognitive) map.

10. Fast-time predictions. Chapter 11 presents a neural network that allows the exploration of an internal model of a maze. The network is capable of internally navigating the maze without modifying its cognitive map. At a given choice point in the maze, animals briefly examine all alternative next places connected to their present place, thereby generating fast-time predictions of the goal. These sequentially generated fast-time predictions are stored in working memories that permit the simultaneous comparison of all the alternative next places. As a result of this comparison, a decision is made and the animal enters the place that best leads to the goal.

Neural networks and neuroscience

Part IV shows how neural network models permit one to simultaneously develop psychological theories and models of the brain. When neural networks are regarded as models of specific brain circuits, they can provide theories that extend to the anatomical and physiological aspects of learning. These theories closely integrate brain mechanisms and psychological function and suggest some basic principles that govern brain–behavior relationships.

Brain–behavior interactions. Most theories of brain function are characterized by (1) assigning a unitary psychological function to a given brain region and (2) proposing that the brain region is involved in a given psychological function. Rejecting the idea that a psychological function can be mapped one to one onto brain regions, Gray et al. (1978, p. 277) suggested that (1) a brain region might participate in different psychological functions and (2) a psychological function might be computed by several brain regions. Supporting this view, Schmajuk (1984a) analyzed multiple theories of hippocampal functions, most of which attributed a fixed psychological function to the hippocampal formation, and found that the approach failed to describe adequately the available experimental data.

If a brain region does not perform any specific psychological function, then its function might be best defined as a transfer function, that is, the relationship between its inputs and outputs (Ranck, 1975). In Marr's terms, the function of a brain structure should be defined at the algorithm level. In agreement with this view, Schmajuk and DiCarlo (1992) suggested that the brain-mapped SD model is distinguished by (1) assigning several computational functions to the hippocampal formation and (2) proposing that the hippocampal formation might process incoming information in different ways. According to Figure 13.2 in Chapter 13, the hippocampal formation receives information about (1) the associations of simple and configural stimuli with the US and (2) the mismatch between the actual and predicted US intensity. Based on these inputs, the hippocampal formation computes and broadcasts (1) a signal proportional to the sum of the associations of simple and configural stimuli with the US (the aggregate prediction of the US) to cerebellar areas in order to control the associations formed with the US, and (2) error signals to association cortex in order to modulate stimulus configuration. At a more molecular level, computation of aggregate prediction signals might be carried out in the entorhinal cortex, whereas computation of error signals might be accomplished in the CA3/CA1 regions. In sum, the brain-mapped SD network strictly defines hippocampal function as the *multiple* transformations (transfer functions) that occur between the hippocampal inputs and outputs, rather than as a unitary psychological function.

If brain areas do not perform specific psychological functions but rather compute certain algorithms, then the relevant aspects of an experimental paradigm are its parameters (e.g., CS duration, CS intensity, and ISI duration) rather than its label (e.g., extinction, trace conditioning). For instance, Schmajuk and DiCarlo (1991b) indicated that, according to the model described in Chapter 4, hippocampal lesions might impair acquisition of classical conditioning according to the combination of variables used in the experimental design rather than the type of paradigm tested.

Brain–behavior dissociations. To further complicate matters, even when a given paradigm with identical experimental parameters is used in an experimental design, different animals (simulated by assuming different initial random weights) might solve the same task using different regions of the neural network. For instance, Schmajuk and DiCarlo (1992) reported an interesting result obtained with the SD network that addresses the relationship between external behavior and internal connections in the brain. In contrast with conventional notions that assume similar behaviors imply identical involvement of different brain regions, simulations with the SD model suggest that even when virtually identical behavior may be displayed by different animals, in each case several brain areas may contribute to the realization of the task in different degrees. Furthermore, the model suggests that although a given behavior can be

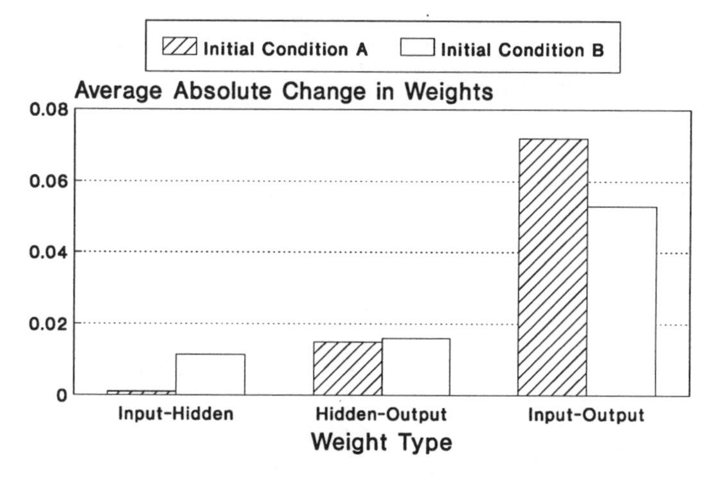

Figure 14.1. Learning distribution during blocking under two different initial conditions. Average of the absolute change in input–hidden units (VH), hidden unit–output (VN), and input–output (VS) weights after training in a blocking paradigm. In both cases, virtually identical behavioral responses were generated. Initial conditions are defined by assigning initial random values to VHs.

observed in the absence of a given brain region in some individuals, that brain region might still participate during the display of behavior in that particular paradigm in other subjects.

Figure 14.1 shows that, even when virtually identical blocking is generated in two simulations using the SD model (Chapter 6) with different initial conditions, various areas of the brain may be differentially engaged in learning. Initial conditions are defined by assigning initial random values to simple stimulus–hidden-unit associations $VH_{i,j}$'s. Figure 14.1 shows the average of the absolute change in input–hidden-unit (cortical) weights $VH_{i,j}$ hidden-unit–output (cerebellar) weights VN_j, and input–output (cerebellar) weights VS_i, after training in a blocking paradigm under two different initial conditions. Under initial conditions *A*, animals establish few configural stimuli in association cortex and many CS_i–US associations in the cerebellum. Under initial conditions *B*, animals show more cortical configural learning and fewer CS_i–US cerebellar associations than under condition *A*. The number of CN_j–US cerebellar associations is equivalent for both conditions. It is important to notice that, according to the model, although blocking can be obtained without cortical involvement, cortical learning may be present during this paradigm.

Also, Figure 14.2 shows that, even when virtually identical negative patterning is obtained in two simulations using the SD model (Chapter 6) with different initial conditions, the cortex and cerebellum may be differently engaged in learning. As in blocking, initial conditions are defined by assigning initial random values to simple stimulus–hidden-unit associations $VH_{i,j}$'s. Figure 14.2 shows the average of the absolute change in input–hidden-unit (cortical) weights $VH_{i,j}$, hidden-unit–output (cerebellar) weights VN_j, and input–output (cerebellar) weights VS_i, after normal animals had been trained in a negative patterning paradigm under two different initial conditions. In both cases, simulations proceeded until the CR to CS_1 and CS_2 was 80 percent, and the CR to CS_1–CS_2 less than 30 percent, of the asymptotic CR in simple acquisition. Under initial conditions *A* and *B*, animals establish a similar number of configural

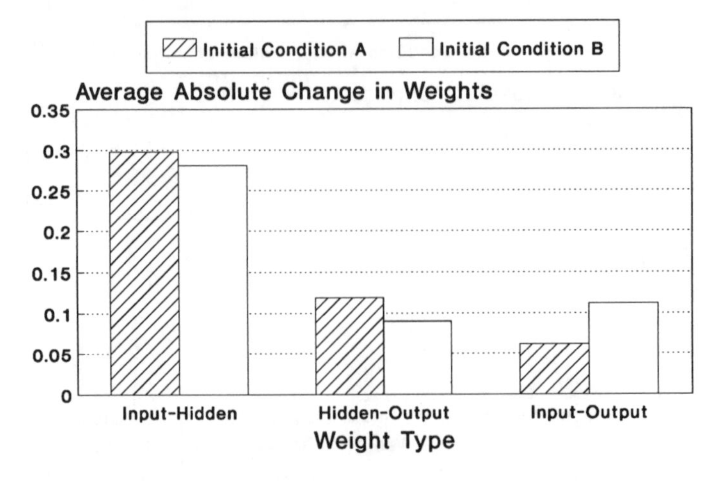

Figure 14.2. Learning distribution during negative patterning under two different initial conditions. Average of the absolute change in input–hidden units (VH), hidden unit–output (VN), and input–output (VS) weights after training in a negative patterning paradigm. In both cases, virtually identical behavioral responses were generated. Initial conditions are defined by assigning initial random values to VHs.

stimuli in association cortex. However, although under condition *A* animals establish more CN_j–US cerebellar associations than under condition *B*, this relation is reversed for the case of CS_i–US cerebellar associations. It is interesting to note that negative patterning always requires cortical learning (Figure 14.2), whereas blocking can be solved with or without cortical associations (Figure 14.1).

Schmajuk and Thieme (1992) reported a result that also addresses the relationship between external behavior and internal connections in the brain. Using the network presented in Chapter 11, they found that even when differently trained groups showed identical performance in a latent learning paradigm, each group had stored different types of information about the environment. Three different groups are used in the latent learning experiment. Group *A* is rewarded at the goal box on the first trial, group *B* on the seventh trial, and group *C* on the thirteenth trial. The model shows latent learning because, when the reward is presented after a period of latency, animals with a preconstructed cognitive map (groups *B* and *C*) display rapid improvement in performance at the same level as group *A*. Figure 14.3 shows average place–view and place–goal associations accrued by groups *A, B,* and *C*. Because place–view associations increase with the number of exploratory movements and this number decreases as the animal is able to encounter the goal, group *C* shows more place–view associations than group *B*, and group *B* more place–view associations than group *A*. In trial 20, the three groups show identical performance, solving the maze in the minimum number of movements, but they differ in the knowledge about the maze that each one has gathered.

Still another example of dissociation between behavior and internal connections is that shown in Chapter 8. During acquisition of avoidance, fear decreases as the animal masters the correct response. Although initially subjects show intense fear to the US, this fear decreases over trials (Kamin, Brimer, and Black, 1963; Solomon and Wynne, 1953; Starr and Mineka, 1977). Subjects

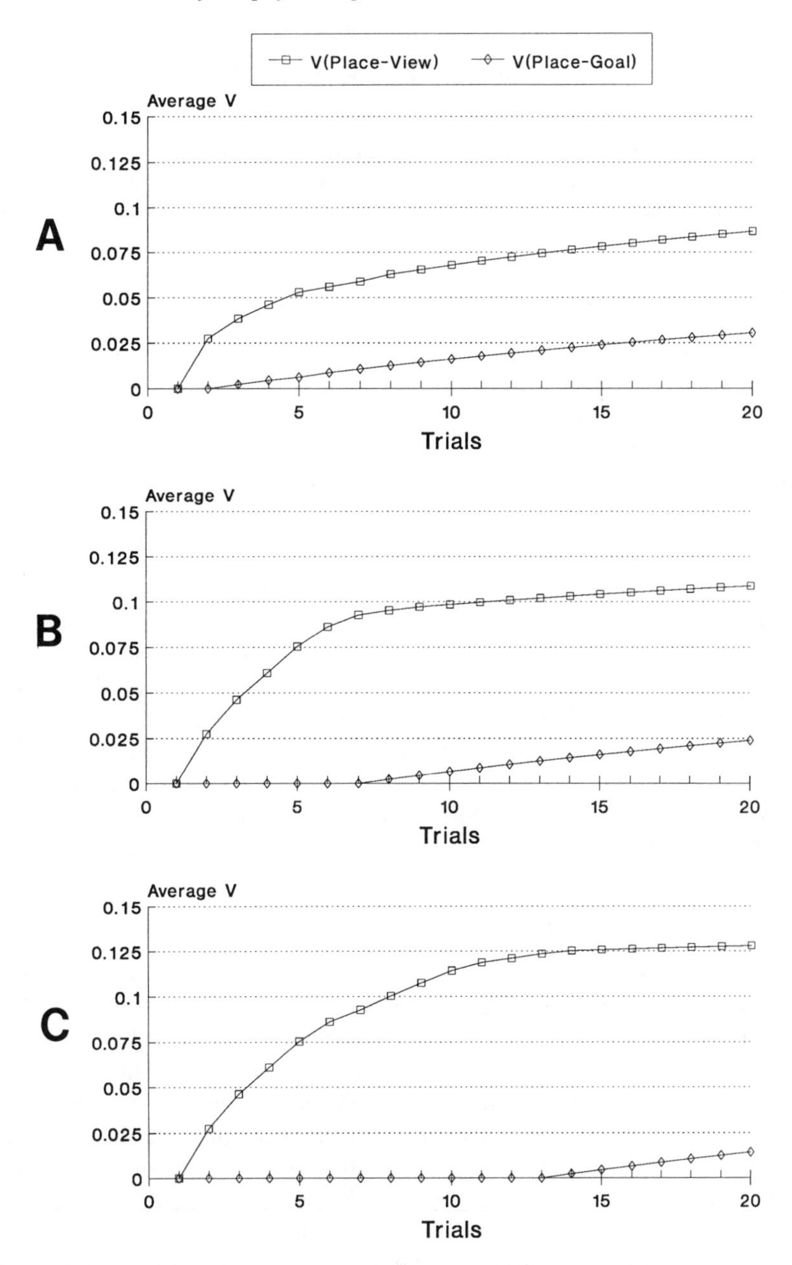

Figure 14.3. Learning distribution during latent learning with different amounts of preexposure to the maze. Average change in place–view and place–goal associations as a number of trials for three different groups. Group *A* is rewarded at the goal box on the first trial, group *B* on the seventh trial, and group *C* on the thirteenth trial.

have been shown to substitute fear responses with stereotyped behavior. This behavior is captured by Equation 8.6 in Chapter 8, which shows that the same operant response can be generated with different combinations of operant and classical associations.

Possibly, the most dramatic example of a dissociation between behavior and internal connections is that given in Chapter 5, which shows that identical be-

BOX 14.2 Some general principles of brain–behavior interactions

1. A brain region is not related one to one with a psychological function:
   (a) A brain region might participate in different psychological functions.
   (b) A psychological function might be computed by several brain regions.
2. The function of a brain region is defined as a transformation between inputs and outputs.
3. A brain structure might have more than one transfer function:
   (a) Different simultaneous outputs.
   (b) Different outputs controlled by other structures.
4. For the brain, there are no behavioral tasks, only a combination of experimental parameters.
5. Identical behaviors might result from differential involvement of different brain regions.

havior can be observed by combining CS–US and CS–Novelty associations of different strengths. For example, identical CRs can be generated by (1) a strong CS–US association and weak CS–Novelty association that follow simple conditioning, or (2) a weak CS–US association and strong CS–Novelty association that follow postconditioning manipulations in a LI paradigm.

Box 14.2 summarizes some general principles of brain–behavior interactions. The practical relevance of these brain-behavioral issues is underscored by the fact that some learning or cognitive abilities might be lost in the presence of some psychopathological disorders.

Evaluation of the models

Box 14.3 summarizes the different levels at which neural networks can be examined. (1) At the behavioral level, simulated behavioral results can be compared with experimental data describing animal behavior. For each of the models introduced in this book, these comparisons are summarized by the tables presented in each chapter. (2) At the anatomical level, interconnections among neural elements in the model are compared with neuroanatomical data. (3) At the neuroanatomical level, model performance is correlated with animal performance after lesioning or at different stages of development. (4) At the computational level, simulated activity of the neural elements of the model is compared with the activity of single neuron or neural population activity. (5) At the neurophysiological level, model performance is correlated with animal performance modifying neural connectivity by inducing LTP (long-term potentiation), LTD (long-term depression), or preventing changes in neural connectivity by blocking LTP. Simultaneous evaluation at behavioral, anatomical, computational, and neurophysiological levels requires the selection of a specific animal preparation, such as the rabbit's nictitating membrane response described in Chapters 12 and 13.

Bunge (1967) analyzed the explanatory power of theories. In general, more advanced theories account for nearly all the facts, but no theory explains them all. Because a number of theories compete to account for a single domain of facts, Bunge suggested that a measure of the actual explanatory efficiency of a

BOX 14.3 Some general principles to evaluate neural networks

1. Comparison of animal and network behavior.
2. Comparison of brain anatomical circuits and network architectural connectivity.
3. Comparison of effects of brain and network lesions (or development).
4. Comparison of neural population activity and network element activity.
5. Comparison of effects of LTP, LTP blockade, and LTD on the brain and modification of connections in the network.

theory is needed. A first measure may be parsimony or economy of thought (explain the most with the least). Bunge suggested that the *explanatory efficiency* of a given theory T can be expressed as $\rho(T) = 1 - n/N$, where n is the number of independent initial assumptions and N the number of independent facts explained by the theory. Explanatory efficiency tends to 1 if the number of independent initial assumptions n is small as compared to the number of independent facts N explained by the theory.

According to Bunge (1967, p. 52), the explanatory power of a theory T is proportional to its range (N, the number of independent facts explained by the theory), accuracy (the truth of the theory's tested consequences), and depth (D, the number of levels to which the theory may be applied). The accuracy of a theory T can be expressed as $A(T) = C/N$, where C represents the number of correct predictions and N the number of testable predictions. According to Bunge, a purely behavioral theory would be assigned depth 1, a theory containing both behavioral and psychological variables depth 2, and a theory incorporating behavioral, psychological, and physiological concepts depth 3. The explanatory power of a theory T is given by $E(T) = A(T)\ D(T)$.

Table 14.2 shows the explanatory efficiency $\rho(T)$ and accuracy $A(T)$ of different classical conditioning models described in Part I. We assume that (1) the range N of all models is the number of classical conditioning paradigms to which all models might be applied (77), (2) paradigms that cannot be described by a model (e.g., latent inhibition) count as incorrect predictions, and (3) all models have depth 3 (they incorporate behavioral, psychological, and physiological concepts).

According to Table 14.2, the real-time version of the delta rule presented in Chapter 3 has the highest parsimony, $\rho(T)$= 0.95, but the lowest accuracy, $A(T) = 0.21$. The cognitive map model presented in Chapter 3 has also a relative high parsimony, $\rho(T)$= 0.92, but still low accuracy, $A(T) = 0.23$. The storage–retrieval (SLG) model presented in Chapter 5 has a lower parsimony, $\rho(T)$= 0.78, with higher accuracy, $A(T) = 0.66$. Finally, the configural (SD) model presented in Chapter 6 has a relative high parsimony, $\rho(T)$= 0.86, and acceptable accuracy, $A(T) = 0.53$. The best combinations of explanatory efficiency and accuracy are those of the storage–retrieval model and the configural model. Table 14.2 suggests that the introduction of additional parameters, from the models presented in Chapter 3 to the model described in Chapter 5, does not seriously reduce parsimony [$\rho(T)$ decreases from 0.95–0.78], but dramatically increases accuracy [$A(T)$ increases from 0.21–0.66].

Table 14.2. *Explanatory efficiency and accuracy of different models.*

Chapter	Model	Correct paradigms C	Variables n	Explanatory Efficiency $\rho(T) = 1 - n/N$	Accuracy $A(T)$
3	Real-time delta rule	16[a]	4	0.95	0.21
3	Cognitive map	18[a]	6	0.92	0.23
5	Storage–Retrieval	20[a] + 31[b]	17	0.78	0.66
6	Configural	25[a] + 16[c]	11	0.86	0.53

Note. C: number of correct predictions; n: number of variables; ρ: explanatory efficiency; N: number of testable predictions (77); A: accuracy; T: theory. [a]: data on classical conditioning, [b]: data on latent inhibition, [c]: data on occasion setting.

The future character of the psychological law

Associationist theories suggest that associations are the basis of learning. Extending the associationist approach, mathematical theories describe animal learning in the context of accurate mathematical associationist models. Some mathematical theories are able to describe behavior as it occurs in real time. Further advancing the ideas fostered by mathematical learning theories, biologically plausible neural network theories combine associationist and connectionist principles in the study of animal learning and cognition.

This book shows that neural network theories that rigorously describe learning and cognitive processes can successfully account for many experimental results. More generally, the book demonstrates that neural network theories provide the necesssary tools to understand the intricate interactions among the numerous temporal and nontemporal parameters controlling the phenomena. Not less important is the fact that the models might become a starting point for detailed neurophysiological analysis of behavior. For all these attractive features, this style of theorizing is likely to become standard in the animal behavioral sciences.

Based on its previous history, where is the future of theorizing in animal learning? An obvious direction is the construction of increasingly accurate quantitative models of behavior. A possible natural evolution of the field lies in the transition from biologically plausible neural networks to biologically realistic neural models. These biologically realistic neural models would display architectures highly isomorphic with neuroanatomical and neurophysiological aspects of a given brain region in a given species. As these theories are developed, a temporary integration of biologically plausible and biologically realistic circuits can be conceived.

Another equally important and likely progression is the development of models with increasing power to describe behavior in real time, independently of their neurophysiological substrates. These advanced neural network theories of animal learning seem likely to have an important impact in the analysis of the adaptive behavior of both animals and autonomous robots (Meyer and Wilson, 1991).

References

Alkon, D. L., Sanches-Andres, J.V., Ito, E., Oka, K., Yoshioka, T., and Collin, C. (1992). Long-term transformation of an inibitory into an excitatory GABAergic synapse response. *Proceedings of the National Academy of Sciences, USA, 89,* 11862–6.

Amsel, A. (1958). The role of frustrative nonreward in noncontinuous reward situations. *Psychological Bulletin, 55,* 102–19.

(1992). *Frustration theory: An analysis of dispositional learning and memory.* New York: Cambridge University Press.

Amsel, A., Rashotte, M. E., and MacKinnon, J. R. (1966). Partial reinforcement effects within subjects and between subjects. *Psychological Monograph: General and Applied, 80* (20, Whole # 628).

Anderson, N. H. and Nakamura, C. I. (1964). Avoidance decrement in avoidance conditioning. *Journal of Comparative and Physiological Psychology, 57,* 196–204.

Andersson, E. and Armstrong, D. M. (1987). Complex spikes in Purkinje cells in the lateral vermis (b zone) of the cat cerebellum during locomotion. *Journal of Physiology (London), 385,* 107–34.

Anger, D. (1956). The dependence of interresponse times upon the relative reinforcement of different interresponse time. *Journal of Experimental Psychology, 52,* 145–61.

Arbib, M. A. and House, D. H. (1987). Depth and detours: An essay on visually guided behavior. In M. A. Arbib and A. R. Hanson (eds.), *Vision, brain and cooperative computation.* Cambridge, MA: MIT Press, pp. 129–63.

Arbib, M. A., Kilmer, W. L., and Spinelli, D. N. (1976). Neural models and memory. In M. R. Rosenzweig and E. L. Bennett (eds.), *Neural mechanisms of learning and memory.* Cambridge, MA: MIT Press, pp. 109–32.

Ayres, J. J. B, Albert, M., and Bombace, J. C. (1987). Extending conditioned stimuli before versus after unconditioned stimuli: Implications for real-time models of conditioning. *Journal of Experimental Psychology: Animal Behavior Processes, 13,* 168–82.

Ayres, J. J. B., Philbin, D., Cassidy, S., and Belling, L. (1992). Some parameters of latent inhibition. *Learning and Motivation, 23,* 269–87.

Baker, A. G. (1974). Conditioned inhibition is not the symmetrical opposite of conditioned excitation: A test of the Rescorla–Wagner model. *Learning and Motivation, 5,* 369–79.

Baker, A. G., Haskins, C. E., and Hall, G. (1990). Stimulus generalization decrement in latent inhibition to a compound following exposure to the elements of the compound. *Animal Learning and Behavior, 18,* 162–70.

Baker, A. G. and Mercier, P. (1982). Extinction of the context and latent inhibition. *Learning and Motivation, 13,* 391–416.

Bakin, J. S., and Weinberger, N. M. (1990). Classical conditioning induces CS–specific receptive field plasticity in the auditory cortex of the guinea pig. *Brain Research, 536,* 271–86.

Balaz, M. A., Capra, S., Kasprow, W. J., and Miller, R. R. (1982). Latent inhibition of the conditioning context: Further evidence of contextual potentiation of retrieval in the absence of appreciable context–US relations. *Animal Learning and Behavior, 10,* 242–8.

Barto, A. G. and Sutton, R. S. (1981). Landmark learning: An illustration of associative search. *Biological Cybernetics, 42,* 1–8.

(1990). Time-derivative models of Pavlovian conditioning. In M. Gabriel and J. Moore (eds.), *Learning and computational neuroscience: Foundations of adaptive networks.* Cambridge, MA: MIT Press, pp. 497–537.

Baum, M. (1966). Rapid extinction of an avoidance response following a period of response prevention in the avoidance apparatus. *Psychological Reports, 18,* 59–64.

(1969). Extinction of avoidance response following response prevention: Some parametric investigations. *Canadian Journal of Psychology, 23,* 1–10.

(1970). Extinction of avoidance responding through response prevention (flooding). *Psychological Bulletin, 74,* 276–84.

(1976). Instrumental learning: Comparative studies. In M. P. Feldman and A. Broadhurst (eds.), *Theoretical and experimental bases of the behavior therapies.* New York: John Wiley & Sons, pp. 113–131.

Beecroft, R. S. (1967). Near goal punishment of avoidance running. *Psychonomics Sciences, 8,* 109–10.

Bellingham, W. P., Gillette-Bellingham, F., and Kehoe, E. J. (1985). Summation and configuration in patterning schedules with the rat and rabbit. *Animal Learning and Behavior, 131,* 152–64.

Berger, T. W. (1984). Long-term potentiation of hippocampal synaptic transmission affects rate of behavioral learning. *Science, 224,* 627–30.

Berger, T. W., Bassett, J. L., and Weikart, C. (1985). Hippocampal–cerebellar interactions during classical conditioning. Paper delivered at the Twenty-sixth Annual Meeting of the Psychonomic Society, Boston, MA, Nov.

Berger, T. W., Clark, G. A., and Thompson, R. F. (1980). Learning-dependent neuronal responses recorded from limbic system brain structures during classical conditioning. *Physiological Psychology, 8,* 155–67.

Berger, T. W. and Orr, W. B. (1983). Hippocampectomy selectively disrupts discrimination reversal conditioning of the rabbit nictitating membrane response. *Behavioral Brain Research, 8,* 49–68.

Berger, T. W., Rinaldi, P. C., Weisz, D. J., and Thompson, R. F. (1983). Single-unit analysis of different hippocampal cell types during classical conditioning of rabbit nictitating membrane response. *Journal of Neurophysiology, 50,* 1197–1219.

Berger, T. W., Swanson, G. W., Milner, T. A., Lynch, G. S., and Thompson, R. F. (1980). Reciprocal anatomical connections between hippocampus and subiculum in the rabbit: Evidence for subicular innervation of regio superior. *Brain Research, 183,* 265–76.

Berger, T. W. and Thompson, R. F. (1978a). Neuronal plasticity in the limbic system during classical conditioning of the rabbit nictitating membrane response. I. The hippocampus. *Brain Research, 145,* 323–46.

(1978b). Neuronal plasticity in the limbic system during classical conditioning of the rabbit nictitating membrane response. II. Septum and mammillary bodies. *Brain Research, 156,* 293–314.

(1982). Hippocampal cellular plasticity during extinction of classically conditioned nictitating membrane behavior. *Behavioral Brain Research, 4,* 63–76.

Berger, T. W., Weikart, C. L., Bassett, J. L., and Orr, W. B. (1986). Lesions of the retrospenial cortex produce deficits in reversal learning of the rabbit nictitating membrane response: Implications for potential interactions between hippocampal and cerebellar brain systems. *Behavioral Neuroscience, 100,* 802–9.

Berry, S. D. and Thompson, R. F. (1979). Medial septal lesions retard classical conditioning of the nictitating membrane response in rabbits. *Science, 205,* 209–11.

Best, M. R., Gemberling, G. A., and Johnson, P. E. (1979). Disrupting the conditioned stimulus preexposure effect in flavor aversion learning: Effects of interoceptive distractor manipulations. *Journal of Experimental Psychology: Animal Behavior Processes, 5,* 321–34.

Best, P. J. and Ranck, J. B. (1982). Reliability of the relationship between hippocampal unit activity and sensory-behavioral events in the rat. *Experimental Neurology, 75,* 652–64.

Bilkey, D. K. and Goddard, G. V. (1985). Medial septal facilitation of hippocampal granule cell activity is mediated by inhibition of inhibitory interneurons. *Brain Research, 361,* 99–106.

Black, R. W. and Black, P. E. (1967). Heart rate conditioning as a function of interstimulus interval in rats. *Psychonomic Science, 8,* 219–20.

Blodgett, H. C. (1929). The effect of introduction of reward upon the maze performance of rats. *University of California Publications in Psychology, 54,* 261–5.

Blough, D. S. (1975). Steady state data and a quantitative model of operant generalization and discrimination. *Journal of Experimental Psychology: Animal Behavior Processes, 104,* 3–21.

Boice, R. and Denny, M. R. (1965). The conditioned licking response in rats as a function of the CS–US interval. *Psychonomic Science, 3,* 93–4.

Bolles, R. C. (1969). Avoidance and escape learning: Simultaneous acquisition of different responses. *Journal of Comparative Psychology and Physiology, 68,* 355–58.

(1970). Species-specific defence reactions and avoidance learning. *Psychological Review, 77,* 32–48.

Bolles, R. C., Moot, S. A., and Grossen, N. E. (1971). The extinction of shuttlebox avoidance. *Learning and Motivation, 2,* 324–33.

Bouton, M. E. (1993). Context, time, and memory retrieval in the interference paradigms of Pavlovian learning. *Psychological Bulletin, 114,* 80–99.

Bower, G. H. (1959). Choice-point behavior. In R. R. Bush and W. K. Estes (eds.), *Studies of mathematical learning.* Stanford, CA: Stanford University Press, pp. 109–24.

Bower, G. H. and Hilgard, E. R. (1981). *Theories of learning.* Englewood Cliffs, NJ: Prentice Hall.

Breese, C. R., Hampson, R. E., and Deadwyler, S. A. (1989). Hippocampal place cells: Stereotypy and plasticity. *The Journal of Neuroscience, 9,* 1097–111.

Brogden, W. J. (1939). Sensory pre-conditioning. *Journal of Experimental Psychology, 25,* 223–32.

Brown, J. S. (1942). The generalization of approach responses as a function of stimulus intensity and strength of motivation. *Journal of Comparative Psychology, 33,* 209–26.

Brown, T. (1820). *Lecture on the philosophy of the human mind,* Vols. 1 and 2. Edinburgh: James Ballantyne.

Brush, E. S. (1957). Traumatic avoidance learning: The effects of conditioned stimulus length in a free responding situation. *Journal of Comparative and Physiological Psychology, 50,* 541–64.

Bryson, A. E. and Ho, Y.-C. (1969). *Applied optimal control.* New York: Blaisdell.

Buchanan, S. L. and Powell, D. A. (1980). Divergences in Pavlovian conditioned heart rate and eyeblink responses produced by hippocampectomy In the rabbit (*Oryctolagus cuniculus*). *Behavioral and Neural Biology, 30,* 20–38.

(1982). Cingulate cortex: Its role in Pavlovian conditioning. *Journal of Comparative and Physiological Psychology, 96,* 755–74.

Bunge, M. (1967). *Scientific research II: The search for truth.* New York: Spinger-Verlag.

Burkhardt, P. E. and Ayres, J. J. B. (1978). CS and US duration effects in one-trial simultaneous fear conditioning as assessed by conditioned suppression of licking in rats. *Animal Learning and Behavior, 6,* 225–30.

Bush, R. R. and Mosteller, F. (1955). *Stochastic models for learning.* New York: John Wiley & Sons.

Carpenter, G. A. and Grossberg, S. (1985). A neural theory of circadian rhythms: Split rhythms, after-effects, and motivational interactions. *Journal of Theoretical Biology, 113,* 163–223.

(1987). A massively parallel architecture for a self-organizing neural pattern recognition machine. *Computer Vision, Graphics, and Image Processing, 37,* 54–115.

Channel, S. and Hall, G. (1981). Facilitation and retardation of discrimination learning after exposure to the stimuli. *Journal of Experimental Psychology: Animal Behavior Processes, 7,* 437–46.

Chantrey, D. (1974). Stimulus preexposure and discrimination learning by domestic chicks: Effects of varying interstimulus time. *Journal of Comparative and Physiological Psychology, 87,* 517–25.

Chorazyna, H. (1962). Some properties of conditioned inhibition. *Acta Biologiae Experimentalis, 22,* 5–13.

Church, R. M. (1984). Properties of the internal clock. In J. Gibbon and L. Allan (eds.), *Timing and time perception.* Annals of the New York Academy of Sciences, Vol. 423, pp. 566–82.

Church, R. M. and Broadbent, H. A. (1991). A connectionist model of timing. In M. Commons, S. Grossberg, and J. E. R. Staddon (eds.), *Neural network models of conditioning and action.* Hillsdale, NJ: Lawrence Erlbaum Associates, pp. 225–40.

Church, R. M., Brush, F. R., and Solomon, R. L. (1956). Traumatic avoidance learning: The effects of CS–US interval with a delayed-conditioning procedure in a free-responding situation. *Journal of Comparative and Physiological Psychology, 49,* 301–8.

Clark, G. A., McCormick, D. A., Lavond, D. G., and Thompson, R. F. (1984). Effects of lesions of cerebellar nuclei on conditioned behavioral and hippocampal neuronal responses. *Brain Research, 291,* 125–36.

Cohen, N. J. and Eichenbaum, H. B. (1993). *Memory, amnesia, and the hippocampal system.* Cambridge, MA: MIT Press.

Cohen, N. J. and Squire, L. R. (1980). Preserved learning and retention of pattern analyzing skill in amnesia: Dissociation of knowing how and knowing that. *Science, 210,* 207–9.

Coons, E. E., Anderson, N. H., and Myers, A. K. (1960). Disappearance of avoidance responding during continued training. *Journal of Comparative and Physiological Psychology, 53,* 290–2.

Cotton, M. M., Goodall, G., and Mackintosh, N. J. (1982). Inhibitory conditioning resulting from a reduction in the magnitude of reinforcement. *Quarterly Journal of Experimental Psychology, 34B,* 163–80.

Crowell, C. R. and Anderson, D. C. (1972). Variations in intensity, interstimulus interval, and interval between preconditioning CS exposures and conditioning with rats. *Journal of Comparative and Physiological Psychology, 79,* 291–8.

Curio, E. (1976). *The ethology of predation.* Berlin: Springer-Verlag.

Cynader, M. and Chernenko, G. (1976). Abolition of directional sensitivity in the visual cortex of the cat. *Science, 193,* 504–5.

Daly, H. B. and Daly, J. T. (1982). A mathematical model of reward and aversive nonreward: Its application in over 30 appetitive learning situations. *Journal of Experimental Psychology: General, 111,* 441–80.

Davenport, D. G. and Olson, R. D. (1968). A reinterpretation of extinction in discriminated avoidance. *Psychonomics Science, 13,* 5–6.

Davidson, T. L., McKernan, M. G., and Jarrard, L. E. (1993). Hippocampal lesions do not impair negative patterning: A challenge to configural association theory. *Behavioral Neuroscience, 107,* 227–34.

Davis, M., Schlesinger, L. S. and Sorenson, C. A. (1989). Temporal specificity of fear conditioning: Effects of different conditioned stimulus–unconditioned stimulus intervals on the fear-startled reflex. *Journal of Experimental Psychology: Animal Behavior Processes, 15,* 295–310.

Denny, M. R. (1971). Relaxation theory and experiments. In F. R. Brush (ed.), *Aversive Conditioning and Learning.* New York: Academic Press, pp. 235–95.

Desmond, J. E. and Moore, J. W. (1988). Adaptive timing in neural networks: The conditioned response. *Biological Cybernetics, 58,* 405–15.

Deutsch, J. A. (1960). *The structural basis of behavior.* Cambridge: Cambridge University Press.

DeVietti, T. L. and Barrett, O. V. (1986). Latent inhibition: No effect of intertrial interval of the preexposure trials. *Bulletin of the Psychonomic Society, 24,* 453–5.

Dickinson, A. (1980). *Contemporary animal learning theory.* Cambridge: Cambridge University Press.

Dickinson, A., Hall, G., and Mackintosh, N. J. (1996). Surprise and the attenuation of blocking. *Journal of Experimental Psychology: Animal Behavior Processes, 2,* 313–53.

Dickinson, A. and Mackintosh, N. J. (1978). Classical conditioning in animals. *Annual Review of Psychology 29,* 587–612.

(1979). Reinforcer specificity in the enhancement of conditioning by posttrial surprise. *Journal of Experimental Psychology: Animal Behavior Processes, 5,* 162–77.

Diez-Chamizo, V., Sterio, D., and Mackintosh, N. J, (1985). Blocking and overshadowing between intra-maze and extra-maze cues: A test of the independence of locale and guidance learning. *The Quarterly Journal of Experimental Psychology, 37B,* 235–53.

DiMattia, B. D. and Kesner, R. P. (1988). Spatial cognitive maps: Differential role of parietal cortex and hippocampal formation. *Behavioral Neuroscience, 102,* 471–80.

Dinsmoor, J. A. (1954). Punishment: I. The avoidance hypothesis. *Psychological Review, 61,* 34–46.

Donahoe, J. W., Burgos, J. E., and Palmer, D. C. (1993). A selectionist approach to reinforcement. *Journal of Experimental Analysis of Behavior, 60,* 17–40.

Douglas, R. (1972). Pavlovian conditioning and the brain. In R. A. Boakes and M. S. Halliday (eds.), *Inhibition and Learning.* London: Academic Press, pp. 529–49.

Douglas, R. and Pribram, K. H. (1966). Learning and limbic lesions. *Neuropsychologia, 4,* 197–220.

Eichenbaum, H. B., Kuperstein, M., Fagan, A., and Nagode, J. (1986). Cue-sampling and goal-approach correlates of hippocampal unit activity in rats performing an odor discrimination task. *Journal of Neuroscience, 7,* 716–32.

Fanselow, M. S. (1990). Factors governing one-trial contextual conditioning. *Animal Learning and Behavior, 18,* 264–70.

Feynman, R. (1967). *The character of physical law.* Cambridge, MA: MIT Press.

Flew, A. (1979). *A dictionary of philosophy.* New York: St. Martin's Press.

Freud, S. (1954 ed.) Project for scientific psychology. *The origins of psycholanalysis, letters to Wilheim Fliess, draft and notes 1887–1902.* New York: Basic Books, Appendix.

Frey, P. W. and Ross, L. E. (1968). Classical conditioning of the rabbit eyelid response as a function of interstimulus interval. *Journal of the Comparative and Physiological Psychology, 65,* 246–50.

Frey, P. W. and Sears, R. J. (1978). Model of conditioning incorporating the Rescorla–Wagner associative axiom, a dynamic attention process, and a catastrophe rule. *Psychological Review, 85,* 321–40.

Gabriel, M. and Schmajuk, N. A. (1991). Neural substrate of avoidance learning in rabbits. In J. W. Moore and M. Gabriel (eds.), *Neurocomputation and learning: Foundations of adaptive networks.* Cambridge, MA: MIT Press, pp. 143–70.

Gaffan, D. (1972). Loss of recognition memory in rats with lesion of the fornix. *Neuropsychologia, 10,* 327–41.

Gallagher, M. and Holland, P. C. (1992). Preserved configural learning and spatial learning impairments in rats with hippocampal damage. *Hippocampus, 2,* 81–8.

Gallistel, C. R. (1980). *The organization of action: A new synthesis.* Hillsdale, NJ: Lawrence Erlbaum Associates.

Garrud, P., Rawlins, J. N. P., Mackintosh, N. J., Goodal, G., Cotton, M. M., and Feldon, J. (1984). Successful overshadowing and blocking in hippocampectomized rats. *Behavioural Brain Research, 12,* 39–53.

Gelperin, A. (1986). Complex associative learning in small neural networks. *Trends in Neuro Sciences, 9,* 323–8.

Gelperin, A., Hopfield, J. J., and Tank, D. W. (1985). The logic of Limax learning. In A. Selverston (ed.), *Model neural networks and behavior.* New York: Plenum Press, pp. 237–61.

Gluck, M. A. and Myers, C. E. (1993). Hippocampal mediation of stimulus representation: A computational theory. *Hippocampus 3,* 491–516.

Gluck, M. A., Reifsnider, E. S., and Thompson, R. F. (1990). Adaptive signal processing and the cerebellum: Models of classical conditioning and VOR adaptation. In M. Gluck and D. E. Rumelhart (eds.), *Neuroscience and Connectionist Theory.* Hillsdale, NJ: Lawrence Erlbaum Associates.

Good, M. and Honey, R. (1993). Selective hippocampus lesions abolish contextual specificity of latent inhibition and conditioning. *Behavioral Neuroscience, 107,* 23–33.

Gordon, W. C. and Weaver, M. S. (1989). Cue-induced transfer of CS preexposure effects across contexts. *Animal Learning and Behavior, 17,* 409–17.

Gormezano, I. (1972). Investigations of defense and reward conditioning in the rabbit. In A. H. Black and W. F. Prokasy (eds.), *Classical conditioning II: Theory and research.* New York: Appleton-Century-Crofts, pp. 151–81.

Gormezano, I. and Coleman, S. R. (1973). The law of effect and CR contingent modification of the UCS. *Conditional Reflex 8,* 41–56.

Gormezano, I., Kehoe, E. J., and Marshall, B. S. (1983). Twenty years of classical conditioning research with the rabbit. *Progress in Psychobiology and Physiological Psychology, 10,* 197–275.

Gormezano, I. and Moore, J. W. (1969). Classical conditioning. In M. H. Marx (ed.), *Learning processes.* New York: Macmillan, pp. 119–203.

Graeff, F. G., Quintero, S., and Gray, J. A. (1980). Median raphe stimulation, hippocampal theta rhythm and threat induced behavioural inhibition. *Physiology and Behavior, 25,* 253–61.

Grahame, N. J., Barnet, R. C., Gunther, L. M., and Miller, R. R. (1994). Latent inhibition as a performance deficit resulting from CS–context associations. *Animal Learning and Behavior, 22,* 395–408.

Grastyan, E., Lissak, K., Madaraz, I., and Donhoffer, H. (1959). Hippocampal electrical activity during the development of conditioned reflexes. *Electroencephalography and Clinical Neurophysiology, 11,* 409–30.

Gray, J. A. (1971). *The psychology of fear and stress.* London: Weidenfeld and Nicholson.

(1975). *Elements of a two-process theory of learning.* London: Academic Press.

(1982). *The neuropsychology of anxiety: An inquiry into the function of the septo-hippocampal system.* New York: Oxford University Press.

Gray, J. A., Feldon, J., Rawlins, J. N. P., Owen, S., and McNaughton, N. (1978). The role of the septo-hippocampal system and its noradrenergic affecrents in behavioural responses to non-reward. In *Functions of the septo-hippocampal system,* Ciba Foundation Symposium 58 (new series). Amsterdam: Elsevier, pp. 275–300.

Gray, J. A. and McNaughton, N. (1983). Comparison between the behavioural effects of septal and hippocampal lesions: A review. *Neuroscience and Biobehavioral Reviews, 7,* 119–88.

Gray, J. A. and Smith, P. T. (1969). An arousal-decision model for partial reinforcement and discrimination learning. In R. Gilbert and N. S. Sutherland (eds.), *Animal discrimination learning.* London:Academic Press, pp. 243–72.

Grossberg, S. (1972a). A neural theory of punishment and avoidance. I: Qualitative theory. *Mathematical Biosciences, 15,* 39–67.

(1972b). A neural theory of punishment and avoidance. II: Quantitative theory. *Mathematical Biosciences, 15,* 253–85.

(1974). Classical and instrumental learning by neural networks. In *Progress in theoretical biology,* Vol. 3. New York: Academic Press, pp. 51–141.

(1975). A neural model of attention, reinforcement, and discrimination learning. *International Review of Neurobiology, 18,* 263–327.

Grossberg, S. and Levine, D. S. (1987). Neural dynamics of attentionally-modulated Pavlovian conditioning: Blocking, inter-stimulus interval, and secondary reinforcement. *Applied Optics, 26,* 5015–30.

Grossberg, S. and Schmajuk, N. A. (1989). Neural dynamics of adaptive timing and temporal discrimination during associative learning. *Neural Networks, 2,* 79–102.

Guthrie, E. R. (1935). *The psychology of learning.* New York: Harper and Row.

Hall, G. (1991). *Perceptual and associative learning.* Oxford: Clarendon Press.

Hall, G. and Channel, S. (1985a). Latent inhibition and conditioning after preexposure to the training context. *Learning and Motivation, 16,* 381–481.

(1985b). Differential effects of contextual change on latent inhibition and on the habituation of an orienting response. *Journal of Experimental Psychology: Animal Behavior Processes, 11,*470–481.

Hall, G. and Minor, H. (1984). A search for context–stimulus associations in latent inhibition. *The Quarterly Journal of Experimental Psychology, 36B,* 145–69.

Hall, G. and Pearce, J. M. (1979). Latent inhibition of a CS during CS–US pairings. *Journal of Experimental Psychology: Animal Behavior Processes, 5,* 31–42.

(1982). Restoring the associability of a preexposed CS by a surprising event. *Quarterly Journal of Experimental Psychology, 34B,* 127–40.

Hall, G. and Schachtman, T. R. (1987). Differential effects of a retention interval on latent inhibition and the habituation of an orienting response. *Animal Learning and Behavior, 15,* 76–82.

Hampson, S. E. (1990). *Connectionistic problem solving.* Boston, MA: Birkhauser.

Haney, G. W. (1931). The effect of familiarity on maze performance of albino rats. *University of California Publications in Psychology, 4,* 319–33.

Hebb, D. (1949). *The organization of behavior.* New York: John Wiley & Sons.

Henneman, E. (1957). Relation between size of neurons and their susceptibility to discharge. *Science, 26,* 1345–47.

(1985). The size-principle: A deterministic output emerges from a set of probabilistic connections. *Journal of Experimental Biology, 115,* 105–12.

Herrnstein, R. J. (1969). Method and theory in the study of avoidance. *Psychological Review, 76,* 49–69.

Herrnstein, R. J. and Hineline, P. N. (1966). Negative reinforcement as shock-frequency reduction. *Journal of the Experimental Analysis of Behavior, 9,* 421–30.

Hinzman, D. L. (1991). Why are formal models useful in psychology? In W. E. Hockley and S. Lewandosky (eds.), *Relating theory and data: Essays on human memory in honor of Bennet B. Murdock.* Hillsdale, NJ: Lawrence Erlbaum Associates, pp. 39–56.

Hirsch, H. V. B. and Spinelli, D. N. (1970). Visual experience modifies distribution of horizontally and vertically oriented receptive fields in cats. *Science, 168,* 869–71.

Hirsh, R. (1974). The hippocampus and contextual retrieval of information from memory: A theory. *Behavioral Biology, 12,* 421–44.

Holland, P. C. (1983). Representation-mediated overshadowing and potentiation of conditioned aversions. *Journal of Experimental Psychology: Animal Behavior Processes, 9,* 1–13.

(1992). Occasion setting in Pavlovian conditioning. In D. Medlin (ed.). *The psychology of learning and motivation,* Vol. 28. San Diego, CA: Academic Press, pp. 69–125.

Holland, P. C. and Forbes, D. R. (1980). Effects of compound or element preexposure on compound flavor aversion conditioning. *Animal Learning and Behavior, 8,* 199–203.

Holland, P. C. and Rescorla, R. A. (1975). The effect of two ways of devaluing the unconditioned stimulus after first- and second-order appetitive conditioning. *Journal of Experimental Psychology: Animal Behavior Processes, 1,* 355–63.

Honey, R. C. and Hall, G. (1988). Overshadowing and blocking procedures in latent inhibition. *The Quarterly Journal of Experimental Psychology, 49B,* 163–86.

(1989). Attenuation of latent inhibition after compound pre-exposure: Associative and perceptual explanations. *The Quarterly Journal of Experimental Psychology, 41B,* 355–68.

Hopfield, J. (1982). Neural networks and physical systems with emergent collective computational properties. *Proceedings of National Academy of Sciences 79:* 2554–8.

Hull, C. L. (1929). A functional interpretation of the conditioned reflex. *Psychological Review, 36,* 498–511.

(1931). Goal attraction and directing ideas conceived as habit phenomena. *Psychological Review, 38,* 487–506.

(1943). *Principles of behavior.* New York: Appleton-Century-Crofts.

(1951). *Essentials of behavior.* Westport, CT: Greenwood Press.

(1952). *A behaviour system.* New Haven, CT: Yale University Press.

Isaacson, R. L. and Woodruff, M. L. (1975). Spontaneous alternation and passive avoidance behavior in rats after hippocampal lesions. In B. L. Hart (ed.), *Experimental Psychobiology.* San Francisco, CA: W. H. Freeman, pp. 103–9.

Ito, M. (1984). *The cerebellum and neural control.* New York: Raven Press.

James, W. (1890). *The principles of psychology.* New York: Holt, Rinehart and Winston.

James, J. D. O., Hardiman, M. J., and Yeo, C. H. (1987). Hippocampal lesions and trace conditioning in the rabbit. *Behavioural Brain Research, 23,* 109–16.

Jarrard, L. E. (1983). Selective hippocampal lesions and behavior: Effects of kainic acid lesions on performance of place and cue tasks. *Behavioral Neuroscience, 97,* 873–89.

(1986). Selective hippocampal lesions and behavior: Implications for current research and theorizing. In R. L. Isaacson and K. H. Pribram (eds.), *The hippocampus,* Vol. 4. New York: Plenum Press, pp. 93–126.

Jarrard, L. E. and Davidson, T. L. (1991). On the hippocampus and learned conditional responding: Effects of aspiration versus ibotenate lesions. *Hippocampus, 1,* 107–17.

Jarrard, L. E. and Meldrum, B. S. (1990). Neurotoxicity of intrahippocampal injections of excitatory amino acids: Protective effects of CPP. *Society for Neuroscience Abstracts, 16,* 429.

Jarrard, L. E., Okaichi, H., Steward, O., and Goldschmidt, R. B. (1984). On the role of hippocampal connections in the performance of place and cue tasks: Comparisons with damage to hippocampus. *Behavioral Neuroscience,98,* 946–54.

Kamin, L. J. (1954). Traumatic avoidance learning: The effects of CS–US interval with a trace-conditioning procedure. *Journal of Comparative and Physiological Psychology, 47,* 65–72.

(1957). The gradient of delay of secondary reward in avoidance learning. *Journal of Comparative and Physiological Psychology, 50,* 445–9.

(1968). "Attention-like" processes in classical conditioning. In M. R. Jones (ed.), *Miami Symposium on the prediction of behavior: Aversive stimulation.* Miami, FL: University of Miami Press, pp. 9–33.

(1969). Predictability, surprise, attention, and conditioning. In B. A. Campbell and R. M. Church (eds.), *Punishment and aversive behavior.* New York: Appleton-Century-Crofts, pp. 279–96.

Kamin, L. J., Brimer, C. J., and Black, A. H. (1963). Conditioned suppression as a monitor of fear of the CS in the course of avoidance training. *Journal of Comparative and Physiological Psychology, 56,* 497–501.

Kasprow, W. J., Cacheiro, H., Balaz, M. A., and Miller, R. R. (1982). Reminder-induced recovery of associations to an overshadowed stimulus. *Learning and Motivation, 13,* 308–18.

Kasprow, W. J., Catterson, D., Schachtman, T. R., and Miller, R. R. (1984). Attenuation of latent inhibition by post-acquisition reminder. *Quarterly Journal of Experimental Psychology, 36B,* 53–63.

Kaye, H. and Pearce, J. M. (1984). The strength of the orienting response during Pavlovian conditioning. *Journal of Experimental Psychology: Animal Behavior Processes, 10,* 90–109.

(1987a). Hippocampal lesions attenuate latent inhibition and the decline of the orienting response in rats. *Quarterly Journal of Experimental Psychology, 39B,* 107–25.

(1987b). Hippocampal lesions attenuate latent inhibition of a CS and of a neutral stimulus. *Psychobiology, 15,* 293–9.

Kehoe, E. J. (1986). Summation and configuration in conditioning of the rabbit's nictitating membrane response to compound stimuli. *Journal of Experimental Psychology: Animal Behavior Processes, 12,* 186–95.

(1988). A layered network model of associative learning: Learning to learn and configuration. *Psychological Review, 95,* 411–33.

Kehoe, E. J. and Graham, P. (1988). Summation and configuration: Stimulus compounding and negative patterning in the rabbit. *Journal of Experimental Psychology: Animal Behavior Processes, 14,* 320–33.

Killeen, P. R. and Amsel, A. (1987). The kinematics of locomotion toward a goal. *Journal of Experimental Psychology: Animal Behavior Processes, 13,* 92–101.

Killeen, P. R. and Weiss, N. A. (1987). Optimal timing and the Weber function. *Psychological Review, 94,* 455–68.

Kimble, D. P. (1968). Hippocampus and internal inhibition. *Psychological Bulletin, 70,* 285–95.

Klein, S. B., Mikulka, P. J., and Hamel, K. (1976). Influence of sucrose preexposure on acquisition of a conditioned response. *Behavioral Biology, 16,* 99–104.

Klopf, A. H. (1988). A neuronal model of classical conditioning. *Psychobiology, 16,* 85–125.

Klopf, A. H., Morgan, J. S., and Weaver, S. E. (1993). A hierarchical network of control systems that learn: Modeling nervous system function during classical and instrumental conditioning. *Adaptive Behavior, 1,* 263–319.

Kohonen, T. (1977). *Associative memory. A system-theoretical approach.* New York: Springer-Verlag.

Kolb, B., Buhrman, K., and McDonald, R. (1989). Dissociation of prefrontal, posterior parietal, and temporal cortical regions to spatial navigation and recognition memory in the rat. *Society for Neuroscience Abstracts, 15,* 607.

Kolb, B., Sutherland, R. J., and Whishaw, I. Q. (1983). A comparison of the contributions of the frontal and parietal association cortex to spatial localization in rats. *Behavioral Neuroscience, 97,* 13–27.

Konorski, J. (1948). *Conditioned reflexes and neuron organization.* London: Cambridge University Press.

Kraemer, P. J., Randall, C. K., and Carbary, T. J. (1991). Release from latent inhibition with delayed testing. *Animal Learning and Behavior, 19,* 139–45.

Kremer, E. F. (1978). The Rescorla–Wagner model: Losses in associative strength in compound conditioned stimuli. *Journal of Experimental Psychology: Animal Behavior Processes, 11,* 15–34.

Krebs, J. R. and Davies, N. B. (1981). *An introduction to behavioral ecology.* Oxford: Blackwell Scientific Publications.

Krnjevic, K., Ropert, N., and Casullo, J. (1988). Septohippocampal disinhibition. *Brain Research, 438,* 182–92.

Lantz, A. E. (1973). Effect of number of trials, interstimulus interval, and dishabituation on subsequent conditioning in a CER paradigm. *Animal Learning and Behavior, 1,* 273–7.

Lawler, E. E. (1965). Secondary reinforcement value of stimuli associated with shock reduction. *Quarterly Journal of Experimental Psychology, 17,* 57–62.

Levine, D. S. (1991). *Neural and cognitive modeling.* Hillsdale, NJ: Lawrence Erlbaum Associates.

Levine, M. W. and Schefner, J. M. (1981). *Fundamentals of Sensation and Perception.* Reading, MA: Addison-Wesley.

Levis, D. J. (1989). The case for a return to a two-factor theory of avoidance: The failure of non-fear interpretations. In Stephen B. Klein and R. R. Mowrer (eds.), *Contemporary learning theories.* Hillsdale, NJ: Lawrence Erlbaum Associates, pp 227–77.

Lewandowsky, S. (1991). Gradual unlearning and catastrophic interference: A comparison of distributed architectures. In W. E. Hockley and S. Lewandosky (eds.), *Relating theory and data: Essays on Human Memory in Honor of Bennet B. Murdock.* Hillsdale, NJ: Lawrence Erlbaum Associates.

Lieblich, I. and Arbib, M. A. (1982). Multiple representations of space underlying behavior. *The Behavioral and Brain Sciences, 5,* 627–59.

Loechner, K. J. and Weisz, D. J. (1987). Hippocampectomy and feature-positive discrimination. *Behavioral Brain Research, 26,* 63–73.

LoLordo, V. M. and Rescorla, R. A. (1966). Protection of the fear-eliciting capacity of a stimulus from extinction. *Acta Biologiae Experimentalis, 26,* 251–8.

Lovibond, P. F., Preston, G. C., and Mackintosh, N. J. (1984). Context specificity of conditioning, extinction, and latent inhibition. *Journal of Experimental Pyschology: Animal Behavior Processes, 10,* 360–75.

Lubow, R. E. (1989). *Latent inhibition and conditioned attention theory.* Cambridge: Cambridge University Press.

Lubow, R. E. and Moore, A. U. (1959). Latent inhibition: The effect of non-reinforced preexposure to the conditional stimulus. *Journal of Comparative and Physiological Psychology, 52,* 415–19.

Lubow, R. E., Rifkin, B., and Alek, M. (1976). The context effect: The relationship between stimulus preexposure and environmental preexposure determines subsequent learning. *Journal of Experimental Psychology: Animal Behavior Processes, 2,* 38–47.

Lubow, R. E., Schnur, P., Rifkin, B. (1976). Latent inhibition and conditioned attention theory. *Journal of Experimental Psychology: Animal Behavior Processes, 2,* 163–74.

Lubow, R. E., Weiner, I., and Schnur, P. (1981). Conditioned attention theory. In G. H. Bower (ed.), *The psychology of learning and motivation,* Vol. 15. New York: Academic Press, pp. 1–49.

Lysle, D. T. and Fowler, H. (1985). Inhibition as a "slave" process: Deactivation of conditioned inhibition through extinction of conditioned excitation. *Journal of Experimental Psychology: Animal Behavior Processes, 11,* 71–94.

Mackintosh, N. J. (1973). Stimulus selection: Learning to ignore stimuli that predict no change in reinforcement. In R. A. Hinde and J. S. Hinde (eds.), *Constraints of learning.* London: Academic Press, pp. 75–96.

(1974). *The psychology of animal learning.* London: Academic Press.

(1975). A theory of attention: Variations in the associability of stimuli with reinforcement. *Psychological Review, 82,* 276–98.

(1983). *Conditioning and associative learning.* Oxford: Clarendon Press.

Maier, N. R. F. (1929). Reasoning in white rats. *Comparative Psychology Monographs, 6,* 29.

Maier, S. F. and Seligman, M. E. P. (1976). Learned helplessness: Theory and evidence. *Journal of Experimental Psychology: General, 105,* 3–46.

Maki, W. S. and Abunawas, A. M. (1991). A connectionist approach to conditional discriminations: Learning, short-term memory, and attention. In M. Commons, S. Grossberg, and J. E. R. Staddon (eds.), *Neural network models of conditioning and action.* Hillsdale, NJ: Lawrence Erlbaum Associates, pp. 241–78.

Marr, D. (1982). *Vision: A computational investigation into the human representation and processing of visual information.* San Francisco, CA: W. H. Freeman, Chap. 4.

Maynard Smith, J. (1978). Optimization theory and evolution. *Annual Review of Ecological Systems, 9,* 31–56.

Mazur, J. E. (1990). *Learning and behavior.* Englewood Cliffs, NJ: Prentice Hall.

McAdam, D., Knott, J. R., and Chiorini, J. (1965). Classical conditioning in the cat as a function of the CS–US interval. *Psychonomic Science, 3,* 89–90.

McCloskey, M. and Cohen, N. J. (1989). Catastrophic interference in connectionist networks: The sequential learning problem. In G. H. Bower (ed.), *The psychology of learning and motivation.* New York: Academic Press, pp. 109–64.

McCormick, D. A., Steinmetz, J. E., and Thompson, R. F. (1985). Lesions of the inferior olivary complex cause extinction of the classically conditioned eyeblink response. *Brain Research, 359,* 120–30.

McCulloch, W. S. and Pitts, W. (1943). A logical calculus of the ideas imminent in nervous activity. *Bulletin of Mathematical Biophysics 5,* 115–33, Chap. 1, 2, 4, 5.

McFarland, D. and Bosser, T. (1993). *Intelligent behavior in animals and robots.* Cambridge, MA: MIT Press.

McLaren, I. P. L., Bennett, C., Plaisted, K., Aitken, M., and Mackintosh, N. J. (1994). Latent inhibition, context specificity, and context familiarity. *The Quarterly Journal of Experimental Psychology, 47B,* 387–400.

McLaren, I. P. L., Kaye, H., and Mackintosh, N. J. (1989). An associative theory of the representation of stimuli: Applications to perceptual learning and latent inhibition. In R. G. M. Morris (ed.), *Parallel distributed processing: Implications for psychology and neurobiology.* Oxford: Clarendon Press.

McNaughton, B. L. (1989). Neuronal mechanisms for spatial computation and information storage. In L. Nadel, L. Cooper, P. Culicover, and R. Harnish (eds.), *Neural connections and mental computations.* New York:Academic Press.

McNaughton, B. L., Barnes, C. A., and O'Keefe, J. (1983). The contributions of position, direction, and velocity to single unit activity in the hippocampus of freely-moving rats. *Experimental Brain Research, 52,* 41–9.

McNaughton, B. L., Leonard, B., and Chen, L. (1989). Cortical-hippocampal interactions and cognitive mapping: A hypothesis based on reintegration of the parietal and inferotemporal pathways for visual processing. *Psychobiology, 17,* 236–76.

McNaughton, B. L. and Nadel, L. (1990). Hebb–Marr networks and the neurobiological representation of action in space. In M. Gluck and D. E. Rumelhart (eds.), *Neuroscience and connectioninst theory.* Hillsdale, NJ: Lawrence Erlbaum Associates.

Meyer, J.-A. and Wilson, S. W. (eds.) (1991). *From animals to animals.* Cambridge, MA: MIT Press.

Micco, D. J. and Schwartz, M. (1972). Effects of hippocampal lesions upon the developments of Pavlovian internal inhibition in rats. *Journal of Comparative and Physiological Psychology, 76,* 371–7.

Millenson, J. R., Kehoe, E. J., and Gormezano, I. (1977). Classical conditioning of the rabbit's nictitating membrane response under fixed and mixed CS–US intervals. *Learning and Motivation, 8,* 351–66.

Miller, G. A., Gallanter, E. H., and Pribram, K. H. (1960). *Plans and the structure of behavior.* New York: Holt, Rinehart & Winston.

Miller, V. M. and Best, P. J. (1980). Spatial correlates of hippocampal unit activity are altered by lesions of the fornix and entorhinal cortex. *Brain Research, 194,* 311–23.

Miller, N. E. (1944). Experimental studies in conflict. In J. McV. Hunt (ed.), *Personality and the behavior disorders.* New York: Ronald Press.

(1948). Studies of fear as an acquirable drive. *Journal of Experimental Psychology, 38,* 89–101.

(1959). Liberalization of basic S–R concepts: Extensions to conflict behavior, motivation, and social learning. In S. Koch (ed.). *Psychology: A study of a science,* Vol. 2. New York: McGraw Hill.

Miller, N. E., Brown, J. S., and Lipovsky, H. (1943). A theoretical and experimental analysis of conflict behavior: III. Approach-avoidance conflict as a function of strength of drive and strength of shock. Cited in Miller (1944). Unpublished manuscript.

Miller, R. R. & Schachtman, T. R. (1985). Conditioning context as an associative baseline: Implications or response generation and the nature of conditioned inhibition. In R. R. Miller and N. E. Spear (eds.), *Information processing in animals: Conditioned inhibition.* Hillsdale, NJ: Lawrence Erlbaum Associates, pp. 51–88.

Milner, P. M. (1960). *Physiological psychology*. New York: Holt, Rinehart, and Winston.

Mineka, S., Cook, M., and Miller, S. (1984). Fear conditioned with escapable and inescapable shock: Effects of a feedback stimulus. *Journal of Experimental Psychology: Animal Behavior Processes, 10,* 307–23.

Mineka, S. and Gino, A. (1980). Disassociation between conditioned emotional response and extended avoidance performance. *Learning and Motivation, 11,* 476–502.

Minsky, M. and Papert, S. (1969). *Perceptrons: An introduction to computational geometry.* Cambridge, MA: MIT Press, Chap. 1, 2, 6.

Mishkin, M. and Petri, H. L. (1984). Memories and habits: Some implications for the analysis of learning and retention. In L. R. Squire and N. Butters (eds.), *Neuropsychology of memory.* New York: Guilford Press, pp. 287–96.

Mizumori, S. J. Y., McNaughton, B. L., Barnes, C. A., and Fox, K. B. (1989). Preserved spatial coding in hippocampal CA1 pyramidal cells during reversible suppression of CA3 output: Evidence for pattern completion in hippocampus. *The Journal of Neuroscience, 9,* 3915–28.

Moore, J. W. (1979a). Brain processes and conditioning. In A. Dickinson and R. A. Boakes (eds.), *Mechanisms of learning and behavior.* Hillsdale, NJ: Lawrence Erlbaum Associates, pp. 111–42.

(1979b). Information processing in space-time by the hippocampus. *Physiological Psychology, 7,* 224–32.

Moore, J. W. and Stickney, K. J. (1980). Formation of attentional-associative networks in real time: Role of the hippocampus and implications for conditioning. *Physiological Psychology, 8,* 207–17.

(1982). Goal tracking in attentional-associative networks: Spatial learning and the hippocampus. *Physiological Psychology, 10,* 202–8.

Moore, J. W., Yeo, C. H., Oakley, D. A., and Russell, I. S. (1980). Conditioned inhibition of the nictitating membrane response in decorticate rabbits. *Behavioural Brain Research, 1,* 397–409.

Morris, R. G. M. (1974). Pavlovian conditioned inhibition of fear during shuttlebox avoidance behavior. *Learning and Motivation, 5,* 424–47.

(1981). Spatial location does not require the presence of local cues. *Learning and Motivation, 12,* 239–60.

Morris, R. G. M., Garrud, P., Rawlins, J. N. P., and O'Keefe, J. (1982). Place navigation impaired in rats with hippocampal lesions. *Nature, 297,* 681–3.

Morris, R. G. M., Schenk, F., Tweedie, F., and Jarrard, L. E. (1990). Ibotenate lesions of hippocampus and/or subiculum: Dissociating components of allocentric spatial learning. *European Journal of Neuroscience, 2,* 1016–28.

Mowrer, O. H. (1947). On the dual nature of learning—a reinterpretation of conditioning and problem solving. *Harvard Educational Review, 17,* 102–48.

Mowrer, O.H. and Lamoreaux, R. R. (1946). Fear as an intervening variable in avoidance conditioning. *Journal of Comparative Psychology, 39,* 29–50.

Moyer, J. R., Deyo, R. A., and Disterhoft, J. F. (1990). Hippocampectomy disrupts associative learning of the trace conditioned eye-blink response in rabbits. *Behavioral Neuroscience, 104,* 243–52.

Muller, R. U. and Kubie, J. L. (1987). The effects of changes in the environment on the spatial firing of hippocampal complex-spike cells. *The Journal of Neuroscience, 7,* 1951–68.

Muller, R. U., Kubie, J. L., Bostock, E. M., Taube, J. S., and Quirk, G. J. (1991). Spatial firing correlates of neurons in the hippocampal formation of freely moving rats. In J. Paillard (ed.), *Brain and space.* New York: Oxford University Press, pp. 296–333.

Muller, R. U., Kubie, J. L., and Ranck, J. B. (1987). Spatial firing patterns of hippocampal, complex-spike cells in a fixed environment. *Journal of Neuroscience, 7,* 1935–50.

North, A. J. and Stimmel, D. T. (1960). Extinction of an instrumental response following a large number of reinforcements. *Psychological Reports, 6,* 227–34.

Oakley, D. A and Russell, I. S. (1972). Neocortical lesions and Pavlovian conditioning. *Physiology and Behavior, 8,* 915–26.

(1975). Role of cortex in Pavlovian discrimination learning. *Physiology and Behavior, 15,* 315–21.

Okaichi, H. (1987). Performance and dominant strategies on place and cue tasks following hippocampal lesions in rats. *Psychobiology, 15,* 58–63.

O'Keefe, J. (1976). Place units in the hippocampus of the freely moving rat. *Experimental Neurology, 51,* 78–109.

(1989). Computations the hippocampus might perform. In L. Nadel, L. A. Cooper, P. Culicover, and R. M. Harnish (eds.), *Neural connections and mental computation.* Cambridge, MA: MIT Press/Brad Books.

O'Keefe J. and Dostrovsky, J. (1971). The hippocampus as a spatial map. Preliminary evidence from unit activity in the freely moving rat. *Brain Research, 34,* 171–5.

O'Keefe, J. and Nadel, L. (1978). *The hippocampus as a cognitive map.* Oxford: Clarendon Press.

O'Keefe, J., Nadel, L., and Wilner, J. (1979). Tuning out irrelevancy? Comments on Solomon's temporal mapping view of the hippocampus. *Psychological Bulletin, 86,* 1280–9.

O'Keefe, J. and Speakman, A. (1987). Single unit activity in the rat hippocampus during a spatial memory task. *Experimental Brain Research, 68,* 1–27.

Olton, D. S. (1978). Characteristics of spatial memory. In H. S. Hulse, H. Fowler, and W. K. Honig (eds.), *Cognitive processes in animal behavior,* New York: John Wiley & Sons, pp. 341–73.

(1986). Hippocampal function and memory for temporal context. In R. L. Isaacson and K. H. Pribram (eds.), *The hippocampus,* Vol. 4. New York: Plenum Press, pp. 281–98.

Olton, D. S., Branch, M., and Best, P. (1978). Spatial correlates of hippocampal unit activity. *Experimental Neurology, 41,* 461–555.

Orr, W. B. and Berger, T. W. (1985). Hippocampectomy disrupts the topography of conditioned nictitating membrane responses during reversal learning. *Journal of Comparative and Physiological Psychology, 99,* 35–45.

Overmier, J. B. and Bull, J. A. (1969). On the independence of stimulus control of avoidance. *Journal of Experimental Psychology, 79,* 464–67.

Overmier, J. B., Bull, J. A., III, and Trapold, M. A. (1971). Discriminative cue properties of different fears and their role in response selection in dogs. *Journal of Comparative Physiological Psychology 76,* 478–82.

Overmier, J. B. and Seligman, M. E. P. (1967). Effects of inescapable shock upon subsequent escape and avoidance responding. *Journal of Comparative and Physiological Psychology, 63,* 28–33.

Page, H. A. and Hall, J. F. (1953). Experimental extinction as a function of the prevention of response. *Journal of Comparative and Physiological Psychology, 46,* 33–4.

Parker, D. B. (1985) Learning-logic. Technical report TR-47, MIT Center for Computational Research in Economics and Management Science.

Pavlov, I. P. (1927). *Conditioned reflexes.* Oxford: Oxford University Press.
Pearce, J. M. and Hall, G. (1980). A model for Pavlovian learning: Variations in the effectiveness of conditioned but not of unconditioned stimuli. *Psychological Review, 87,* 532–52.
Pearce, J. M., Kaye, H., and Hall, G. (1983). Predictive accuracy and stimulus associability: Development of a model of Pavlovian learning. In M. Commons, R. J. Herrnstein, and A. R. Wagner (eds.), *Quantitative analyses of behavior,* Vol. 3. Cambridge, MA: Ballinger, pp. 241–55.
Penick, S. and Solomon, P. R. (1991). Hippocampus, context, and conditioning. *Behavioral Neuroscience, 105,* 611–7.
Platt, J. R., Kuch, D. O., and Bitgood, S. C. (1973). Rats lever-press duration as psychophysical judgments of time. *Journal of the Experimental Analysis of Behavior, 19,* 239–50.
Port, R. L., Mikhail, A. A., and Patterson, M. M. (1985). Differential effects of hippocampectomy on classically conditioned rabbit nictitating membrane response related to interstimulus interval. *Behavioral Neuroscience, 99,* 200–8.
Port, R. L. and Patterson, M. M. (1984). Fimbrial lesions and sensory preconditioning. *Behavioral Neuroscience, 98,* 584–9.
Port, R. L., Romano, A. G., and Patterson, M. M. (1986). Stimulus duration discrimination in the rabbit: Effects of hippocampectomy on discrimination and reversal learning. *Physiological Psychology, 14,* 124–9.
Port, R. L., Romano, A. G., Steinmetz, J. E., Mikhail, A. A., and Patterson, M. M. (1986). Retention and acquisition of classical trace conditioned responses by rabbits with hippocampal lesions. *Behavioral Neuroscience, 100,* 745–52.
Pribram, K. H. and Isaacson, R. L. (1975). Summary. In R. L. Isaacson and K. H. Pribram (eds.), *The hippocampus,* Vol. 2. New York: Plenum Press, pp. 429–41.
Quirk, G. J., Muller, R., and Kubie, J. L. (1990). The firing of hippocampal place cells in the dark depends on the rats' recent experience. *Journal of Neuroscience, 10,* 2008–17.
Ranck, J. B. (1975). Behavioral correlates and firing repertoires of neurons in the dorsal hippocampal formation and septum of unrestrained rats. In R. L. Isaacson and K. H. Pribram (eds.). *The hippocampus,* Vol. 2. New York: Plenum Press, pp. 207–46.
Rashotte, M. E., Griffin, R. W., and Sisk, C. L. (1977). Second-order conditioning of the pigeon's keypeck. *Animal Learning and Behavior, 5,* 25–38.
Ratcliff, R. (1990). Connectionist models of recognition memory: Constraints imposed by learning and forgetting functions. *Psychological Review, 97*: 285–308.
Rawlins, J. N. P. (1985). Associations across time: The hippocampus as a temporary memory store. *The Behavioral and Brain Sciences, 8,* 479–96.
Reiss, S. and Wagner, A. R. (1972). CS habituation produces a "latent inhibition effect" but no active "conditioned inhibition." *Learning and Motivation, 3,* 237–45.
Rescorla, R. A. (1967). Inhibition of delay in Pavlovian fear conditioning. *Journal of Comparative and Physiological Psychology, 64,* 114–20.
(1968). Probability of shock in the presence and absence of CS in fear conditioning. *Journal of Comparative and Physiological Psychology, 66,* 1–5.
(1971a). Summation and retardation tests of latent inhibition. *Journal of Comparative and Physiological Psychology, 75,* 77–81.
(1971b). Variation in the effectiveness of reinforcement and nonreinforcement following prior inhibitory conditioning. *Learning and Motivation, 2,* 113–23.
(1973). Effect of US habituation following conditioning. *Journal of Comparative Physiological Psychology, 82,* 137–43.
(1976a). Pavlovian excitatory and inhibitory conditioning. In W. K. Estes, (ed.), *Handbook of learning and cognitive processes,* Vol. 2. Hillsdale, NJ: Lawrence Erlbaum Associates, pp. 7–35.
(1976b). Stimulus generalization: Some predictions from a model of Pavlovian conditioning. *Journal of Experimental Psychology: Animal Behavior Processes, 2:* 88–96.
(1978). Some implications of a cognitive perspective on Pavlovian conditioning. In S. H. Hulse, H. Fowler, and W. Honig, (eds.), *Cognitive processes in animal behavior.* Hillsdale, NJ: Lawrence Erlbaum Associates, pp. 15–50.
(1980). Simultaneous and successive associations in sensory preconditioning. *Journal of Experimental Psychology: Animal Behavior Processes, 6,* 207–16.
Rescorla, R. A. and LoLordo, V. M. (1965). Inhibition of avoidance behavior. *Journal of Comparative and Physiological Psychology, 59,* 406–12.
Rescorla, R. A. and Wagner, A. R. (1972). A theory of Pavlovian conditioning: Variations in the effectiveness of reinforcement and non-reinforcement. In A. H. Black and W. F. Prokasy

(eds.), *Classical conditioning II: Theory and research.* New York: Appleton-Century-Crofts, pp. 64–99.

Revusky, S. (1971). The role of interference in association over delay. In W. K. Honig and P. H. R. James, (eds.), *Animal Memory.* New York: Academic Press, pp. 155–213.

Rickert, E. J., Bent, T. L., Lane, P., and French, J. (1978). Hippocampectomy and the attenuation of blocking. *Behavioral Biology, 22,* 147–60.

Rickert, E. J., Lorden, J. F., Dawson, R., Smyly, E., and Callahan, M. F. (1979). Stimulus processing and stimulus selection in rats with hippocampal lesions. *Behavioral and Neural Biology, 27,* 454–65.

Rizley, R. C. and Rescorla, R. A. (1972). Associations in second-order conditioning and sensory preconditioning. *Journal of Comparative Physiological Psychology 81,* 1–11.

Roberts, S. (1981). Isolation of an internal clock. *Journal of Experimental Psychology: Animal Behavior Processes, 7,* 242–68.

Robinson, G. B. (1986). Enhanced long-term potentiation induced in rat dentate gyrus by coactivation of septal and entorhinal inputs. *Brain Research, 379,* 56–62.

Robinson, G. B., Port, R. L., and Berger, T. W. (1989). Kindling facilitates acquisition of discriminative responding but disrupts reversal learning of the rabbit nictitating membrane response. *Behavioural Brain Research, 31,* 279–83.

Robinson, G. B. and Racine, R. J. (1986). Interactions between septal and entorhinal inputs to the rat dentate gyrus: Facilitation effects. *Brain Research, 379,* 63–7.

Rosellini, R. A., De Cola, J. P., and Warren, D. A. (1986). The effect of feedback stimuli on contextual fear depends upon the length of the minimum ITI. *Learning and Motivation, 17,* 229–42.

Rosenblatt, F. (1962). *Principles of neurodynamics.* Washington, DC: Spartan Books, Chap. 1, 2, 3, and 6.

Ross, R. T. and Holland, P. C. (1981). Conditioning of simultaneous and serial feature-positive discriminations. *Animal learning and Behavior, 9,* 293–303.

Ross, R. T., Orr, W. B., Holland, P. C., and Berger, T. W. (1984). Hippocampectomy disrupts acquisition and retention of learned conditional responding. *Behavioral Neuroscience, 98,* 211–25.

Rozin, P. and Kalat, J. W. (1971). Specific hungers and poisoning as adaptive specializations of learning. *Psychological Review, 78,* 459–86.

Rudell, A. P., Fox, S. E., and Ranck, J. B. (1980). Hippocampal excitability phase-lock to theta rhythm in walking rats. *Experimental Neurology, 68,* 87–96.

Rudy, J. W., Krauter, E. E., and Gaffuri, A. (1976). Attenuation of latent inhibition by prior preexposure another stimuli. *Journal of Experimental Psychology: Animal Behavior Processes, 2,* 235–47.

Rudy, J. W., Rosenberg, L., and Sandell, J. H. (1977). Disruption of taste familiarity effect by novel exteroceptive stimulation. *Journal of Experimental Psychology: Animal Behavior Processes, 88,* 665–9.

Rudy, J. W. and Sutherland, R. J. (1989). The hippocampal formation is necessary for rats to learn and remember configural discriminations. *Behavioural Brain Research, 34,* 97–109.

Rumelhart, D. E., Hinton, G. E., and Williams, G. E. (1986). Learning internal representations by error propagation. In D. E. Rumelhart and J. L. McClelland (eds.), *Parallel distributed processing: Explorations in the microstructure of cognition, Vol. 1: Foundations.* Cambridge, MA: Bradford Books, MIT Press; pp. 318-362

Rumelhart, D. E. and McClelland, J. L. (1986). PDP models and general issues in cognitive science. In D. E. Rumelhart and J. L. McClelland (eds.), *Parallel distributed processing: Explorations in the microstructure of cognition, Vol. 1: Foundations.* Cambridge, MA: Bradford Books, MIT Press, pp. 110–49.

Salvatierra, A. T. and Berry, S. D. (1989). Scopolamine disruption of septo-hippocampal activity and classical conditioning. *Behavioral Neuroscience, 103,* 715–21.

Schenk, F. and Morris, R. G. M. (1985). Dissociation between components of spatial memory in rats after recovery from the effects of retrohippocampal lesions. *Experimental Brain Research, 58,* 11–28.

Schmajuk, N.A. (1984a). Psychological theories of hippocampal function. *Physiological Psychology, 12,* 166–83.

(1984b). A model for the effects of the hippocampal lesions on Pavlovian conditioning. *Abstracts of the Society for Neuroscience, 10,* 124.

(1986). Real-time attentional models for classical conditioning and the hippocampus. Unpublished Ph.D. dissertation, University of Massachusetts.

(1987). SEAS: A dual memory architecture for computational cognitive mapping. In *Proceedings of the Ninth Annual Conference of the Cognitive Science Society.* Hillsdale, NJ: Lawrence Erlbaum Associates, pp. 644–54.
(1989). The hippocampus and the control of information storage in the brain. In M. Arbib and S. I. Amari (eds.), *Dynamic interactions in neural networks: Models and data.* New York: Springer-Verlag, pp. 53–72.
(1990). Role of the hippocampus in temporal and spatial navigation: An adaptive neural network. *Behavioral Brain Research, 39,* 205–29.
Schmajuk, N. A. (1994). Behavioral dynamics of escape and avoidance: A neural network approach. In D. Cliff, P. Husbands, J.-A. Meyer, and S. Wilson (eds.), *From Animals to Animats 3,* Cambridge, MA: MIT Press, pp. 118–27.
Schmajuk, N. A. and Axelrad, E. Communication and consciousness: A neural network conjecture. *Behavioural and Brain Sciences* (in press).
Schmajuk, N. A. and Blair, H. T. (1993). Stimulus configuration, place learning, and the hippocampus. *Behavioral Brain Research, 59,* 103–17.
Schmajuk, N. A. and DiCarlo, J. J. (1991a). Neural dynamics of hippocampal modulation of classical conditioning. In M. Commons, S. Grossberg, and J. E. R. Staddon (eds.), *Neural network models of conditioning and action.* Hillsdale, NJ: Lawrence Erlbaum Associates, pp. 149–80.
(1991b). A neural network approach to hippocampal function in classical conditioning. *Behavioral Neuroscience, 105,* 82–110.
(1992). Stimulus configuration, classical conditioning, and the hippocampus. *Psychological Review, 99,* 268–305.
Schmajuk, N. A. and Christiansen, B. A. (1990). Eyeblink conditioning in rats. *Physiology and Behavior, 48,* 755–8.
Schmajuk, N. A., Lam, Y. W., and Gray, J. A. (1996) Latent inhibition: A neural network approach. *Journal of Experimental Psychology: Animal Behavior Processes, 22*: 321–49.
Schmajuk, N. A., Lamoureux, J., and Holland, P. (1997). Occasion setting and stimulus configuration: A neural network approach. *Psychological Review,* in press.
Schmajuk, N. A. and Moore, J. W. (1985). Real-time attentional models for classical conditioning and the hippocampus. *Physiological Psychology, 13,* 278–90.
(1986). A real-time attentional-associative model for classical conditioning of the rabbit's nictitating membrane response. In *Proceedings of the Eighth Annual Conference of the Cognitive Science Society.* Hillsdale, NJ: Lawrence Erlbaum Associates, pp. 794–807.
(1988). The hippocampus and the classically conditioned nictitating membrane response: A real-time attentional-associative model. *Psychobiology, 46,* 20–35.
(1989). Effects of hippocampal manipulations on the classically conditioned nictitating membrane response: Simulations by an attentional-associative model. *Behavioral Brain Research, 32,* 173–89.
Schmajuk, N. A. and Segura, E. T. (1980). A motivational and learning system. In G. E. Lasker (ed.), *Applied systems and cybernetics,* Vol. 2. New York: Pergamon Press, pp. 770–4.
Schmajuk, N. A., Spear, N. E., and Isaacson, R. L. (1983). Absence of overshadowing in rats with hippocampal lesions. *Physiological Psychology, 11,* 59–62.
Schmajuk, N. A. and Thieme, A. D. (1992). Purposive behavior and cognitive mapping: An adaptive neural network. *Biological Cybernetics, 67,* 165–74.
Schmajuk, N. A., Thieme, A. D., and Blair, H. T. (1993). Maps, routes, and the hippocampus: A neural network approach. *Hippocampus, 3,* 387–400.
Schmajuk, N. A., Urry, D., and Zanutto, B. S. The frightening complexity of avoidance: An adaptive neural network. In J. E. R. Staddon and C. Wynne (eds.), *Models of action.* Hillsdale, NJ: Lawrence Erlbaum Associates (in press).
Schmaltz, L. W. and Theios, J. (1972). Acquisition and extinction of a classically conditioned response in hippocampectomized rabbits (*Oryctolagus cuniculus*). *Journal of Comparative and Physiological Psychology, 79,* 328–33.
Schneiderman, N. (1966). Interstimulus interval function of the nictitating membrane response of the rabbit under delay versus trace conditioning. *Journal of Comparative and Physiological Psychology, 62,* 397–402.
(1972). Response system divergences in aversive classical conditioning. In A. H. Black and W. F. Prokasy (eds.), *Classical conditioning, II: Current research and theory.* New York: Appleton-Century-Crofts, pp. 341–76.
Schneiderman, N. and Gormezano, I. (1964). Conditioning of the nicitating membrane of the rabbit as a function of CS–US interval. *Journal of Comparative Physiological Psychology, 57,* 188–95.

Schnur, P. and Lubow, R. E. (1976). Latent inhibition: The effects of ITI and CS intensity during preexposure. *Learning and Motivation, 7,* 540–50.

Schoenfeld, W. N. (1950). An experimental approach to anxiety, escape, and avoidance behavior. In P. H. Hoch and J. Zubin (eds.), *Anxiety.* New York: Grune and Stratton, pp. 70-99.

Sears, L. L. and Steinmetz, J. E. (1990). Acquisition of classical conditioned-related activity in the hippocampus is affected by lesions of the cerebellar interpositus nucleus. *Journal of Neuroscience Research, 27,* 681–92.

Seligman, M. E. P. (1970). On the generality of the laws of learning. *Psychologyical Review, 77,* 406–18.

(1975). *Helplessness: On depression, development, and death.* San Francisco, CA: W. H. Freeman & Co.

Seligman, M. E. P. and Campbell, B. A. (1965). Effects of intensity and duration of punishment on extinction and avoidance response. *Journal of Comparative and Physiological Psychology, 59,* 295–7.

Seligman, M. E. P. and Johnston, J. C. (1973). A cognitive theory of avoidance learning. In F. J. McGuigan and D. B. Lumsden (eds.), *Contemporary approaches to conditioning and learning.* Washington, DC: V. H. Winston, pp. 69–110.

Seligman, M. E. P. and Maier, S. F. (1967). Failure to escape traumatic shock. *Journal of Experimental Psychology, 74,* 1–9.

Semple-Rowland, S. L., Bassett, J. L., and Berger, T. W. (1981). Subicular projections to retrospenial cortex in the rabbit. *Society of Neurosciences Abstracts, 7,* 886.

Seward, J. P. and Levy, H. (1949). Latent extinction: Sign learning as a factor in extinction. *Journal of Experimental Psychology, 39,* 660–8.

Sidman, M. (1953). Two temporal parameters of the maintenance of avoidance behavior by the white rat. *Journal of Comparative and Physiological Psychology, 46,* 253–61.

Siegel, S. and Domjan, M. (1971). Backwards conditioning as an inhibitory procedure. *Learning and Motivation 2,* 1–11.

Skelton, R. W. (1988). Bilateral cerebellar lesions disrupt conditioned eyelid responses in unrestrained rats. *Behavioral Neuroscience, 102,* 586–90.

Skinner, B. F. (1938). *The behavior of organisms.* New York: Appleton-Century-Crofts, p. 227.

Skrandies, W. (1984). Scalp potential fields evoked by grating stimuli: Effects of spatial frequency and orientation. *Electroencephalology and Clinical Neurophysiology, 58,* 325–32.

Smith, M. C. (1968). CS–US interval and US intensity in classical conditioning of the rabbit's nictitating membrane response. *Journal of Comparative and Physiological Psychology, 66,* 679–87.

Smith, M. and Gormezano, I. (1965). Effects of alternating classical conditioning and extinction sessions on the conditioned nictitating membrane response of the rabbit. *Psychonomic Science, 3,* 91–2.

Smythies, J. R. (1966). *Brain mechanisms and behavior.* New York: Academic Press.

Sokolov, E. N. (1960). Neuronal models and the orienting reflex. In M. A. B. Brazier (ed.), *The central nervous system and behavior.* New York: Macy Foundation, pp. 31-36.

Solomon, P. R. (1977). Role of the hippocampus in blocking and conditioned inhibition of rabbit's nictitating membrane response. *Journal of Comparative and Physiological Psychology, 91,* 407–17.

(1979). Temporal versus spatial information processing theories of the hippocampal function. *Psychological Bulletin, 86,* 1272–9.

Solomon, P. R., Brennan, G., and Moore, J. W. (1974). Latent inhibition of the rabbit's nictitating membrane response as a function of CS intensity. *Bulletin of the Psychonomic Society, 4,* 445–8.

Solomon, R. L., Kamin, L. J., and Wynne, L. C. (1953). Traumatic avoidance learning: The outcomes of several extinction procedures with dogs. *Journal of Abnormal and Social Psychology, 48,* 291–302.

Solomon, P. R. and Moore, J. W. (1975). Latent inhibition and stimulus generalization of the classically conditioned nictitating membrane response in rabbits (*Oryctolagus cuniculus*) following dorsal hippocampal ablation. *Journal of Comparative and Physiological Psychology, 89,* 1192–1203.

Solomon, P. R., Vander Schaaf, E. R., Thompson, R. F., and Weisz, D. J. (1986). Hippocampus and trace conditioning of the rabbit's classically conditioned nictitating membrane response. *Behavioral Neuroscience, 100,* 729–44.

Solomon, R. L. and Wynne, L. C. (1953). Traumatic avoidance learning: Acquisition in normal dogs. *Psychological Monographs, 67,* 354.

(1954). Traumatic avoidance learning: The principles of anxiety conservation and partial reversibility. *Psychological Review, 61,* 353–85.
Soltysik, S. (1960). Studies on avoidance conditioning: II. Differentiation and extinction of avoidance responses. *Acta Biologiae Experimentalis, 20,* 171–82.
(1964). Inhibitory feedback in avoidance conditioning. In A. Escobar (ed.), *Feedback systems controlling nervous activity.* Mexico: Sociedad Mexicana de Ciencias Fisiologicas, pp. 316–31.
(1985). Protection from extinction: New data and a hypothesis of several varieties of conditioned inhibition. In R. R. Miller and N. E. Spear (eds.), *Information processing in animals: conditioned inhibition.* Hillsdale, NJ: Lawrence Erlbaum Associates, pp. 369–94.
Spear, N. E. (1981). Extending the domain of memory retrieval. In R. R. Miller and N. E. Spear (eds.), *Information processing in animals: Memory mechanisms.* Hillsdale, NJ: Lawrence Erlbaum Associates, pp. 341–78.
Spear, N. E., Miller, J. S., and Jagielo, J. A. (1990). Animal memory and learning. *Annual Review of Psychology, 41,* 169–211.
Spence, K. W. (1952). The nature of the response in discrimination learning. *Psychological Review 59,* 89–93.
Squire, L. R., Shimamura, A. P., and Amaral, D. G. (1989). Memory and the hippocampus. In J. H. Byrne and W. Berry (eds.), *Neural models of plasticity.* New York: Academic Press, pp. 208–39.
Staddon, J. E. R. and Zhang, Y. (1991). On the assignment-of-credit problem in operant learning. In M. Commons, S. Grossberg, and J. E. R. Staddon (eds.), *Neural network models of conditioning and action.* Hillsdale, NJ: Lawrence Erlbaum Associates, pp. 279–93.
Starr, M. D. and Mineka, S. (1977). Determinants of fear over the cause of avoidance learning. *Learning and Motivation, 8,* 332–50.
Steinbuch, K. (1961). Die Lernmatrix. *Kybernetik, 1,* 36–45.
Steinmetz, J. E., Logan, C.G., and Thompson, R. F. (1988). Essential involvement of mossy fibers in projecting the CS to the cerebellum during classical conditioning. In C. Woody, D. Alkon, and J. McGaugh (eds.), *Cellular mechanisms of conditioning and behavioral plasticity.* New York: Plenum Press, pp. 143–8.
Stewart, M. and Fox, S. E. (1990). Do septal neurons pace the hippocampal theta rhythm? *Trends in Neuroscience, 13,* 163–68.
Sutherland, R. J., Kolb, B., and Whishaw, I. Q. (1982). Spatial mapping: Definitive disruption by hippocampal or medial frontal cortical damage in the rat. *Neuroscience Letters, 31,*271–6.
Sutherland, N. S. and Mackintosh, N. J. (1971). *Mechanisms of animal discrimination learning.* New York: Academic Press.
Sutherland, R. J., Wishaw, I. Q., and Kolb, B. (1988). Contributions of cingulate cortex to two forms of spatial learning and memory. *The Journal of Neuroscience, 8,* 1863–72.
Sutton, R. S. (1991). Reinforcement learning architectures for animals. In J. Meyer and S. Wilson (eds.), *From animals to animals.* Cambridge, MA: MIT Press, pp. 288–96.
Sutton, R. S. and Barto, A. G. (1981a). An adaptive network that constructs and uses an internal model of its world. *Cognition and Brain Theory, 4,* 217–46.
(1981b). Toward a modern theory of adaptive networks: Expectation and prediction. *Psychological Review, 88,* 135–70.
Tesauro, G. (1990). Neural models of classical conditioning: A theoretical viewpoint. In S. J. Hanson and C. R. Olson (eds.), *Connectionist modelling and brain function: The developing interface.* Cambridge, MA: MIT Press, pp. 74–104.
Testa, T. J. (1974). Causal relationships and the acquisition of the avoidance response. *Psychological Review, 81,* 491–505.
Testa, T. J. and Ternes, J. W. (1977). Specificity of conditioning mechanisms in the modification of food preferences. In L. M. Barker, M. R. Best, and M. Domjan (eds.), *Learning mechanisms in food selection.* Waco, TX: Baylor University Press, pp. 229–53.
Thompson, R., Harmon, D., and Yu, J. (1984). Detour problem-solving behavior in rats with eocortical and hippocampal lesions: A study of response flexibility. *Physiological Psychology, 2,* 116–24.
Thompson, R. F. (1986). The neurobiology of learning and memory. *Science, 233,* 941–7.
Thorndike, E. L. (1898). Animal intelligence: An experimental study of the associative processes in animals. *Psychological Monographs,* 2 (4, Whole No. 8).
Tipler, P. A. (1982). Physics. New York: Worth Publishers.
Toates, F. (1986). *Motivational systems.* Cambridge: Cambridge University Press.

Tolman, E. C. (1932a). *Purposive behavior in animals and men.* New York: Irvington Publishers.
(1932b). Cognitive maps in rats and men. *Psychological Review, 55,* 189–208.
Tolman, E. C. and Honzik, C. H. (1930). "Insight" in rats. *University of California Publications in Psychology, 4,* 215–32.
Trobalon, J. B., Chamizo, V. D., and Mackintosh, N. J. (1992). Role of context in perceptual learning in maze discriminations. *The Quarterly Journal of Experimental Psychology, 44B,* 57–73.
Vertes, R. P. (1982). Brain stem generation of hippocampal EEG. *Progress in Neurobiology, 19,* 159–86.
Vinogradova, O. S. (1975). Functional organization of the limbic system in the process of registration of information: Facts and hypothesis. In R. L. Isaacson and K. H. Pribram (eds.), *The hippocampus.* New York: Plenum Press, pp. 3–69.
Vinogradova, O. S., Brazhnik, E. S., Karanov, A. M., and Zhadina, S. D. (1980). Neuronal activity of the septum following various types of deafferentation. *Brain Research, 187,* 353–68.
Wagner, A. R. (1961). Effects of amount and percentage of reinforcement and number of acquisition trials on conditioning and extinction. *Journal of Experimental Psychology, 62,* 234–42.
(1978). Expectancies and the priming of STM. In S. H. Hulse, H. Fowler, and W. K. Honig (eds.), *Cognitive processes in animal behavior.* Hillsdale, NJ: Lawrence Erlbaum Associates, pp. 177–209.
(1979). Habituation and memory. In A. Dickinson and R. A. Boakes (eds.), *Mechanisms of learning and motivation.* Hillsdale, NJ: Lawrence Erlbaum Associates, pp. 53–82.
(1981). SOP: A model of automatic memory processing in animal behavior. In N. E. Spear and R. R. Miller (eds.), *Information processing in animals:* Memory mechanisms. Hillsdale, NJ: Lawrence Erlbaum Associates, pp. 5–47.
Wagner, A. R., Logan, F. A., Haberlandt, K., and Price, T. (1968). Stimulus selection in animal discrimination learning. *Journal of Experimental Psychology, 76,* 171–80.
Waldrop, M. M. (1992). *Complexity.* New York: Simon & Schuster.
Walter, W. G. (1951). *The living brain.* London: Gerald Duckworth.
Weikart, C. and Berger, T. W. (1986). Hippocampal lesions disrupt classical conditioning of cross-modality reversal learning of the rabbit nictitating membrane response. *Behavioural Brain Research, 22,* 85–90.
Weisendanger, R. and Weisendanger, M. (1982). The corticopontine system in the rat. *Journal of Comparative Neurology, 208,* 227–38.
Weisman, R. G. and Litner, J. S. (1969). Positive conditioned reinforcement of Sidman avoidance in rats. *Journal of Comparative and Physiological Psychology, 68,* 597–603.
(1972). The role of Pavlovian events in avoidance training. In R. A. Boakes and M. S. Halliday (eds.), *Inhibition and learning.* London: Academic Press, pp. 253–70.
Weisz, D. J., Clark, G. A., and Thompson, R. F. (1984). Increased responsivity of dentate granule cells during nictitating membrane response conditioning in rabbit. *Behavioral Brain Research, 12,* 145–54.
Weiss, C., Houk, J. C., and Gibson, A. R. (1990). Inhibition of sensory responses of cat inferior olive neurons produced by stimulation of red nucleus. *Journal of Neuropshysiology, 64,* 1170–85.
Weiss, M. and Pellet, J. (1982a). Raphe–cerebellum interactions. I. Effects of cerebellar stimulation and harmaline administration on single unit activity. *Experimental Brain Research, 48,* 163–70.
(1982b). Raphe–cerebellum interactions. II. Effects of midbrain raphe stimulation and harmaline administration on single unit activity of cerebellar cortical cells in the rat. *Experimental Brain Research, 48,* 171–6.
Welsh, J. P. and Harvey, J. A. (1989). Cerebellar lesions and the nictitating membrane reflex: Performance deficits of the conditioned and unconditioned response. *Journal of Neuroscience, 9,* 299–311.
Werbos, P. J. (1974). Beyond regression: New tools for prediction and analysis in the behavioral sciences. Unpublished Ph.D. dissertation, Harvard University, Chap. 2, 3, and 6.
Westbrook, R. F., Bond, N. W., amd Feyer, A.-M. (1981). Short- and long-term decrements in toxicosis-induced odor-aversion learning: the role of duration of exposure to an odor. *Journal of Experimental Psychology: Animal Behavior Processes, 7,* 362–81.
Whishaw, I. Q. (1991). Latent learning in a swimming pool place task by rats: Evidence for the use of associative and not cognitive mapping processes. *The Quarterly Journal of Experimental Psychology, 43B,* 83–103.

Whishaw, I. Q. and Kolb, B. (1984). Decortication abolishes place but not cue learning in rats. *Behavioural Brain Research, 11,* 124–34.

Whishaw, I. Q. and Tomie, J. (1991). Acquisition and retention by hippocampal rats of simple, conditional and configural tasks using tactile and olfactory cues: Implications for hippocampal function. *Behavioral Neuroscience, 105,* 787–97.

Wible, C. G., Findling, R. L., Shapiro, M. W., Lang, E. J., Crane, S., and Olton, D. S. (1986). Mnemonic correlates of unit activity in the hippocampus. *Brain Research, 399,* 97–110.

Wickelgren, W. A. (1979). Chunking and consolidation: A theoretical synthesis of semantic networks, configuring in conditioning, S–R versus cognitive learning, normal forgetting, the amnesic syndrome, and the hippocampal arousal system. *Psychological Review, 86,* 44–60.

Wickens, C., Tuber, D. S., and Wickens, D. D. (1983). Memory for the conditioned response: The proactive effect of preexposure to potential conditioning stimuli and context change. *Journal of Experimental Psychology: General, 112,* 41–57.

Widrow, B. amd Hoff, M. E. (1960). Adaptive switching circuits. In *1960 IRE WESCON Convention Record,* pp. 96–104.

Wilkie, D. M. (1987). Stimulus intensity affects pigeon's timing behavior: Implications for an internal clock model. *Animal Learning and Behavior, 15,* 35–9.

Wilkie, D. M. and Palfrey, R. (1987). A computer simulation model of rat's place navigation in the Morris water maze. *Behavior Research Methods, Instruments, and Computers, 19,* 400–3.

Wilklund, L., Björklund, A., and Sjölund, B. (1977). The indolaminergic innervation of the inferior olive. 1. Convergence with the direct spinal afferents in the areas projecting to the cerebellar anterior lobe. *Brain Research, 131,* 1–21.

Willshaw, J. W., Buneman, O. P., Longuet-Higgins, H. C. (1969). Non-holographic associative memory. *Nature, 222,* 960–2.

Winston, P. H. (1977). *Artificial intelligence.* Reading, MA: Addison-Wesley.

Wyss, J. M. and Sripanidkulchai, K. (1984). The topography of the mesencephalic and pontine projections from the cingulate cortex. *Brain Research, 293,* 1–15.

Yeo, C. H., Hardiman, M. J., Moore, J. W., and Russell, I. S. (1984). Trace conditioning of the nictitating membrane response in decorticate rabbits. *Behavioural Brain Research, 11,* 85–8.

Young, G. A. (1976). Electrical activity of the dorsal hippocampus in rats operantly trained to lever press and to lick. *Journal of Comparative and Physiological Psychology, 90,* 78–90.

Zimmer-Hart, C. L. and Rescorla, R. A. (1974). Extinction of Pavlovian conditioned inhibition. *Journal of Comparative and Physiological Psychology, 86,* 837–45.

Zipser, D. (1985). A computational model of hippocampal place fields. *Behavioral Neuroscience, 99,* 10006–18.

(1986). Biologically plausible models of place recognition and goal location. In J. L. McClelland, D. E. Rumelhart, and the PDP Research Group, *Parallel distributed processing. Explorations in the microstructure of cognition. Vol. 2: Psychological and biological models.* Cambridge, MA: The MIT Press, pp. 432–70.

Zipser, D. and Rumelhart, D. E. (1990). Neurobiological significance of new learning models. In E. Schwartz (ed.), *Computational neuroscience.* Cambridge, MA: MIT Press, pp. 192–200.

Companion diskette

This book comes with a companion **diskette** with a version of the original program for the Schmajuk and DiCarlo (1992) model, written for DOS computers. The program was written in collaboration with Patrick Lam, Chris Lai, and Tad Blair.

The diskette permits the simulation on a personal computer of the paradigms described in Chapter 2 and different brain manipulations described in Chapter 13. Perhaps more interesting, it allows users to simulate their experimental designs and examine the model's predictions. This is an exciting way of communicating theory that might increase the interaction between theorists and experimenters in the field.

To run

From the diskette

1. Insert the diskette in a drive.
2. Type the drive letter followed by a colon (e.g., A: or B:). Press Enter.
3. Type HIPPO. Press Enter.
4. Make a selection from the MENU.

From the hard disk

1. Insert the diskette in a drive.
2. Type MKDIR [Directory name]. Press Enter.
3. Type CD [Directory Name]. Press Enter.
4. Type COPY A:*.* C:\[Directory name]. Press Enter.
5. Type HIPPO. Press Enter
6. Make a selection from the MENU.

0. DESIGN YOUR OWN
Enter trial definitions

1. DELAY CONDITIONING ACQUISITION & EXTINCTION
10 (CS1)+, 10 (CS1)-

2. TRACE CONDITIONING ACQUISITION & EXTINCTION
10 (CS1)+, 10 (CS1)-

3. BLOCKING
10(CS1)+, 10 (CS1,CS2)+, 1 (CS1)-, 1 (CS2)-

4. OVERSHADOWING
10 (CS1,CS2)+, 1 (CS1)-, 1 (CS2)-

5. CONDITIONED INHIBITION
40 alternating (CS1)+ & 1 (CS1,CS2)-

6. DISCRIMINATION ACQUISITION & REVERSAL
40 alternating (CS1)+ & (CS2)-, 14 alternating (CS1)- & (CS2)+

7. ACQUISITION/EXTINCTION SERIES
Acquisition & extinction to criterion 5 times

8. NEGATIVE PATTERNING (XOR)
120 alternating (CS1)+ & (CS2)+ & (CS1,CS2)-, 1 (CS1)-, 1 (CS2)-

9. POSITIVE PATTERNING (XOR)
66 alternating (CS1,CS2)+ & (CS1)- & (CS2)-, 1 (CS1,CS2)-

10. SIMULTANEOUS FEATURE POSITIVE PATTERNING
40 alternating (CS1,CS2)+ & (CS2)-, 1 (CS1,CS2)-, 1 (CS1)-

11. SERIAL FEATURE POSITIVE PATTERNING
74 alternating (CS1,CS2)+ & (CS2)-, 1 (CS1,CS2)-, 1 (CS1)-

PLEASE SELECT A CONDITIONING PARADIGM

Figure 1. Menu of different classical conditioning paradigms. The menu includes the "design your own" case.

The menu

Figure 1 shows the menu of different classical paradigms ready to be simulated. In addition, Option 0 lets the users design their own experiment. Users can decide the number of CSs to be used, CS duration, US duration, ISI, types of trials, and order of trial presentation.

Once a paradigm has been selected, or designed, the type of lesion to be simulated should be decided.

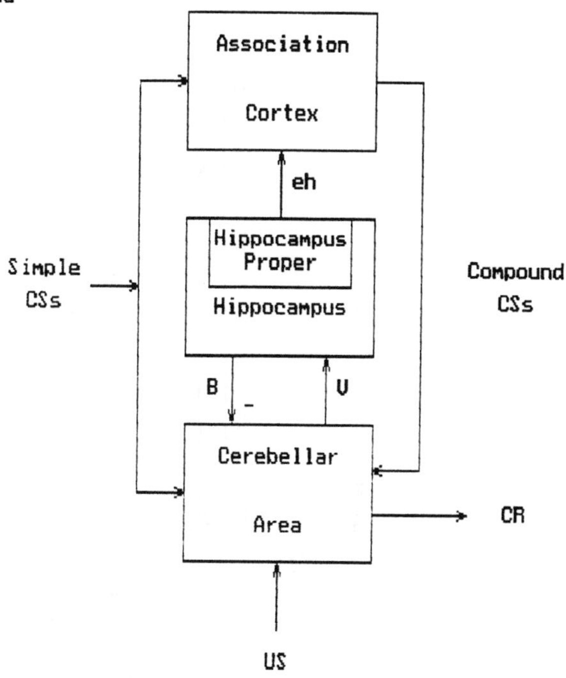

Figure 2. Block diagram of the brain-mapped S–D model. When a lesion condition is selected, the "lesioned" block is dashed.

The block diagram

After all selections have been made, the next screen shows a diagram of the model indicating the blocks that might be affected by the selected lesions (Figure 2).

The graphic simulation

Next, the screen shows the simulations of the selected paradigm under the preferred brain condition (Figure 3). The left panels display real-time simulated conditioned and unconditioned responses in different trials color-coded according to the type of trial that is being simulated. Vertical dashed lines indicate CS onset and offset. The vertical point line indicates US onset. Trial 1 is always represented at the bottom of the panel. The right panels show (1) peak CR: peak CR as a function of trials, (2) hidden weights: average VHs as a function of trials, and (3) output weights: average VSs and VNs as a function of trials.

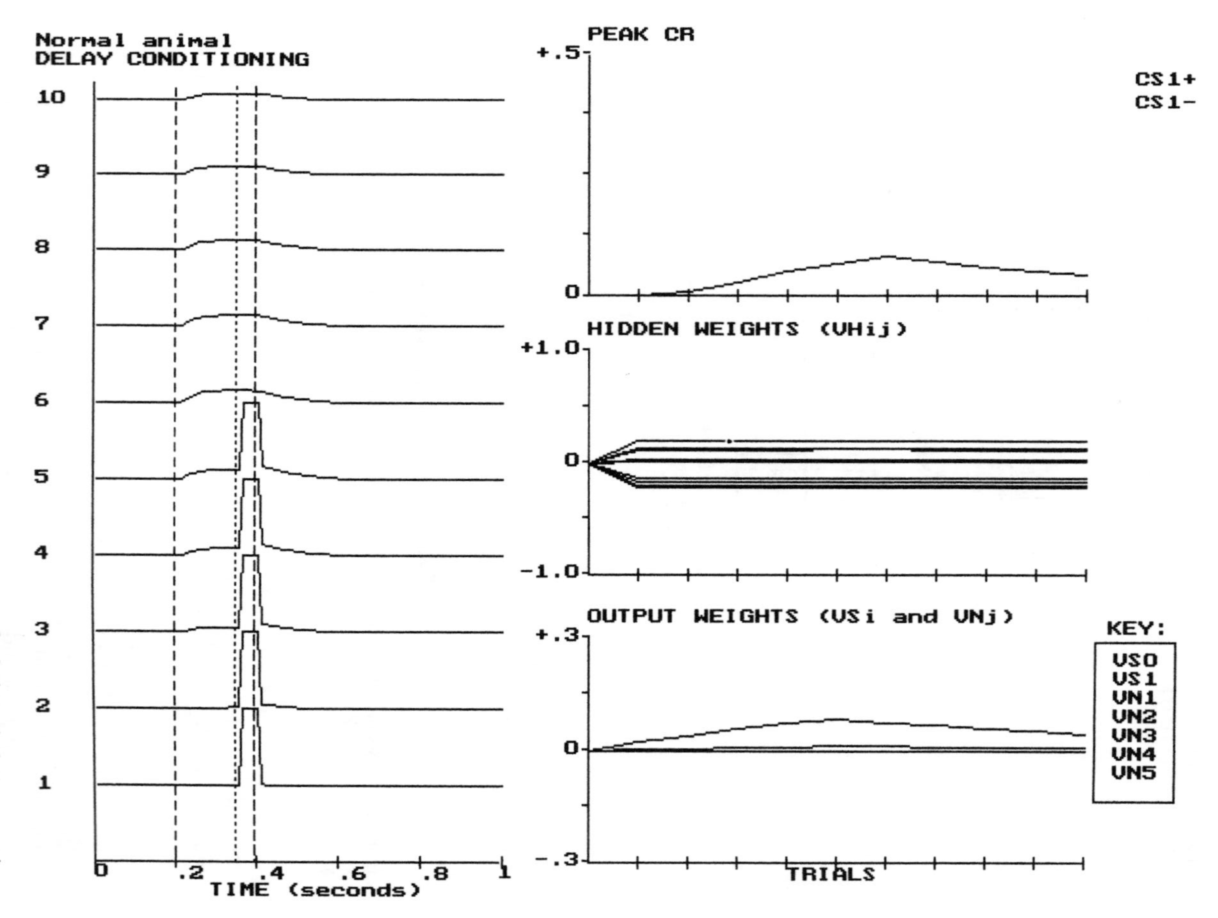

Figure 3. Real-time simulations of acquisition and extinction of classical conditioning. Left panels. Real-time simulated conditioned and unconditioned response in trials 1–10 (1–5 acquisition, 6–10 extinction). Vertical dashed line indicates CS onset and offset, vertical point line indicates US onset. Trial 1 is represented at the bottom of the panel. Right Panels. Peak CR: Peak CR as a function of trials. Hidden weights: average VHs as a function of trials; output weights: average VSs and VNs as a function of trials.

The numerical results

After the graphic simulation is completed, the screen shows numerical values of some of the simulated variables that might be compared with experimental results. Because the DOS program is not the original program with which the simulations presented in the book were obtained, some variations can be expected.

Technical questions about running or using the program should be addressed to the author.

Author Index

Subject Index

Italic page numbers indicate definitions; **boldface** page numbers indicate boxes.

WITHDRAWN